T0141140

Of Sheep, Oranges, and Yeast

Cary Wolfe, Series Editor

40 OF SHEEP, ORANGES, AND YEAST: A MULTISPECIES IMPRESSION
Julian Yates

39 FUEL: A SPECULATIVE DICTIONARY
Karen Pinkus

38 WHAT WOULD ANIMALS SAY IF WE ASKED THE RIGHT QUESTIONS?
Vinciane Despret

37 MANIFESTLY HARAWAY
Donna J. Haraway

36 NEOFINALISM
Raymond Ruyer

35 INANIMATION: THEORIES OF INORGANIC LIFE
David Wills

34 ALL THOUGHTS ARE EQUAL: LARUELLE AND NONHUMAN PHILOSOPHY
John Ó Maoilearca

33 NECROMEDIA
Marcel O'Gorman

32 THE INTELLECTIVE SPACE: THINKING BEYOND COGNITION
Laurent Dubreuil

31 LARUELLE: AGAINST THE DIGITAL
Alexander R. Galloway

30 THE UNIVERSE OF THINGS: ON SPECULATIVE REALISM
Steven Shaviro

(continued on page 367)

Of Sheep, Oranges, and Yeast

A MULTISPECIES IMPRESSION

Julian Yates

posthumanities 40

UNIVERSITY OF MINNESOTA PRESS
Minneapolis · London

The University of Minnesota Press gratefully acknowledges the generous assistance provided for the publication of this book by the Center for Material Culture Studies and Department of English at the University of Delaware.

Portions of chapter 1 appeared in "Counting Sheep: Dolly Does Utopia (Again)," *Rhizomes* 8 (2004), http://www.rhizomes.net/issue8/yates2.htm, and in "Sheep Tracks," in *Animal, Vegetable, Mineral: Ethics and Objects,* ed. Jeffrey Jerome Cohen, 173–209 (Washington, D.C.: Oliphaunt Books, 2012). Portions of chapter 2 appeared in "What Was Pastoral (Again)? More Versions," in *The Return of Theory in Early Modern English Studies,* ed. Paul Cefalu and Bryan Reynolds, 93–118 (New York: Palgrave Macmillan, 2011); reprinted with permission. Portions of chapter 3 appeared in "Towards a Theory of Agentive Drift: Or, a Particular Fondness for Oranges in 1597," *Parallax* 8, no. 1 (2002): 47–58, http://www.tandfonline.com/doi/abs/10.1080/13534640110119614?journalCode=tpar20. Portions of chapter 4 appeared in "Orange," in *Prismatic Ecology: Ecotheory beyond Green,* ed. Jeffrey Jerome Cohen, 85–103 (Minneapolis: University of Minnesota Press, 2013). Portions of the Conclusion appeared in "Oves et Singulatim: A Multispecies Impression," in *Renaissance Posthumanisms,* ed. Joseph Campana and Scott Maisano, 167–94 (New York: Fordham University Press, 2016).

Copyright 2017 by the Regents of the University of Minnesota

All rights reserved. No part of this publication may be reproduced, stored in a retrieval system, or transmitted, in any form or by any means, electronic, mechanical, photocopying, recording, or otherwise, without the prior written permission of the publisher.

Published by the University of Minnesota Press
111 Third Avenue South, Suite 290
Minneapolis, MN 55401-2520
http://www.upress.umn.edu

Printed in the United States of America on acid-free paper

The University of Minnesota is an equal-opportunity educator and employer.

22 21 20 19 18 17 10 9 8 7 6 5 4 3 2 1

Library of Congress Cataloging-in-Publication Data
Names: Yates, Julian, author.
Title: Of sheep, oranges, and yeast : a multispecies impression / Julian Yates.
Description: Minneapolis : University of Minnesota Press, 2017. |
Series: Posthumanities; 40 | Includes bibliographical references and index.
Identifiers: LCCN 2016023162 | ISBN 978-1-5179-0066-3 (hc) |
ISBN 978-1-5179-0067-0 (pb)
Subjects: LCSH: Literature—Philosophy. | Literature, Experimental. |
American literature—21st century. | Sheep in literature.
Classification: LCC PN45 .Y38 2017 | DDC 801—dc23
LC record available at https://lccn.loc.gov/2016023162

For Noah and Naomi

Contents

Impression 1

PART I. SHEEP

1. Counting Sheep in the Belly of the Wolf 35

2. What Was Pastoral (Again)? More Versions (*Otium* for Sheep) 91

PART II. ORANGES

3. Invisible Inc. (Time for Oranges) 135

4. Gold You Can Eat (On Theft) 179

PART III. YEAST

5. Bread and Stones (On Bubbles) 223

Erasure 271

Acknowledgments 275

Notes 279

Index 345

Impression

Moments of disorientation are vital.

—SARA AHMED, *Queer Phenomenology*

My favorite trope for dog tales is "metaplasm." Metaplasm means a change in word, for example by adding, omitting, inverting, or transposing its letters, syllables, or sounds. The term is from the Greek *metaplasmos,* meaning remodeling or remolding. Metaplasm is a generic term for almost any kind of alteration of a word, intentional or unintentional. I use metaplasm to mean the remodeling of dog and human flesh, remolding the codes of life, in the history of companion-species relating.

Compare and contrast "protoplasm," "cytoplasm," "neoplasm," and "germplasm." There is a biological taste to "metaplasm"—just what I like in words about words. Flesh and signifier, bodies and words, stories and worlds: these are joined in nature cultures. Metaplasm can signify a mistake, a stumbling, a troping that makes a fleshly difference. . . . Woof!

—DONNA HARAWAY, *The Companion Species Manifesto*

I am going to begin with a memory of skin. Not my memory. Not your memory. Not any one person's memory exactly. It belongs to the thing we call a "character" in a play. Like all memories, it comes from without even as it seems to well up from within. The trigger? A parchment—writing material manufactured from the skin of a sheep, a goat, or a cow. Jack Cade from William Shakespeare's *Henry VI, Part 2* (1590/1591) touches a parchment and finds himself involuntarily "touched" in return. Worse still, this parchment hurts. It forces him to recall his encounters with the law. He registers this pain as a "sting." And this sting leads him to bemoan the fate of the lamb from which this parchment was made.

I begin with Jack's skin memory, a memory that sympathetically transfers the pain of the knife that flays the lamb to human skin that is stung by a seal, because it makes legible the anthropo-zoo-genetic bases to the worlds we live. This virtual pain, pain that went unfelt by the lamb, dead already, endures or dwells within the parchment as a potential that Jack realizes. A different and differentiating order of pain—not the suffering of the lamb as it was killed, nor of Jack as he signs his life away, but still another form of pain that registers their join, the parchment itself a literal and figural passage between them. This parchment encounter or skin memory reveals the co-making or cowriting of human, other animal, plant, and mineral presents that constitutes what I call a multispecies impression. Crucially for my purposes, Jack's encounter suggests also that the folding together of different beings at the level of flesh comes interlaced with the possibility of retrieving those ligatures, of unfolding or recutting them, so as to make visible, sensible, and accessible the archival function of flesh as the substrate on and with which biopower "writes." Throughout this book, Jack's questions remain my questions: How to distribute that production of pain differently? How to incline that flesh toward the production of less lethal pleasures? What might an inquiry into the way beings are joined or folded together offer to something I hesitatingly name "us"?

SKIN MEMORIES

Meet Jack. Jack Cade, rebel leader and self-proclaimed "parliament" of the Commons. Dick the Butcher is his friend and constant companion. The two are embarked on what usually plays as grisly comedy or parody, an antiwriting utopia that almost immediately turns violent. The whole thing fizzles in the matter of a few short scenes at the mere mention of Henry V. "The name of Henry the Fifth hales them to an hundred mischiefs," moans Jack, "and makes them leave me desolate."[1] It's Dick, of course, who speaks the line that everyone remembers: "The first thing we do, let's kill all the lawyers" (4.2.71). And Jack agrees, "Nay, that I mean to do" (4.2.72). But he pauses and adds these more open-ended lines that momentarily retard the proceedings:

> Is not this a lamentable thing, that of the skin of an innocent lamb should be made parchment; that parchment, being scribbled o'er, should undo a man? Some say the bee stings, but I say, 'tis the bee's wax; for I did but seal once to a thing, and I was never mine own man since. How now? Who's there? (4.2.72–76)

It is not clear what happens on stage at this moment. Does Jack hold the "this" to which he refers aloft, calling all eyes to a parchment? What exactly does he see, touch, feel, or hear as he remarks on the "lamentable" gathering of resources (the lambs and labor) necessary to its production? Uncannily, earnestly, half-jokingly (the script is radically unstable), Jack rewrites the truths of common experience: bees don't sting, but wax that bears the imprint of a seal does. The joke's on him, though, for once upon a time, he took upon himself the singular act of putting his name to a legal document and has "never [been his] own man since." Later, we learn that he probably did not sign but made his mark, "like an honest plain-dealing man" (4.2.94–95).

Jack's skin memory is a common trope in the period, a turning or reordering of surfaces that discloses the splicing together of different kinds of matter and different states of animation (living and dead) that make up the built world.[2] The zoomorphic play, the disorienting dose of reference, the palpable *thisness* of the encounter, produces a leveling effect. Jack looks into the parchment, touches it, and recovers the metonymic chain of variously animate and variously manufactured remains that enables a lamb to sting like a bee. The synesthesic play of the lines registers the phenomenological "feel" of the rhetorical zoomorphism he employs. It registers the presence of a general flesh of being in the reduced forms we routinely accept and put to use: in this case, parchment, wax, and the apparatus of writing but also the deprivations certain human subjects endure so that their social betters do not.[3] When Jack arrives at the seal—the impress of a metallic-backed sign in wax that authorizes the document—he registers its mark as a "sting" and momentarily transforms the parchment back into the living, breathing skin of the lamb or lambs from which it was made. As he does so, he becomes a living relay that offers the lamentation that previously had not been voiced, a lamentation that alliteratively recalls the word we use to name the lamb, its suffering transducted, vibrating within the sound box that is the actor or "shadow," as he was called in the period, who bodies forth the now equally defunct Jack. Perhaps, if we listen very carefully, we shall hear even the faint buzzing of the bees that made the wax, and beyond that, more faintly still, the sounds of the smelting and working of the metal that formed the seal, or even the quiet of unmined metal ores.

Flickering uncertainly between the literal and the figural, Jack Cade's skin memory offers something like a rubric for the mode of description I aim to craft in this book. His handling or performance of the parchment thematizes a politics of the trope or the figure, an understanding of historical process that regards interventions in the writing machine or the figural life of "things" as one of the most important or durable modes of political action. In using the words "writing machine," I seek to remind readers that language remains the first inhuman technology, which both in spoken and written forms serves as a tool for rhetorical persuasion and as an external device for installing memories in individuals and collectives.[4] Deriving from the Greek word "to turn, to change, to alter" *(trepein),* tropes designate a set of relays or switches whose turning or performance—always more and less than we wish for, always subject to chance, to error, and so to the gravity of other things as they pull upon us—choreographs our relations with other beings. By their repetition, tropes spin off scripts, partial forms, figures that keep our worlds the same or enable them to change. By their turning, tropes seed our discourses with possibilities for imagining our worlds differently, possibilities that manifest, sometimes, as dissident, heterotopic inversions, botched dead ends, cast off with a shrug, for they seem to constitute mere nonsense or not-yet-sense.

In act 4, scene 2 of *Henry VI, Part 2,* Jack tropes parchment. In the place of either looking or reading, or by the oscillation of the parchment between the two, the material-semiotic "flesh" of Jack's present, the carving up of the world into differently articulated beings, becomes knowable. In his hands, parchment becomes mobile, plastic, knowable precisely as a point of convergence between matter and metaphor and so as a field of affective intensity that momentarily takes hold of him. Jack time travels. He reverses the mode of production so that the dead-alive animal, plant, and mineral remainders the parchment collates become animate once more. He transforms the indexical and so unremarkable remains of the sheep and bees and mineral ores into partial beings. He registers these remainders, polities that once were, or are, perhaps, merely patterns, with the only substrate he has to hand, by and through the way the parchment impresses itself upon his senses. Perhaps his fingers trace out the remnants of the sheep's wool on the "hair" (grain or skin) side of the parchment, or he is surprised by the smoothness to the "gut" (flesh) side. Parchment pleasures cohabit with parchment pains.[5]

The parchment does not disappear. It does not serve as some static-free conduit to the beyond, falling away to reveal the lamb or lambs, to say nothing of the bees. Handling parchment, holding on to parchment, "reading" into as opposed to reading off of, touching as opposed to tracing your finger along the script to keep your place, not putting the parchment to use, render the world made by parchment objectively knowable, less durable, less effective, and therefore more affecting. The durability of parchment as backing for writing derives, in part, from the wetness that comes with its wear at the hands of the animals we name human. Fail to absorb the parchment into your habit world and you deny its surface the stability that comes from the "aldehyde tannage, the oxidation of fats and oils (from the fingers) . . . which are applied (unknowingly) as" you touch it and it touches you.[6] Jack's open-ended inquiry suspends or retards this use, interferes with the medium's material-semiotic efficacy. Instead, he offers its constellation of matter his voice, vocalizes its components, refuses to treat it as a derived and so naturalized set of use values. Its surfaces vibrate through him, stitching the skin that was, that is, to his as he performs this dense archive of sensation. Biosemiotic relay to parchment—in another vocabulary, its "wetware"—Jack shall deny it the fatty moisture that comes from his fingers and instead weep and wail. But that weeping and wailing is not his own even as he remains the point of enunciation. Neither does it belong to the lamb or lambs. It belongs to the world that parchment makes.

What is Jack in this moment of joining with parchment? Not human. Not any one thing. Bodied forth by an actor on stage or screen, vocalized by you and me as we read his lines or they are read to us, he makes manifest the fictive process of giving voice or face to things. Jack designates one instance of a generalized *prosopopoeia,* the trope that means to give voice or face, the trope of apostrophe, which also "implies that the original face can be missing or non-existent," that it exists only because of its being figured or because of the program of figuration itself.[7] Given the historical moment from which Jack hails, it is tempting to be more specific still and write that he embodies a schoolboy exercise in *ethopoeia*—learning the nuts and bolts of rhetorical performance by impersonating a literary "character," though, here, the *ethos* (character) is not "human" but a collation of different beings.[8] If parchment exists as a marshaling of terrestrial resources to craft a durable archive that takes animals and plants as a substrate to acts of human writing, Jack momentarily reverses the arrangement. He becomes the sensing substrate to the *thing* that is parchment, the matter that registers its impress. He poses parchment as a question, both real and rhetorical. And in doing so, he offers his voice as nothing more than a biotechnical instrument that emits sound or noise as it amplifies stimulus, connection, and loss. His voice testifies not to his existence as some self-identical being with an essence so much as to the brute facticity of his state of animation and his presence to the parchment, *then* and *there* or the successive *here*s and *now*s of the play's performance. I breathe and therefore I shall sing or wail on behalf of this *thing,* this gathering, into which I inquire and which, depending on your answer, I shall contest absolutely.

Nothing is restored. The lamb or lambs do not live again. Jack and his fellows do not suddenly find themselves "free." Jack's skin memory remarks the relationships between different creatures that inhere to parchment, but it does not redeem them. There is recognition but no identification. On the contrary, the differently articulated skins, of Jack and the lamb or lambs, serve as successive orders of media that merely host or "back" one another. The lamb is dead. Its skin remains in tooled form as the backing to one form of writing. So, now, Jack's voice (also dead, bodied forth by the actor) reanimates the lamb in translated form. The nested structure of nonequivalent relations or parasitic effects in which beings serve as a backing or screen for one another remains even as it is acknowledged. Jack remarks on the lamb's suffering, and that suffering is understood to stand in some approximate relation to that of Jack and his fellows, but it may not be reduced to a simple identity. Parchment connects radically different scales of being. The biblical coding to the lamb's "innocence" (Agnus Dei), for example, manifests here both as an historically particular animal, tooled into a writing surface, and also as a sign of a particular order of Christian charity that the carving up of the world by parchment-backed property rights violates, pitting Christian against Christian. The putative universalism of one theological settlement finds itself deployed to object to its particularizing exclusions or, worse, its excluded middles: illiterate lambs, differently literate and so differentiated groups of human animals. Despite their ostensibly shared beliefs, Jack feels more kinship with the tooled skin of the unlamented lamb than with his Christian neighbor who lords over the land upon which he lives. Thus Jack, through this transversal relation to the lamb, by and through the lamb (or is it the flock?), decides to become a subject and a historical actor. Thereafter, Jack and those who flock to him seek to recut the flesh that constitutes their being, opening this infrastructure to other contracts (sacral, natural, pastoral, legal, economic, ecological) and polities than those to which he has had to "seal" his name. We watch as they marshal their hastily weaponized tools into the instruments of an insurgent writing machine that attempts to rewrite or overwrite the world backed and maintained by parchment.[9]

In the wool-dependent medieval and Renaissance England from which Jack hails, the sheepy lexicon that Jack inhabits derived from the shifting relations between variously "human" persons, sheep, cows, goats, dogs, wolves, grass, and other plant actors as large swaths of land were transformed from arable, agricultural use or tillage to pasturage, recoding labor relations, land use, and status in the process. Rival polities of human and nonhuman entities (variously articulated "persons" and "sheep") fought over this coding of labor and land that went by the name of enclosure. In *Henry VI, Part 2,* we watch the metaphorical afterlife of these scripts and the zoomorphic figures they slough off play out. The usually "sheepy" or "sheepish" Commons turns wolfish as it reclothes itself with the skin of its predatory betters. Jack speaks from within this process. "Jack Cade the clothier," says First Rebel, who announces Jack's entrance, "means to dress the Commonwealth, and turn it, and set a new nap upon it" (4.2.4–5)—invert its social hierarchies, that is, and put a new "wooly skin" on it. The word *nap* derives from the language of sheep shearing.[10] The Commons shall be young again, reclothed by the shearing of their betters. Second Rebel has the lingo down pat too. "So he had need, for 'tis threadbare" is his winking comeback.

Jack speaks from within the process. He theorizes the rebels' techniques. He enters the play in act 3 as the skilled mimetic operator, "John Cade of Ashford," whom the Duke of York boasts has "seduced" into a performance of the now defunct heir to the throne, John Mortimer, whose "face," "gait," and "speech" he "resemble[s]" (3.1.371–73). York recalls seeing the "headstrong Kentishman's" performance in Ireland, where, outnumbered by a "troop of kerns," "thighs peppered with darts," he turned "sharp-quilled porcupine" (3.1.363) and then, when rescued, "caper[ed] upright like a wild Morisco, / Shaking the bloody darts as he his bells" (3.1.364–65). Jack's skin-stretching performance (he's a porcupine become hybrid converted Muslim or Morris dancer, depending on which sense of Morisco you hear) turns heads—York's among them—and so he turns intelligencer or spy, betraying first the Irish and now, hopes York, the Commons.[11] The play begins by voiding the historical "Jack Cade," whoever he may have been, in favor of a spectacular series of surfaces or "Cade effects" that York appears to manipulate. Indeed, in York's speech, Cade is recognized and valued as an unruly mimetic agent that parasitically eats its host, obliterating all sense of the original. Fittingly, York's characterization of Jack might describe also the multitemporal palimpsesting or splicing together of words and occasions that creates the script Jack speaks. By turns, he summons phrases attributed to John Ball, Jack Straw, and Wat Tyler's uprising of 1381; Cade's Kentish uprising of 1450; the 1517 uprising by xenophobic apprentices; the Hackett rebellion in July 1591; and the felt makers' revolt of June 1592. Jack condenses a long story of popular protest; he vocalizes an archive of captured speech.[12]

When we meet Jack in the flesh in act 4, he is about to embark on his "becoming Mortimer," just as York had said was the plan. I like to picture York's remote, off-stage satisfaction. Come act 4: he plans to "reap the harvest which that coistrel [groom/rascal] sowed" (3.2.381). But Jack's no mere groomsman to York's horse. Nor are his fellows. Instead, just as Dick commands silence and Jack begins to speak as if a king, deploying a royal "we" that shall enable him to absorb the Commons, Dick and friends carve him up. They subject his name and words to metaplasmic refolding:

> CADE: We, John Cade, so termed of our supposed father—
> BUTCHER: Or rather of stealing a cade of herrings.
> CADE: For our enemies shall fall before us, inspired with the spirit of putting down kings and princes—command silence.
> BUTCHER: Silence.
> CADE: My father was a Mortimer—
> BUTCHER: He was an honest man and a good bricklayer.
> CADE: My mother was a Plantagenet—
> BUTCHER: I knew her well, she was a midwife.
> CADE: My wife descended from the Lacys—
> BUTCHER: She was indeed a pedlar's daughter and sold many laces.
> WEAVER: But now of late, not able to travel with her furred pack, she washes bucks here at home.
> CADE: Therefore am I of an honourable house.
> BUTCHER: Ay, by my faith, the field is honourable, and there was he born, under a hedge; for his father had never a house but the cage.
> (4.2.31–49)

Punning on the alternative sense of "cade" as "barrel," Dick reveals Jack to be a "thief"; chipping away at the mortar in Mortimer, he renders him a "bricklayer"; rearing a "jennet" (feisty, young female horse) from Plantagenet, he renders his mother a "midwife" (a euphemism); unraveling "Lacey" into "laces," he restrings his mother's claim to noble birth into a common trade; transposing a *d* into a *g*, he puts Jack's father in the "cage." Dick the Butcher and company's fractal refolding of Jack produces another "we" as the voices of the rebels make his skin ripple or bubble, emitting a multitude or polity of other voices that will not be silent, that refuse to disappear into this unroyal, nonsovereign "we."[13]

In the breakneck violence that follows, the open-ended idiocy to Jack's inquiry and this crafting of a partial, fragmentary, horizontal, self-ruining "we" get lost or transposed into other registers. The tropes turn. The now encrypted or only partially retrievable forms of insurgent writing come back as the grisly forms of a general violence. The scene descends into something like a gruesome knock knock joke. "How now? Who's there?" asks Jack. And in walks the clerk of Chatham, right on cue: Emmanuel, Emmanuel who "can read and write and cast account" (4.2.79). He carries a "book in his pocket with red letters in't" (4.2.83). The "red letters" refer to the red ink used to rubricate or mark up a text. Emmanuel knows how to write several hands, signs his name as opposed to making a mark. And, as Dick explains, Emmanuel happens to be the phrase "they use to write . . . on the top of letters," deeds, and legal documents (4.2.91). It means "god be with us." Kill all the lawyers? Look, here comes one now: Emmanuel, the walking personification of institutional writing. They lead him offstage, where they promise to string him up with "his pen and inkhorn about his neck" (4.2.100–101). Emmanuel is only the first in a succession of killings all aimed at different forms of literacy, Latinity, and the printed book. Dick the Butcher's knife "razes" (erases) written forms along with all those who can write or sign their names. His knife does double duty. It functions both as a tool for correcting mistakes on parchment by scraping away the surface on which the offending words appear and as a weapon for killing and cutting.[14] Within the play's lexicon, such an equivalence or condensation becomes increasingly explicit as Dick's employment as a butcher comes to metaphorize (or not) his acts of killing. "They fell before thee like sheep and oxen," says Jack of their betters, "and thou behaved'st thyself as if thou hadst been in thine own slaughterhouse" (4.3.3–5), which now comprehends all London.[15]

Come the end, however, this same sheepy lexicon bites back at the rebels. Vanquished by "the name of Henry the Fifth" (4.8.56), Jack takes to his heels. We meet him soon after, on the run, "ready to famish" (4.10.2), as an uncertainly ovine or bovine Jack has climbed "o'er a brick wall . . . into this garden to see if [he] can eat grass or pick a sallet" (4.10.7–8). He is quickly discovered and dispatched by its owner, who just happens to be out for a walk. Jack dies with famine on his lips—"O I am slain! Famine and no other hath slain me" (4.10.59) and "I, that never feared any, am vanquished by famine" (4.10.74). By the end of the scene, we will, in effect, have been watching the retraining of Jack's mouth: no longer the self-predicating "parliament" of the land, the mouth of this sheep turned wolf is denied flesh as he is forced to eat grass. Jack gets to narrate his transformation, his becoming "cattel" (or herd animal) on the way to a becoming "soile" or dirt, the figural process by which he and his fellows are rendered or processed as if sheep.[16]

Want to imagine something different, something better, it is to be hoped? Then, you may need to take Jack at his word and inquire into the lot of sheep as well as the host of entities you come into being with. The material-semiotic routines that render sheep "sheep" and you and I theoretically "human persons" rely on a co-constitutive anthropo-zoomorphism whose early modern forms I have been reading. Tinker with these relays—imagine sheep differently, for example, as something other than an expressed series of use values or a figure for the divine gift—and you may begin to imagine people differently also. Leave one or the other of the terms intact and the figural or material-semiotic passages between them, their ligatures, shall cause you to loop back upon yourself as the usual routines reassert themselves. You have to take care; you have to remember that Jack's question—"Is not this a lamentable thing?"—functions pedagogically as a moment of retarding inquiry. It is not an end in itself. Jack provides no answers. Instead, he narrates the successive troping of the animals we name "human" and "sheep," the "ontological choreography" this lexicon sets in motion. All he does is suggest an opening, the possibility for some other form of Commonwealth or world in common to emerge by and through the actions of his followers.[17]

MULTISPECIES IMPRESSIONS

Welcome to *Of Sheep, Oranges, and Yeast*. Welcome to an orientation that takes for granted that what we call "humanity derives," in Donna Haraway's terms, "from a spatial and temporal web of interspecies dependencies" and that this web holds true at the level of the most evolved of metaphors as well as at that of the genome.[18] I am interested in how modeling our archives as the reduced form of a general or generative text, an unreduced biosemiotic or multispecies impression, might alter our protocols for reading and crafting narratives, for telling stories about those traces and texts that we name "past." What happens if we assume that our texts play host to or are crowded with other forms of writing or marking (human and not), traces that our own acts erase, obliterate, but also render sensible, knowable, precisely by taking so many others as a substrate? Let's imagine that our archives are full of other forms of expression not reducible to the physiology of a human sensorium, modes of inscription that do not allow themselves to be linearized or readily processed but that necessarily we receive by and as they press or impress themselves upon us. They may manifest, as in Jack's skin memory, in the form of animal and plant remainders, as in the nightmares of conservators, in the form of unwanted fungal growths that obliterate the writing on a parchment. Fragments of "bacteria, fungi, protists and such" dally in our archives as in our genome, rendering them a multispecies impression, sometimes a "symphony," in Haraway's terms, but also a cacophony of "benign and dangerous symbionts."[19] Such remainders are keyed to the use differently animated beings make of one another, keyed, in other words, to the processes that build a world. Here the phrase "built world" might be heard to include the reproductive technologies of plants, the camouflage rhetoric of insects, all manner of creaturely display, as well as the mineral efflorescence of rock formations or the movements of a glacier.[20] As Michel Serres observes, "we aren't the only ones to write and read, to code, to decipher the codes of others, to understand, mutate, invent, communicate, exchange signals, process information, encounter one another . . . to thus win our lives. Everything in the world does it." The challenge, then, lies in finding those modes of translation that permit "the invasive order [to] become a reciprocal dialogue" and so pass from parasitism to symbiosis together.[21]

My aim in this book is to imagine a series of scripts for literary and historical study that attempt to own and proceed on the basis of this common relation to coding that knows no stable ontological differences between animal, plant, fungal, microbial, viral, mineral, or chemical actors. I assume that our archives are marked by multiple modes of finitude whose traces remain. I do so rooted in a conviction that tropes *matter* and that the medial loops in which beings come or are habitually required to show up contain within them a fund of other unrealized or latent possibilities that might enable us to proceed on the basis that "to become one is to become with many."[22] Tuning in to the likes of Jack's skin memory,

I trace a series of anthropo-zoo-genetic figures or switches, points of contact or exchange between beings, that make or spin off differently configured groups of animals (human and otherwise), plants, and fungi. These tropes or figures constitute something like an infragenre or switchboard that subsists within and without texts of all kinds, whose surfaces they anchor, interrupt, deform, or cause to ripple in the metaplasmic sense that Haraway offers and that Jack, Dick, and company perform. "Metaplasm is a generic term for almost any kind of alteration of a word," Haraway writes, "intentional or unintentional." Metaplasm "can signify a mistake, a stumbling, a troping that makes a fleshly difference." The word provides no shelter from the fact that the world is populated by inhuman and frequently indifferent agencies not our own.[23] What Haraway likes about *metaplasm* is the "biological taste" to the word. "Compare and contrast 'protoplasm,' 'cytoplasm,' 'neoplasm,' and 'germplasm,'" she suggests, and you will come to appreciate the crosscutting or splicing together of "flesh and signifier, bodies and words, stories and worlds" that the word offers. *Metaplasm* names the technique by which a biological as well as semiotic archive is successively performed to create all manner of beings, extending our sense of the figure into the world of biology, and so expanding our sense of the figure as a tropic actor whose job it is to splice together words and things, signs and matter, to constitute worlds and maintain them.

My method consists of a type of a tracing or unfolding of tropes as they move between and among registers, discourses, and disciplines—a modest supplement or footnote, perhaps, to Erich Auerbach's *Mimesis*. The difference is that I seem no longer to know the difference between types of beings or between foreground and background. Moreover, the figures I trace do not always manage to generate fully or, subjected to a more capacious sense of writing or marking, find themselves strangely disfigured, gone mobile across or athwart the lines of so-called species difference or kingdom. The scenes I stage seem overwhelmed by their staffage or inundated by otherwise than human or "nonconventional entities."[24] Figures *(figurae)*, as a category, still matter. But they count for more than the bodying forth of absent things in media (representations) or even the effects of that bodying forth on a reader or viewer (a memory, an affective response). They designate privileged material–semiotic zones *(topoi),* scenes of writing or marking, from which forms of life issue. These figures are time bound. They produce time effects. But those effects are keyed more to questions of performance or the timing of their activation than to fidelity to a historical period. Tropology orients itself to chronology by way of this spacing, by the way figures organize, outline, schematize, or splice. More importantly, as a trope turns, as a figure generates, it becomes subject to mutation or change. Its performance both over- and underproduces, generating a mimetic excess, fracture, or residue that must be managed. It is to these residues, leftovers, or after-images; their management; and the possibilities they offer that this book attends.

In what follows, it might help if you think of me as hunting or beach-combing, ferreting out this or that fugitive figural turn or trope that enables us to turn our discourses differently. Sometimes my inventorying may constitute a kind of theft or shoplifting, a kleptomaniacal impulse that refuses to let go of something—such as Jack's question. The reason is that none of these figures is ever finished or at an end. Each tropic performance remains tuned to its respective historical moment but also and always turns toward an aesthetic (sensory) domain whose temporality and consistency remain all its own and that offers a potential or energetics regardless of the constraints of time and place. *Poiesis,* performance, the cascade of action, constitutes the wild card in the deck, sporting unintended and sometimes excised forms of being that hover on the edge of sense.[25]

My approach stands surety with a flat ontology that models the world in terms of actor-networks, assemblages, ecologies, or some otherwise associative or additive grid, knot, or mesh. Such models enable us to question the primacy of human language as anything other than a subset of larger systems or codes of reaction and response (olfactory, visual, auditory, and so on) broadening access to the privilege accorded to humans by the order of finitude bestowed by language to include nonhumans (animals, plants, fungi, stones, stars).[26] Nevertheless, I remain interested in what might be gained, even as we provincialize the "human," by maintaining, as Cary Wolfe suggests, that part of what it means to "be," for us, entails owning or being owned by "the radically ahuman technicity and mechanicity of language (understood in the broadest sense as a semiotic system through which creatures 'react' and 'respond' to each other)."[27] When, for example, Jacques Derrida stages "the history of life, of what . . . [he calls] . . . differance—as the history of the *grammè*" in *Of Grammatology,* he begins with the writing event of "genetic inscription" and "'short programmatic chains' regulating the behavior of the amoeba or the annelid up to the passage beyond alphabetic writing to the orders of the *logos* and of a certain *homo sapiens.*"[28] Derrida asserts the governing function of a program that amounts to what he calls, following anthropologist André Leroi-Gourhan, "a 'liberation of memory,' of an exteriorization always already begun but always larger than the trace which, beginning from the elementary programs of so-called 'instinctive' behavior up to the constitution of electronic card indexes and reading machines enlarges difference and the possibility of putting in reserve." It is this orientation to the re-marking or redoubling of the trace, to the constitution of an archive, an ongoing archivalization that both remembers and forgets, that constitutes or renders the "human" as this prosthetizing movement, a reaching beside or within that metaplasmically folds the outside in—the human, as Derrida offers, in another context, always an "infra-human."[29]

Human-oriented writing manifests as one subset to a general question of coding, "and that's why," even as "human" "ex-appropriation is radically different" from other forms of life, "it requires a thinking of différance and not of opposition . . . in the case of what one calls 'non-living,' the 'vegetal,' the 'animal,' 'man,' 'God.'"[30] "Writing," a relation to the archival per se, to the production of an archive, may produce the set of effects or "being there" we designate as and by "human," but there is nothing "human" about it. And with regard to these other forms of life or states of animation with which we come into being, Derrida insists, as Wolfe offers, "not on one line [of demarcation] . . . but many. But not 'no line' either," though this requires admitting, as a starting point, that "the material processes—some organic, some not—that give rise to different ways of responding to the world for different living beings are radically asynchronous, moving at different speeds, from the glacial pace of evolutionary adaptations and mutations to the fast dynamics of learning and communication that, through neurophysiological plasticity, literally rewire biological wetware."[31] Thus, as Wolfe argues, Derrida's refusal of "biological continuism" paradoxically "makes possible a more robust naturalistic account of the processes that give rise to that which cannot be reduced to the biological alone—or even more radically still, to the organic *per se.*" The key difference lies in that the "relation to technicity and temporality," to the movement of the trace, may no longer be reserved to the "human." We have to reckon instead with multiple orders of finitude as they mark us.[32] Human *poiesis* or making represents but one form of a generalized zoo-bio-biblio-processing, the packaging and production of beings that live within and through others that they acknowledge, disavow, ignore, come to love, come to hate, or mundanely never know, even as they take them still for and as a substrate. Thus what we call "writing" or "coding" comes to constitute not a stable, identifiable structure to the world (a referent) so much as a zone or horizon of emergence from which our concepts of life, death, animation, and categories of being (animal, vegetable, mineral, etc.) emerge.

It makes little sense, then, as Wolfe ventures, to exclude other animals or forms of life from this process of exteriorization and so an archive. "Animal behaviors and forms of communication," he writes, "are 'already-there,' forming an exteriority, an 'elsewhere,' that enables some animals more than others to 'differentiate' and 'individuate' their existence."[33] It might be ventured, then, that the question as to the threshold or internally divided and marked line between a supposedly automatic reaction and an eventful, organic response, to what Derrida calls an "abyss" or "limitrophy," will tend to present to us and is allied to what comes to count as an archive, and that the question will, in one sense or another, be decided by the way in which we decide this archive's limits. Far from a simple repository, as Derrida reminds in *Archive Fever* and Jack's tuning in to the sovereign media ecology of his day recovers, the archive remains tied always to an "anarchivic" and "anarchiviolithic" violence or erasure, even as it orients itself toward the future.[34] The archive, as it were, names the site of exchange between the worlds of flesh and discourse, a contact zone from which different orders of life and death effects, states of animation, emerge, as well as traces of the acts of writing or coding of the host of others we come into being with.

In modeling our archives as coterminous with the articulation of different forms of flesh, I take up Wolfe's invitation in *Before the Law* to think "something like a distinction between *bios* (so-called 'human' life or the life of the social) and *zoë* (the facticity of life) that obtains within the domain of domesticated animals itself." I am interested in the ways different beings or forms of life derive from the crosscutting or passing back and forth between *bios* and *zoë,* within and across classificatory lines such as the discourse of species and kingdom and so to the possibility of modeling a more "highly differentiated biopolitical [or zoopolitical] field."[35] "What we need to remember," he elaborates, "is that biopolitics acts fundamentally not on the 'person' or the 'individual,' nor even, finally, on 'the body,' but rather at the more elemental level called 'flesh'" through "which 'the [human] body' is both sustained and threatened." This material-semiotic "flesh" constitutes the "communal substrate shared by humans with other life forms" as they are parceled out in an ongoing biopolitical articulation and writing of the world that crosscuts between them.[36] "It makes little or no sense," for example, he continues, "to lump together in the same category the chimpanzee who endures biomedical research, the dog who lives in your home and receives chemotherapy, and the pig who languishes in the factory farm." The *bêtise* or asininity of "*the* animal" as an undifferentiated category "masks . . . the transversal relations in which animals, and our relations are caught under biopolitical life." And, for Wolfe, this insight offers a way of explaining and approaching the curious circumstances that lead, on one hand, to the "billions of animals in factory farms, many whom are very near to or indeed exceed cats and dogs and other companion animals in the capacities we take to be relevant . . . (the ability to experience pain and suffering, anticipatory dread, emotional bonds and complex social interactions and so on)," and, on the other, to the advent of health insurance for companion animals, who "are felt to be members of our families and our communities, regardless of their species."[37] The issue here is parsed most carefully by Nicole Shukin, whose *Animal Capital* takes as its burden the project of demonstrating the way the "discourses and technologies of biopower hinge on the species divide" so as to enable the "zoo-ontological production of species difference as a strategically ambivalent rather than absolute line, allowing for the contradictory power to both dissolve and reinscribe borders between humans and animals."[38] We are all, then, mixed beings, sharing characteristics that belong to one another, characteristics that wander across the lines of species, kingdom, and kind.

Different polities of humans and other animals, plants, fungal, microbial, and mineral actors, form intersecting and rival multispecies whose attempts to "write" the world mark our discourses as their variously backed impressions or forms of writing are folded together. There seems little reason, then, to limit the purchase of biopolitics, or now zoopolitics, the crosscutting between *bios* and *zoë,* to the flesh of animals, even as extending its reach to plants, fungi, and beyond may strain the limits of what we may be said to include within a human household or collective and cause the "limitrophy" between "reaction" and "response" to bloat to still more alarming proportions, "complicating, thickening, delinearizing, folding, and dividing that line precisely by making it increase and multiply."[39] Why not assume that what we call biopolitics and biopower manifest merely as historically particular forms whose anthropic orientation, if not a misnomer, is not part of their definition? Do other forms of life practice biopolitics or zoopolitics? Plants assuredly do: their very forms (cut and cuttable fruits and flowers) address themselves to the animate or at least ambulatory beings we call "animals." It would seem inhospitable, then, to begin on the basis of a reduced sense of the biopolitical even as such reductions might need to be made or have determinate, historical circumstances that require their reduction. Accordingly, I take as my focus three differently scaled and differently animated, material–semiotic and rhetorical actors (sheep, oranges, and yeast) whose biomatter marks the archives we inherit and the trace-chains we live at multiple points. The chapters that follow take the form of a series of cascades that trace the remains of variously "sheepy," "orangey," and "yeasty" archives, figures which refer to partially retrievable routines or recipes for making the human animals we name persons via a process that makes also those entities we name "sheep," "oranges," and "yeast." Along with all manner of differently configured "sheep," "oranges," and "yeast," the cast includes a host of variously "sheepy," "orangey," and "yeasty" persons.

Why sheep, oranges, and yeast and not some other constellation of actors? The answer remains, in one sense, arbitrary. Choose other entities and you shall tell other stories, find yourself charting different courses. That said, focusing on three differently scaled actors from different biological kingdoms (animal, plant, and fungus) allows me to explore the way in which the scaling of our relation to these different types of entities produces differently configured biosemiotic archives (bodies and texts). Differences in scale canalize the modes of relation between proximal forms of being and decide the extent to which we may be said to share (or not) a common world, as Jakob von Uexküll's famous modeling of what seems to us the highly delimited relation of a tick to its environment *(Umwelt)* makes plain.[40] Differences between beings traditionally understood to derive from ontology (animal, vegetable, mineral, and so on) cease to have any necessary explanatory or classificatory function once all beings are understood to be differently scaled encodings of information (technical, bioinformatic, social). Singular "difference" plays out instead as plural "differences"—differences in scale, form, connectivity, and mobility—that derive from how these beings are distributed and differentiated, in relation to each other, within themselves, and within their own groups. The differing morphologies of sheep, oranges, and yeast make possible but also limit the transversal relations that form between beings even as they offer resources for possible identifications. Still other modes of relation exist or are configured at levels of articulation that may or may not be recognized or registered consciously—via the sweetness of oranges to us; the fragrance of their trees; the stench of their rot; or via the effervescence of a beer; the smell of yeast as it ferments; the texture of the dough you knead; the bread you eat; or, for that matter, the burning itch of a yeast infection.

Some of these forms produce hospitable, because reversible, symbiotic joining, mutual porting, weird identifications, transversal movements, hybrid hospitalities. Still others orphan whole species in the worst kinds of one-way extinctions and genocides. No guarantees can be made as to the progressive or even positive cast to the modes of association that an orientation to the multispecies might yield—even as the term itself has within it a pull or vector toward a "becoming with" or to what Vinciane Despret regards as self-conscious modes of "anthropo-zoo-genetic practice" rooted in the cultivation of trust or respect. The process remains, as in Serres's model of the parasite, avowedly neutral, even as he longs for symbiosis or merely some modality that may prove less lethal.[41] There exist and will continue to exist positive and negative forms, hospitable and abusive multispecies configurations. "Man," as the adage goes, "is a wolf to man," but sometimes the teeth that do the biting, as Jack knows, belong to sheep. Biopolitics serves as the reduced form of an ecological or zoo/bio/political production of worlds that articulates multiple forms of "life" that crisscross the lines of species, pitting different alliances of human, other animal, plant, and other actors against one another.[42]

Here it seems important to own the way, for Haraway, the "multi" or "companion species" and the "mess-mate" do not refer to stable configurations of beings so much as they serve to trope the discourse of species itself to offer a mode of figuration that proceeds on the basis that "to be one is to become with many."[43] The word functions as a *catachresis,* a misuse or even abuse of terms, that gestures forth to designate something but not to delimit it, leaving the *thing* it designates or gathers open to name itself by and through our encounters with it. As Haraway tracks the word *species* through different registers, from logic to biology or zoology, she finds that it "contains its own opposite in the most promising way," referring both to "the relentlessly 'specific' or particular and to a class of individuals with the same characteristics."[44] The word figures a convergence between a hyperattention to particulars (an inventory of differences, and the differences such differences might make) and the possibility of a universalizing type or typing of forms. Debates as to whether such forms are "taxonomic conveniences" or refer to "earthly organic entities" enact the mutually coextensive or associative logic at work, such that "species is about a dance linking kin and kind," about the worlds made possible by different models, dance steps, routines, or inputs. Categories emerge from the process of positing species; ontology manifests as an ongoing, necessarily unfinished or aporetic product subject always to renegotiation. Not quite "under erasure," as the saying goes, still, the successive paw prints or otherwise impressed tracks of all the beings the term *multispecies* gathers shall, hopes Haraway, erase its previous iterations. Projective or compositional in its cast, species designates not a mutually agreed upon or closed set of categories but an ongoing "matter of concern." The oscillation between the drive toward particulars and the desire for universals (however botched) marks less a contradiction than a signal that the discourse of species consists of an ongoing poetic process of making and manufacture, a dividing up and folding over of a general flesh to produce differently configured beings whose limits remain in play and are subject to debate.

Following Haraway, the positing of a multispecies does not represent an always already compromised anthropocentric endeavor but, in this altered form, an open system or inquiry, the posing of a set of questions to the various beings we encounter as to what counts for them, whether, indeed, there is something that might be classified as a "they" or, for that matter, a "we." In so doing, the figure of the multispecies comes to look a lot like Jack and friends as he winks at his fellows and deploys an absorptive royal "we" that is successively recut by Dick and company's voices, though we may have to change what we understand by "voice," "face," and any number of terms that tie us to a metaphysics of presence as opposed to one rooted in performance. "Jack," as I have emphasized, remains a character, something personed by an actor who speaks certain sentences and cultivates certain poses.

And, still, the structure of Jack's retarding question, "Is not this a lamentable thing," obtains. That question shall be answered by whatever happens next. The period of questioning it opens does not endure. Instead, it indicates a time-bound period of possible inquiry. The syntax it leaves open will close and yield an answer that decides the limits as to whom and what shall be lamented, *here* and *now,* and whom and what Dick's knife shall "cut." The work of redescription, assuming that to be for us is to cohabit with other forms of finitude, does nothing to lessen our obligations.

HOSPITABLE GRAFTS

The temptation, then, will be to refuse the "cut" of Dick's knife entirely. We might even panic. Traumatized by all the calls *(phone)* that we receive from other beings that we once thought were at a distance *(tele)* but that now resonate within us *(infra),* the "human" might find itself (like Jack) transformed into some imperfect but potential receiver for all manner of signals from a world that thus far has been processed as noise or static. There may, after all, be as many as twelve thousand years' worth of dropped or blocked calls from other forms of life, depending on how you care to date the beginning of the Holocene and of agricultural practices. Such an infinite receptivity might constitute an order of consoling *askesis* as we somehow try to make good on an infinite obligation that would render us whole, if not Holy.[45] We might find ourselves embarked on a never to be completed inventory of all the variously animated beings whose remainders populate our archives, our bodies, and our psyches—something on the order of a self-contracted, or auto-archiving, fever, a will to document proof of life *(zoë)* everywhere, as if the term comes to possess some infinitely translatable exchange value that equates to the good/s. We would amass further and further redescriptions of our practices in the hope of securing some realizable technique or technology of witness by which we might make good on an absolute or immanent mode of hospitality—the symbiosis after which Serres inclines. Or we might attempt to outrun the cut entirely by crafting some technique that aims to reach into *poiesis* itself and to adjust our acts of making as we make, seeing off the errancy that haunts the cascade it sets in motion and thereby the production of abusive worlds. We could attempt, as it were, to archive our acts of making before they happen in some exhaustive management and evacuation of risk. In such a way, all our inventions, our attempts to make a world, would never have to surrender their prospects and promise to the wild card phenomenalization of their projects.[46]

But what if the "cut" of Dick's knife is in fact constitutive and necessary? What if belatedness is simply our *milieu,* our particular order of finitude, and with that comes an inexhaustible burden or debt as well as a host of opportunities? As Wolfe writes in the closing cast to *Before the Law,* "discrimination, selection, self-reference, and exclusion cannot be avoided." Decisions (cuts), in all their madness, shall get made, indeed, have already been made. Thus, like Jack, we shall be left, after the fact, wondering or having to inquire into whether a historically specific *this* as it presents itself to us *now* constitutes a "lamentable thing."[47] For despite all our efforts, still, in a moment of madness and blindness, we shall have to act, to call for the "cut" of Dick's knife, even if we choose to process that "cut" as the scraping away of one surface or layer that we now regret. For "hospitality to be hospitality, to be real," continues Wolfe, it "must be something 'determinate' and 'conditioned'; my laws will not protect you if they aren't."[48] If figures of hospitality stand hostage to an either–or structure, "*either* unconditionally embrace all forms of life as subjects of an immunitary protection, *or* suffer the autoimmune consequences that follow," then what gets missed, he argues, is the way the drawing down of biopolitical boundaries, deciding the boundaries of citizenship or belonging, "is precisely the condition of possibility for any possible affirmation, thus opening the community to its others, potentially all its others, wherein reside the inseparable possibilities of both promise and threat."[49] This is precisely not a "cop out" that defaults to human primacy even as it may constitute an *alibi,* a defense that explains why we were not where we were supposed to be, why we failed to acknowledge the presence of so many others that we came into being with. To put things very directly, like it or not, "we *must* choose, and by definition we *cannot* choose everyone and everything at once. But this is precisely what ensures that, *in the future,* we *will have been wrong*" and shall have to begin all over again. The very instability in knowing the difference between an automatic, machine-like "reaction" and an apparently organic, vital, or living "response" means that the "act of selection and discrimination, in its contingency and finitude . . . can never be juridical," and we find ourselves returned to what might be termed a foundational *aporia* keyed to our own particular order of finitude.

Ethics and politics find themselves (as they always were) rooted through technics and media, through the various switchboards that connect and disconnect calls and so demarcate the limits of answerability—the routinized distinction between a "who" and a "what." There's no scandal here. Such is merely the result of the fact that Being comes hardwired. Decisions, as the unmoored sense of Wolfe's governing "we" indicates, shall be made. Indeed, it is the business of our infrastructures to routinize them, to create quasi-automatic "cuts" that eventuate the continuum of flesh. Decision ends up spliced with our poetic acts, both in the literal sense of making and manufacture and the conceptual territories with which they are allied such that merely to make something is to unmake something or someone else.[50] Hence the proliferation of altered regimes of variously posthuman description that explicitly question the criteria by which this "we" might be configured, such as Bruno Latour's actor-networks; Haraway's "cyborg," "companion," "multispecies," and kin-making inquiries; Jane Bennett's vibrant or "vital materialism"; Stacy Alaimo's "transcorporeality"; Timothy Morton's burgeoning lexicon of dark ecology and hyperobjects; the "*zoë* egalitarianism" of Rosi Braidotti; and the "agential cuts" and intra-action of Karen Barad—or any number of inspiring projects that take up the burden of exploring which modes of imaging, visualizing, sonifying, or animating an object (and thereby also its analysts) create "ethically," which is to say also, as Isabelle Stengers observes, "technically" well-modeled experimental subjects.[51] These modes of description essentially replay Jack's retarding language of Commons and Commonwealth, of utopia become "cosmopolitics" or "ecology," as we attempt to craft translation tools that inquire into the way our collective flesh gets written. When the likes of Bruno Latour, for example, issues a call for the crafting of new "speech impedimenta" or ways of speaking that might animate a potential citizen if not subject in his parliament of things, or when Haraway asks us to think about the mediatizing of entities by way of critter-cams, duct tape, or agility sports for the dog–person companion species, we are being invited to try out new rhetorical and technical means by which to transform noise into news of an other, to tinker with the relays in our collective writing machines.[52]

Welcome, we might say, by extension, to a model for the university campus or intellectual commons of a reconfigured posthumanities that reorganizes itself so that its various disciplines, two cultures or not, are understood to represent different skill sets that each analyzes a segment in the life cycle of some *thing*. Questions of metaphysics as they have traditionally been posed within the humanities turn out, all along, to have been a reduced form of inquiry into a general *physis* that knows no stable boundaries between kinds—animal, plant, fungus—or between differing states of animation: organic and inorganic; the living and the dead. Likewise, questions that play out in the life sciences as matters of technique or investigative protocol turn out to have within them moral philosophical and metaphysical scripts, blatant or concealed, that continue to shape our encounters with the world. It is here that the "thing" we name a "literary" or "cultural critic" might be productively retasked or reunderstood. I argue that we find ourselves reterritorialized in questions of form, rhetoric, genre, and translation, understood now as ways of moving, ferrying, or shifting things (persons, concepts, plants, animals) between and among different spheres of reference, leading us to focus critical energies on our ways of making things speak to and of "us" and, so, in a reunderstood or renewed sense of aesthetics.[53]

For those of us housed in the humanities, in the semiotic or rhetorical charnel house of the collective, and who are trained, like the actor who plays Jack, to rake through the bones or to reinflate dead skins and make them speak, and so to produce effects of liveliness in our variously timed "presents," it seems important to take ownership of the ideological function to our discourses and to the function of humanities-based research within a larger ecology of practices or disciplines and to render that function not in terms of the "human" as some retrievable and viable content but as the effect generated by a succession of different routines (rhetorical and semiotic) and by different media ecologies that foster (or do not) worlds hospitable to the host of others we come into being with. Paradoxically, then, even as human language might be understood to constitute merely one instance of a generalized question concerning coding, its privilege deterritorialized across the boundaries of species and kingdom, I argue that we find ourselves reterritorialized precisely in questions of media, of form, genre, and trope. Such, for me, would be the expertise of something named a "literary critic" in the intellectual commons that the posthumanities might be said to convoke.

A strategic difference, however, between Latour's positing of a "parliament of things" and the projects of many of us housed in the humanities resides in the way we find ourselves oriented to our objects of study. Tuned to things past, to the fragments of chains of making long since severed or attenuated, partially interrupted, and so to actor networks that have dropped actants as they have added new ones, we are obliged to deal with the fractured objects that result from these dropped connections. It is these texts or traces, these partial connections, that we take as our points of departure. We serve, in Latour's terms, as "avatars" of the freeze-frame or, to speak an allied language, "vicars of [lost] causations" or causations gone missing.[54] Our object remains always the archive of a practice, the remnants of some thing, which, by our joining, we only partially re/activate, alive to the ways the figure of the archive itself as actor-network enables certain modes of joining and disables others and so makes certain worlds or prospects un/thinkable. Our expertise remains "generic" or "rhetorical," then, in the best sense of the terms, "virtual," a matter of the archive, of translation, of the forms ideas take as they travel. What we have called the "human," then, becomes one set of screens, merely, a translational relay, or a node that registers the world our acts of making render. The human as a concept idles; and the labor of thinking, of feeling, and so of reworking our discourses, begins again, all over again, from the beginning. Such is the idiotic hiatus or pause that I see Jack introduce into Dick's cut, a proroguing or begging of the question that Dick's knife shall nevertheless settle. What have you unmade by your acts of making? And how must we begin again?[55]

It is to this project of redescription, of having to begin, in Wolfe's terms, all over again, that *Of Sheep, Oranges, and Yeast* seeks to contribute, preserving the form of Jack's question as I move across supposed divides between differently animated beings, charting their crosscutting or interlacing through the general flesh that we receive parceled out into different archival forms: Jack's bleating parchment world, the fruit twinned with a color that we name "orange," the yeasty rising of processed grains that we name "bread." I can promise little by way of consolation for such "cuts" that come. I shall not escape the syntax of Jack's question as I search out emergent, time-bound forms of an immanent hospitality as they mark our archives. The contents of this book remain meagerly insufficient to the task for, like Jack, I become little more than an occasion for such remainders that haunt our archives to presence. I comb the archive searching for hospitable grafts, allowing the particularity of sheep, oranges, and yeast to dislocate the tropes that co-make us. Such moments of possibility that I find prove small even as their effects may resonate. But you have to tread carefully. Tropes turn, but they also turn on you, spinning off effects that improve the conditions of one of their players at the expense of still others.

I am under no illusion that I shall speak for sheep, oranges, and yeast. That is not the point. We know that the "other" is always rendered voluble, despite its silences, made to speak, forced to speak, summoned, or variously rendered. The material question remains tied to the sites of enunciation—the media-specific forms by which this news of possible other voices arrives. But neither am I quite ready to surrender the question of what sheep, oranges, or yeast might be said to "want" or "lack," even as I do not have even a fleeting glimmer of the syntax that would allow me to pose that question.[56] Instead, I retain that question of voice as an unruly, empty set, a shifting, multiplying set of lacunae or tubercular growths within my flesh that taxes my feats of description to remind me that we actively constitute a world by and in our successive acts of writing. One day this empty set may be recognized as full or otherwise phenomenalized—sheep, oranges, yeast, may, in fact, already be asking for things, but in ways I do not yet comprehend—and when it is, our discourses, the "conditions of production of knowledge" and the "conditions of existence" for us (all), will be irrevocably changed.[57] It is for these reasons that I emphasize that, for now, my object remains tied to our particular order of finitude, to the archive, understood broadly as a multispecies contact zone, the material-semiotic and rhetorical "flesh" on and with which biopower writes to produce different "backed" forms of life that crisscross the lines of species and kingdom.[58]

It might be best, then, if you consider this book a kind of wide-eyed theft, stealing the material-semiotic-rhetorical chains we name "sheep," "orange," or "yeast" to see what other ways of being we might spin off from them.[59] Is it possible, for example, to find instances of multispecies being, "writing," or *technē,* "whose effects," as Timothy Campbell writes, "cannot be measured solely in mastery"? Is it possible to imagine multispecies writing machines or anthropo-zoo-genetic practices whose cultivation of trust and mutual respect unfolds as practices of "attention and play"?[60] Accordingly, I proceed on the assumption that nested within our discourses remain caches of possibility, small as they may be, that might provide resources for such an endeavor. Of course, thief that I am, whatever I learn, whatever feats of *ethopoeia* I undertake, athwart the lines of kingdom and kind, I remain, inhospitably, in receipt of stolen goods. Such is what it means, for me, to be, to eat, and to read.

If, as I go, it seems that the scenes I juxtapose straddle too great a historical distance or decouple questions of biopolitics from their usual moorings, then consider that I have been directed by my attempted fidelity to sheep, oranges, and yeast. Sheep, oranges, and yeast operate at different scales and speeds than human animals. They obey different chronologies. And their respective physiology or ways of being distribute them differently with regard to the biopolitical as more and less readily fungible to the animal we name "human." It seems only fair, then, to let them outface me, to allow them to direct me down what may appear to be sidetracks, cul-de-sacs, dead ends, that might still constitute alternate paths. I make this point not to assuage concerns as to the historical specificity of my claims but to announce that to travel by trope or by figure means that distances accentuated by chronologies tied to human finitude (politics, economics, epistemic categories, sexuality, and more) are calibrated differently for sheep, oranges, and yeast. Historically, the distance between two points may seem immense. Fold things differently, however, figuratively, according to other scales, and that distance disappears or reveals itself to operate differently.

Part I, "Sheep," takes up the biopolitical quotient of Jack's skin memory to disclose the undergirding *oves* (sheep) to the *omnes et singulatim* (all and one by one) logic of pastoral power. Chapter 1, "Counting Sheep in the Belly of the Wolf," unfolds Jack's skin memory by tracking the figure of the talking or animated sheep summoned to tell the truth about human labor relations back to its origins in the utopian discourses that subtend the worlds of pastoral and pastoral care. Inhabiting the defining scene of our sheepy shepherding, of biopolitics, the chapter unfolds as a counting, a marking off and enumeration, not of sheep exactly, but of hybrid sheep–human figures. Is it possible, I ask, to mine these figures for other ways of being (sheepy and not)? Chapter 2, "What Was Pastoral (Again)? More Versions (*Otium* for Sheep)," attempts to alter this scene of counting, of logistical calculation, and the articulation of sheep as living stock, by digging in to our discourses to locate those moments that inquire into what might be said to "count" for sheep. Key here is the way the articulation of sheep and human animals as coeval multiplicities produces a set of conceptual resources for plotting the relation between leisure and labor, play and work, that comes routed through scenes of pastoral deactivation or *otium.* With *otium,* I argue, we encounter a zoo-anthropo-genetic machine that coproduces "sheep" and "human" animals by and through their relation to a grounding plantlike *phusis*. *Otium* names or designates a figure of pure growth to which all creaturely life might be expanded (or reduced) but is programmed also to differentiate that "life" into different forms. Enter the capital fellow that is the apparently animate, agentive, human subject leading a parade of apparently abject, sheepish sheep, sloughed off from this counting that projects and then eventuates the vegetal equivalence it posits. Combing through the idylls of pastoral and its competing definitions, I offer a reading of ethologist Thelma Rowell's studies of, or with, sheep as the latest chapter in the long history of pastoral. I argue that with Rowell's studies, we encounter a radically altered ecology of practice in which classical themes and techniques find themselves rezoned to produce a writing machine that attempts to own its anthropo-zoo-genetic function. I argue that Rowell invites sheep to write their own forms of pastoral or, perhaps, epic. And this rewriting allows the drama of subjectification to play out across the lines of flesh as it is parceled out in animal and plant forms.

Part II, "Oranges," shifts focus to the differently scaled world of plants and their reproductive technologies as those may be said to mark our discourses. In chapter 3, "Invisible Inc. (Time for Oranges)," I inhabit the archive of letters, memoirs, bills, reports on security, and more generated by a prison escape from the Tower of London tuned to the way in which these texts are calibrated by the presence of oranges that were used to make the invisible ink that allowed the inmates to circulate letters planning the escape. My aim is to produce something on the order of a multispecies monograph or singularity writing tuned to the distributed remains of oranges that populate this repertoire of texts, inflecting the relations between ostensibly human players. Whole and cut, cut and pieced, whole again, oranges—the individual orange always morphing back into a multiplicity or stream of oranges, differentiated only in the moment of their consumption or use—punctuate the story. Discursively, the chemical properties or efficacy of orange juice as a medium for secret writing come to fund all manner of revelation effects, as in the bringing out of secret writing, in ways that demarcate the lines of competing polities of human actors. Reading oranges back into the stories told to explain the escape allows me to apprehend the multispecies basis to a struggle most often parsed in terms of confessional difference or as an episode in the early life of the prison system. Oranges serve to differentiate these competing polities whose interests, existence, and political efficacy rest on their ability to deploy the orange in a variety of roles. Lurking within the story, however, remains the chimerical figure of a warder who, so we are led to understand, merely likes the fruit. The chapter ends by inquiring into this liking, this fetish, or addiction for oranges, or, as one of the escapees has it, "golden apples." Chapter 4, "Gold You Can Eat (On Theft)," charts the arrival of oranges or "golden apples" in Western Europe as a formalizing moment in the life of citrus that collides with the emergence of the commodity form. Oranges, golden apples, gold, as it was frequently remarked, you can eat, interrupt exchange. Orange desire, orange addiction, the allure of orange and oranges, both traumatize and inspire modes of association between human subjects and litter our discourses with excessive orange remainders—such as the orange writing I discuss in the previous chapter. Structured as a *florilegium,* or book of flowers, the chapter constitutes an archival heterospace keyed to the ways "orange," the dispersal strategy of a particular genus of plant, by and through its recruitment of human animals, comes to interrupt acts of exchange, inclining them toward an economy of the gift or accusations of theft. Charting these thefts, or their tropic deployment, offers a way of approaching the existence of small-scale, erring, time-bound polities, whose mode of association becomes *thing*-like, unfolding by and through orange, in ways that reroute the flows of matter that sustain their governing infrastructures.

Part III, "Yeast," shifts scale again to focus on the invisible or only partially visible world of yeast as a fungal actor in bread production. In chapter 5, "Bread and Stones (On Bubbles)," I examine the alliance that formed between yeast and early modern civic authorities to ensure that the daily bread for which all Christians pray showed up at the appointed intervals. Keyed to the suprasign of the Eucharist, the figure of a quotidian daily bread was premised on the animation of a yeasty ferment that enabled cities to regulate the price of bread in an economy in which the price of wheat and other commodities might fluctuate depending on the conditions of the weather, the relative supply of wheat for flour, and a host of other factors. I explore the affective relays that formed based on the growing assumption of the availability of bread such that, by the advent of modernity, bread becomes twinned with the figure of the paving stone as two things that remain the same wherever we go. With yeast, with things that bubble, ferment, and inflate, we encounter processes of serial repeatability that produce moments of apparent stillness and so pass "as if" stone. What drives this chapter, then, is the way our sense of infrastructure, of shelter, of a convoking series of relays that constitute our built worlds, derive from alliances of animal, plant, fungal, and mineral actors. What, I ask, might happen if this convoking function were rezoned, pried loose from its sacramental theology, and understood, instead, to be a space of encounter with the multispecies basis of our lives, here, with yeast, as we embark on one mode of microbiopolitics?

The book ends by returning to Jack's question and to the seeming inevitability to the "cut" of Dick's knife. This set of erasures returns to the matter of hospitality that haunts our discourses as we attempt to imagine the contours and limits of the posthumanities. I do so by way of a table at which lambs, oranges, yeast, and human subjects once all sat or found themselves "convoked." It remains, of course, a table riven by its losses, a table marked by a set of partial and fading presences, prints, or impressions. The project of this book is to render this humanist table urgent, a scene of possible transformation, of multispecies table talk.

I
SHEEP

1

Counting Sheep in the Belly of the Wolf

At any moment I shall be able to throw off with a single shake the coat (half donkey hide and half sheepskin) which grows over one's true and natural skin under the influence of the environment. . . . This winter I went for almost three months without seeing the sun, except in some distant reflection. The light which gets through to the cell is halfway between the light of a cellar and that of an aquarium.

—ANTONIO GRAMSCI TO GIULIA, Milan Prison, February 27, 1928

Ideas do not die. Not that they survive simply as archaisms. At a given moment they may reach a scientific stage, and then lose that status or emigrate to other sciences. Their application and status, even their form and content, may change; yet they retain something essential throughout the process, across the displacement, in the distribution of a new domain. Ideas are always reusable.

—GILLES DELEUZE AND FÉLIX GUATTARI, *A Thousand Plateaus*

The model of knowledge imagined by Gilles Deleuze and Félix Guattari in these lines represents a dispersed and distributed field. Ideas slide, shifting or mixing as they pass, but they retain something of their essence in the passage. A displacement from chemistry to popular culture, for example, from physics to hypnosis, marks a fundamental change in state, reanimating an idea now as a hybrid of interfering times and places, at once of the "future" even as it is marked by the "past." Forms matter. Matter moves.

Speaking a different but allied language, Michel Serres remarks this dispersal with a language of thunderous acceleration. "Time," he writes, "does not always flow according to a line . . . nor according to a plan but, rather, according to an extraordinarily complex mixture, as though it reflected stopping points, ruptures, deep wells, chimneys . . . rendings, gaps."[1] "Once you accept this," he adds, "it's not hard to accept the fact that . . . things that are very close can exist in culture, but the line [of chronological time] makes them appear very distant from one another."

My aim in this chapter is to chart the passage of one "idea" that has traveled from the texts of Renaissance humanist Thomas More into a series of sites in popular culture, political theory, and scientific practice. The "idea," or, more correctly, trope, is of a variously animated sheep, individuated from the flock, summoned to the table talk of its animal masters and made to speak the truth about human labor relations. You had your first taste of this trope when I introduced you to Jack Cade, rebel leader of the Commons in *Henry VI, Part 2*. Jack's encounter with parchment provoked a retarding question about the folding together and parceling out of different forms of flesh to produce the Commons, their betters, their world. "Is not this a lamentable thing," asks Jack, "that of the skin of an innocent lamb should be made parchment; that parchment, being scribbled o'er, should undo a man?"[2] He pauses and begins to unfold the various beings he finds there: the dead lamb, tooled into a writing surface, tattooed with bee's wax impressed by a metallic seal that "stings."

Why sheep? Why sheep "talking"? This animal summoning may seem an esoteric choice were it not for the fact that we are surrounded by sheep, sheep made to talk, and to talk about us. In *Capital,* Karl Marx explains the "value-relation" of a coat to the linen from which it was made by analogy to the way a Christian's "resemblance to the Lamb of God" is shown in his "sheep-like nature."[3] Marx is being mean, obviously, but he alludes to the way the conjoined history of pastoral and pastoral care as allied discourses in figuring human labor in the West is predicated on a series of sheepy metaphors that manages successive multiplicities. The purchase of biopower proceeds on the basis of an articulation of human subjects in terms of population, a reckoning and risk management of the collective flesh predicated on the statistical tracking of birth and death rates and now all manner of data points that recalls, at some distance, the scene of shepherds counting their sheep. We live and die, then, as if or in reference to "sheep," even as many of us also play the part of literal or figurative "shepherds," shepherds to sheep and "shepherds" to human not "sheep." Charlie Chaplin captures the point vividly when he deploys the recalcitrance of a lone black sheep against the accelerated and attenuated temporality of the assembly line in *Modern Times* (1936).[4]

The film opens with the image of a flock of sheep (Figure 1a) that almost immediately dissolves into the image of commuters heading up from the subway (Figure 1b), the two images syncopated by the use of the same down-to-business, percussive sound track. The sheep never reappear. But this wrapping of images and beings that march to the same beat provides the governing basis or ground for all that follows. Indeed, when we meet the Little Tramp, his movements calibrated by the machine of the assembly line, that assembly line figures an abusive iteration of the flock, the articulation of sheep and their not-sheep cousins as a fungible set of resources, the animate relays of our machines, a laboring-power that might also be alienated as so much live or living "stock."[5] It's this same latent "sheepiness," however, that enables us to remark the Little Tramp in the first place, to pick him out of the flock or the crowd. The appearance of difference among the ostensibly not-sheep conglomeration of commuters, the irregular black banding of certain fedoras and not others in a sea of otherwise white and tan hats, condenses into a binary black-and-white difference that births our hero, for whom we wait, for whom we shall care, for whom we care a great deal already—he is, after all, Charlie Chaplin. Birthed by and through the turning of this sheepy trope, marked out as that heroic black sheep, the Little Tramp shall grind the assembly line to a halt and seek his own forms of utopian cancellation, living on the edges or within the interstices of a city that his gestures appear to cancel.

Shakespeare plays the same sort of trick as Chaplin, of course, at the end of *Henry VI, Part 2*. As you remember, Jack and Dick the Butcher set out to recut their world, to alter its relays, but their crafting of a purely horizontal, nonhierarchical "we," an unroyal, nonsovereign, collective historical subject, doesn't get very far. The state fleeces them with its

a

b

Figure 1. Image captures from the opening sequence of Charlie Chaplin's *Modern Times* (1936), featuring commuting factory workers "dissolving" into sheep and then returning to their human selves.

shiny ideological lures ("Henry the Fifth") and the Commons return to their figural sheepy lives. Jack exits the play on the run, hungry, dying of hunger, forced to eat grass, narrating his transformation and registering thereby his subjection to the "paradoxically distributive side of the Christian pastorate" that, even as it attends to the salvation of each and every sheep, accepts "the sacrifice of a sheep that could compromise the whole." "The sheep that is the cause of scandal," Michel Foucault explains, "or whose corruption is in danger of corrupting the whole flock, must be abandoned, possibly excluded, chased away, and so forth."[6] Scandal that he is, black sheep, or wolf in sheep's clothing that the state claims him to be, Jack dies as if a sheep, descending the food chain as first he eats grass and then endures the successive "wet" deaths of vermiculation (being eaten by worms) only then to be consumed by plants, his name then bodied forth by theatrical revival (reinflated by the breath of the wormy wetware we name actors and which they named "shadows").

Chaplin's and Shakespeare's sheepy dissolves prove unstable. They cut both ways. The double articulation of sheep and human not-sheep as at once facialized, singular creatures and as morphing multiplicity sports a series of reversible passages. The lexicon of terms and repertoire of tropes that result derive from this sheepy "historical a priori" or governing "positivity," which decides which statements are thinkable and writable but which also, necessarily, permits still other statements to be made and thoughts to be had—even as they may constitute mere nonsense or not-yet-sense.[7] Accordingly, from time to time, the animals we name human morph into sheep, and sheep find themselves emphatically singularized and rendered talkative.[8] The mechanism at work is not a simple form of anthropomorphism or even a reciprocal zoomorphism but, instead, a set of figural possibilities generated by the mutual constitution of "sheep" and not-sheep "humans." Chaplin runs the constitutive metaphor backward to provoke a critique of industrial capitalism and to induce us to misbehave. He invites us to stray or idle with the Little Tramp as he fails to habituate his movements to the world around him or as that habituation produces an erring feedback or static that interrupts the machine. We take pleasure in what Foucault might call this "misconduct" *(mal conduire),* much as some of us might in Jack and Dick's clowning.[9]

Such instances stand as one semiotic fine edge of the multispecies basis to our lives, the way in which, in this case, what it means for us to labor comes routed through a set of concerns to do with live or living "stock," variously articulated "pools" of labor power as they are produced in different historical moments. In the figural passages between sheep and their not-sheep humans, in the turning of the tropes by which the two are coproduced, we begin to access the procedures by which the energetics of "flesh" are marshaled and configured to do different kinds of "work." Here it seems important to recall the way for Foucault the articulation of this flesh upon and with which biopower "writes" becomes a way to eventuate a biological continuum and so to "introduce[e] . . . a break into the domain of life that is under power's control: the break between what must live and what must die."[10] Famously, he goes on to remark the way race, "the distinction between races," serves to establish "a biological-type caesura within a population" that allows "power to treat that population as a mixture of races, or to be more accurate, to treat the species, to subdivide the species it controls, into the subspecies known precisely as races." These breaks come rooted through our figural correlations to sheep and cows, to all those animals corralled under the name of "cattle" (as capital) and so articulated, at base, as some fungible biomass.

My aim in this and the next chapter is twofold. I begin by trying to inhabit one of these caesuras, the point of contact and separation between sheep and not-sheep humans. I am interested in the way the articulation of these coeval multiplicities produces differing states of animation that render sheep "sheepish" (which is to say plant-like) and certain (but by no means all) not-sheep humans "human." I do so by counting or tracing dissonant ovine figures that return from time to time. And this counting leads me to take up Donna Haraway's circumspect insight that "plumbing the category of labor more than the category of rights" might "nurture responsibility with and for other animals" alive to the fact that "relations of use" are "almost never symmetrical ('equal' or calculable)."[11] What potential resources inhere to our sheepy anthropo-zoo-genesis? What might it mean to try to think the relations that form (and do not) between herd animals and their attendant entities as partly open, partly foreclosed, forms of multispecies relating? Might these relations yield a different order of concepts for personhood and sociality than we have received thus far? Posing these questions requires that we understand that the biopolitical quotient to pastoral power remains almost entirely neutral, sporting progressive along with abusive relations. Hence the string of paradoxes that Foucault takes pains to store up as he attempts to chart its forms.[12] The individualizing power, for example, that "cuts" certain individuals from the flock merely realizes a different application of the same differentiating ethic of care and concern offered to each and every member of flock for the general health of the flock. "The salvation of a single sheep," writes Foucault, "calls for as much care from the pastor as does the whole flock; there is no sheep for which he must not suspend all his other responsibilities and occupations, abandon the flock, and try to bring it back."[13] Individualized expulsion and violence unfold from the same imperative that directs the care, cure, or caress that another individual may receive.

So it is that the distributive process that "cuts" Jack from the flesh of the flock for the good of the flock relies on the same order of individualizing power that enables us to pick the black sheep or every-sheep-to-be Little Tramp out from the otherwise undifferentiated crowd of hats as the commuting "hands" of the factory head to work. Jack's momentary self-promotion up the hierarchy of animal figures in the bestiary of political fable corresponds to Chaplin's singling out of the Little Tramp as that most dangerous of things, a sheep with ideas, whose very being, whose physiology, disassembles the assembly line. The *dispositifs* or apparatuses of personhood that render us individuals prove coterminous with a molar interchangeability as we disappear back into the crowd.

"*Omnes et singulatim,*" we might say with Foucault—all together and one by one. All shall count and all shall be counted, even as we may discount *you*:

> The shepherd counts the sheep; he counts them in the morning when he leads them to pasture, and he counts them in the evening to see that they are all there, and he looks after each of them individually. He does everything for the totality of the flock, but he does everything also for each sheep of the flock. And it is here that we come to the famous paradox of the shepherd which takes two forms. On the one hand, the

> shepherd must keep his eye on all and on each, *omnes et singulatim,* which will be the great problem both of the techniques of power in Christian pastorship, and of . . . let's say, modern techniques of power deployed in . . . technologies of population. *Omnes et Singulatim.* And, then, in an even more intense manner, the second form taken by the paradox of the shepherd is the problem of the sacrifice of the shepherd for his flock, the sacrifice of himself for the whole of his flock, and the sacrifice of the whole of his flock for each of the sheep.[14]

Such is the essence of a pastoral power "exercised on a multiplicity rather than on a territory."[15] Its forms demand everything of its sheep and its shepherds, which it binds together. We remain shepherds, then, even as we are also sheep, oscillating between these roles as we navigate our worlds. Occasionally, we are produced as a goat, dog, wolf, or donkey, less often, a llama. The ovine figure predominates. It remains our lot, then, to play out these variously encrypted metaphors and encounter their paradoxical involutions, our coats changing as we do. By all means, cultivate those modes of "misconduct," "resistance," "attack," or "counterattack" that Foucault finds "*within* the field of the pastorate" when and as they appear.[16] For goodness' sake, "let's misbehave." But take care as you do so, for it may prove perilous to trust too much or for too long in their efficacy. The tropes keep turning. "They are sheep and calves," as Hamlet says dismissively to Horatio, "which seek out assurance in that."[17]

Might be best not to resist exactly but instead to join in. Allow the tropes to swallow you whole and inhabit them. That's what Antonio Gramsci attempts, even if it is not something he necessarily recommends. We join him in his prison cell in Milan in 1928 by way of a letter he wrote home, one in a long series of letters written during his imprisonment that offer an affective, experiential supplement to the more famous *Prison Notebooks.*

TROPE 1. "BECOMING SHEEP": MILAN, 1928

Early on in his prison letters, letters he wrote largely to his wife, Giulia, sons Delio and Carlo, and sister-in-law Tania, Gramsci expresses a "horror of being reduced to" what he calls "conventional letter writing, or, what's worse . . . conventional prison letter writing."[18] Thereafter, his letters remain deeply personal but take on a documentary quality as he seeks to archive his experience of the restricted object-world and altered relationship to time and space that characterized Italy's prison ecology in the 1920s and 1930s.

Confined first on the island of Ustica, where six hundred of the island's thirteen hundred inhabitants are political prisoners, he describes a penal colony in which no work is done. Instead, prisoners experience an overpopulated and so desolated *otium* or freedom from work: too much time, too little to do. They take walks and spend their days buying provisions with their four *lira* per diem; locked into barracks at 5:00 P.M., they gamble away what's left in order to spend their evening. Time runs backward. Always more of an "idea" than a reality, "Italy" unmakes itself as they sort themselves into groups by region. Gramsci feels that he has been delivered back into a premodern world, and his letters start to "rough out . . . vignette[s] of peasant life"—describing, for example, an animal trial that he witnesses of a pig "which was found having an unlawful feed in the village street and was led off to prison like a common malefactor" (38). Things take a turn for the worse when he is transferred to the penitentiary in Milan. There, Gramsci passes his days in a ten by fifteen by twelve foot cell lit by a "so-called *boco di lupo* [wolf's mouth]"—a tiny window with bars on the inside through which "one can see . . . a stretch of sky" but through which "one cannot look onto the courtyard . . . nor see anything to the left or right" (50).

Deep in the belly of the wolf, Gramsci suffers a breakdown. In the letter dated February 27, 1928, he rallies or performs a rallying. "I don't think I was ever actually disoriented," he writes, "what was happening was this: I was going through a series of crises of resistance to the new mode of life . . . with its rules, its routine, its privations and its necessities—an enormous complex of little things which follow each other mechanically day after day, month after month, year after year, like grains of sand in a gigantic hour glass" (77). He registers this implacable imposition of an environment as an involuntary becoming donkey spliced with a becoming sheep.[19] He tried to resist the process, he says, but now regards his opposition as "inefficacious and inept." A mistake. It drained what vital reserves he had left. But he reassures his "darling Giulia" that his "real skin" subsists somewhere beneath or within the hide or fleece that grows over or through it. He "shall be able to throw [it] off with a single shake" (77), he writes, when or if the opportunity arises.

Until then he has made "the calm decision not to oppose" the overwriting of his body, his *sensorium,* by the prison and its routines but instead to "dominate and control the process with the help of a certain ironic spirit." He refuses to accept that the prison cell constitutes a kinetic dead zone. Instead, it serves as an occasion for an altered set of questions and regime of description. "You mustn't think that my life drags by," he offers. "Once you have got used to this aquarium existence, your sense organs are attuned to the reception of dull twilight impressions which flow quivering through it" (77–78). He's being brave, he knows, admits or declares, reaffirming thereby his decision that "I'm still looking at things from an ironic standpoint" (78). For by this adaptation, by this attunement to the altered scale of his curtailed environment, he perceives that "a whole world begins to swarm around you, with its own particular vivacity, its own peculiar laws, its own essential being." He compares the process to "look[ing] at an old tree trunk":

> First of all you take in nothing but moist fungus, and maybe a snail oozing its hesitant slimy way forward. Then you become aware all of a sudden of whole colonies of little insects busying themselves with their affairs, making the same efforts and following the same route time and time again. If one maintains one's own independent position, if one doesn't allow oneself to become a snail or ant, the whole thing ends up by becoming very fascinating. (78)

Look closely and whole polities of actors will reveal themselves to you in what you had taken for a singular entity—be it a tree trunk, a prison cell, the prison itself, the world writ large. Such an orientation is not exactly pleasant. It hurts and yet still it has to be cultivated, willed, and even desired. You have to allow yourself to feel the new hide or hairs work their way through the one you had taken to be your "real skin." You cannot allow it simply to be imposed from without. You cannot allow yourself to succumb to the being-docile of the prison that would render you the literal donkey or sheep that finds itself written as beast of burden or livestock, a commodity backed by a living being, *for* human animals. Instead, you have to conserve an "ironic standpoint," a mode of being there, which we might gloss as an ability still, despite everything, to pose your existence in the form of a question, even if you put it only to yourself.

In this letter, Gramsci generates a succession of zoomorphic forms none of which quite bears the weight of what he hopes to describe. Never just a donkey but also a sheep, never simply a donkey-sheep but also a snail, an ant, and so on, his metaphors mix. They run on or run together as he reaches after a form that might render the process in which he finds himself enmeshed. Crisscrossed by a different order or economy of *bios* and *zoë*, recoded as a form of what his political and philosophical inheritors shall call "bare life"—though, given his metaphors, the Italian *vita nuda* seems more appropriate—Gramsci produces a miasmic or chimera-like beast fable that reveals the coarticulation of different forms of life across the supposed divides of species or kind as well as media (air to water).[20] The incarcerated human, from an "ironic standpoint," manifests as a donkey-sheep-ant-man to men; less a fish than a man underwater, underground, his senses traumatized and tranquilized by the light-scarce environs of his bathypelagic prison cell-aquarium. So it is that Gramsci decides to live on as if a donkey-sheep-human, as if a snail or ant, for that best captures the subterranean or subaquatic form of life that is the penitentiary. As he notes, that life has its compensations, however ironic their register. Swallowed whole by the state become wolf to his donkey-sheep, Gramsci continues to set down marks. He continues to count. His letters, letters that he sends out, never to see again, provide a point of reference, an orientation. The paper that bears his script stands in relation to his "true" but increasingly overgrown skin. There is a cost, he knows, a cost he registers half-jokingly, that given the lack of sun, he "shan't be able to restore . . . [his] famous smoked look." Bleached by his deep-sea confinement, he translates the "look" of his flesh to paper. If only his family could cultivate the same "ironic standpoint," they might all be happier. If only they would write more. But for now, "it's clear that as far as this [his family] is concerned," he concludes, "I must be pretty disoriented. But that's inevitable, I suppose. I embrace you tenderly."

Following Gramsci, who manages this feat of description under harrowing circumstances, it seems only right to begin by owning up to our constitutive "sheepiness" and our sometimes shepherding and attempt to inhabit the "ironic standpoint" he pries loose from his confinement. I am a not-sheep "sheep" and not-sheep shepherd. Every day I wake to find myself a sometimes sheep-shepherd-dog-goat-wolf-donkey-llama. The tropes are simply too powerful, too old, to slough off or disavow. So, instead, let's allow our "half donkey hide and half sheepskins" to presence and chart the forms of expression or order of statements they occasion. Such an endeavor amounts to something on the order of a proactive or predatory mimesis, an acting out and up that produces a different order of "kine" or "cattel" aesthetics that owns the co-making of "sheep" and ostensibly "human" persons along with their dog, wolf, goat, donkey, and llama alliances. Not resistance, then, in Gramsci's sense of things, in this chapter I cultivate and so attempt to steal the trope. Inhabiting the material–semiotic–rhetorical chain, I rework its terms as I seek after some syntax that might parse these beings differently and so alter what it means to manifest as "sheep" and not-sheep "human."[21] As you read, and as I count, you can picture me falling asleep, if you want, like all the West's shepherds, a not entirely "human" Bo Peep, dreaming of all the woolly growths that erupt through our discourses.

If this all sounds a bit utopian, that's perfectly fine. Indeed, as you shall see, that's almost exactly right.

TROPE 2. DOLLY THE SHEEP IS DEAD: ROSLIN, 2003

On February 18, 2003, any casual grazer of a news feed might have chanced upon a link to the following announcement in the online "News and Comment" section of the British journal *Nature*: "Celebrity Clone Dies of Drug Overdose."[22] "For over six years, every bleat of the world's most famous sheep has been analysed for biological significance and hints of decrepitude," begins the obituary. "No longer: Dolly was put down by lethal injection last Friday [February 14]. She was six and a half years old, and suffering from lung cancer caused by a virus." "In 20 years time, Dolly won't be remembered for the practical applications that she led to," observes Dr. Harry Griffin, then assistant director of the Roslin Institute in Edinburgh, "but for opening our eyes to the idea that the cells in our bodies are much more flexible than we had thought.... Preliminary post-mortem results show that apart from the cancer and well publicized arthritis... there were no other signs of aging"—putting paid, apparently, to the speculation that Dolly's woes may have been caused by irregularities in her chromosome indicating that her "biological age might equal that of her and her mother combined." Dorset Finns can usually expect to live ten to twelve years. Dolly only made it to six. "Spookily," observed an earlier review of her condition in January, in the now defunct popular life sciences magazine *Acumen,* "early death seems to be the fate of animal clones the world over."[23] "Are clones," it continued, "like the replicants in *Bladerunner,* doomed to early extinction?" But Harry Griffin has no truck with such sci-fi speculations. According to him, Dolly's "celebrity may partly have been to blame." "Early in life she had a weight problem," he observes, "she was fed a lot of excess food to get her to perform for the cameras."

Who or what was Dolly? She appeared at the time as a contested pathology (clones do or do not age prematurely); as a shift in *epistēmē* (our genes are subject to micromanipulation); as a clash of genres: obituary meets parody, science fiction, *TV Guide*; and as a temporal riddle. Dolly's uncertain age (was she six or eighteen?) created an ontologically variable signal, testifying either to the future success and routinization of cloning, splicing together asexual with sexual reproduction, to realize the malleability of the flock and its flesh at the level of the genome, or a prematurely decayed or atrophied future (the fate of all utopias)? This uncertainty multiplied to produce a hyperattention to temporality in the bleating of her human commentators. She was haunted, on one hand, by Philip K. Dick and Ridley Scott and their dystopian futures and, on the other, by her matronym, Dolly Parton, and written into the text of stardom, of tragic overeating and addiction. It was the cameras, finally, that killed Dolly, a star that burned too brightly. That was the hype, anyway.

Survived by four of her lambs, Dolly is now on display at the Museum of Scotland. Taxidermy preserves her as an icon to be variously stuffed, animated by whichever human voice is agreed to speak as her best witness. She endures as a privileged node or experimental object, a yet-to-be-decided witness to the ways in which technoscience synchs up to pastoral, articulating the "animal" as if a "plant," to produce the fiction of the multiplicity in one, not quite part for whole, so much as a splicing, or, as in Sarah Franklin's felicitous term, a "dolly mixture," that turns back the clock from parent to embryo, "grow[ing] animal cells in a dish, as if they were bacteria or cultured plant cells" that may be "transformed *en masse.*"[24] Strange flesh this, Dolly the sheep become plant-animal, presaging the way for the micromanipulation of our own flesh or revealing only the latest chapter in this story.

TROPE 3. "PERFECT METAPLASTS": LONDON, 1667–68

Rewind to London 1676 and the possible miscibility of sheep and not-sheep; human flesh was a matter of some speculation, regret, and no little hilarity. Throughout the month of May, Thomas Shadwell's *The Virtuoso* (1676) played nightly at Dorset Garden, and there, on stage, you might encounter, among other sensations, report of a series of experiments conducted by the Royal Society transfusing the blood of a young sheep into a man. Shadwell was out to satirize the society, and in act two, scene two, we hear Sir Nicholas Gimcrack, chief experimenter and titular virtuoso, proclaim the virtues of the procedure, recounting how, following the transfusion of "sixty-four ounces" of blood into a madman,

> the patient from being maniacal or raging became wholly ovine or sheepish: he bleated perpetually and chewed the cud; he had wool growing on him in great quantities and a Northamptonshire sheep's tail did soon emerge or arise from his anus or human fundament.[25]

Without missing a bleat, Shadwell has Sir Nicholas go on to write what, had he succeeded, might have proved the final chapter of England's land enclosure movement. No longer requiring land for pasture such that he must convert arable land to pasturage, Sir Nicholas plans to transfuse sheep blood into yet more likely candidates and become self-sufficient in wool. Premised on an exchange of bodily humors across the lines of kind

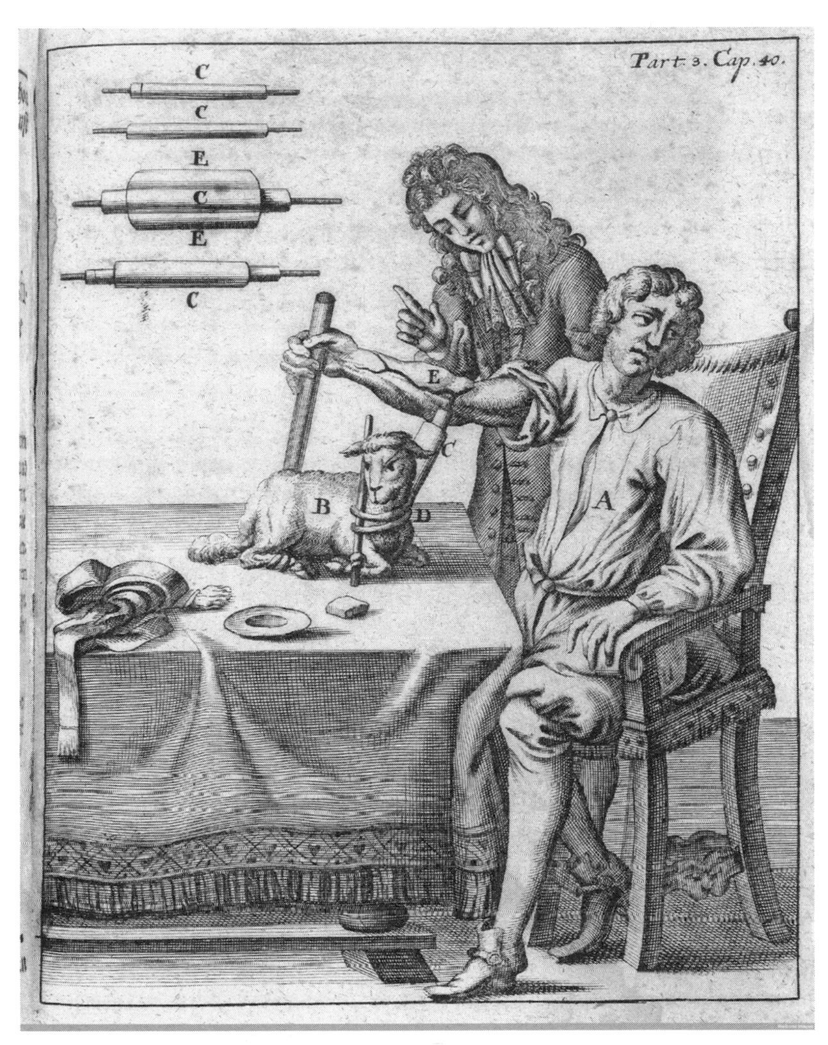

Figure 2. An early blood transfusion from lamb to man. Matthias Gottfried Purmann, *Grosser und gantz neugewundener Lorbeer-Krantz, oder Wund Artzney… Zum andern Mahl vermehrt heraus gegeben* (Frankfurt: Widow and Heirs of M. Rohrlach, Leignitz, 1705). Wellcome Library, London.

and so on the fungibility of ovine and human flesh, he hopes to render still more of his patients "sheepish" and, in fact, sheep. He shall have his tailor make all his suits from the wool of these transformed creatures, he says, and boasts that he will "shortly have a flock of 'em and make all my clothes of 'em," for their wool is "finer than beaver" (2.2.210).

Snarl, "an old pettish fellow, a great admirer of the last age," will have none of it, but Bruce and Longvil, "gentlemen of wit and sense"—well, not exactly—agree, offering that Snarl "is in the wrong in abusing transfusion, for excellent experiments may be made in changing one creature into the nature of the other," and that the process may yet be "improved" so that we may "alter the flesh of creatures that we eat, as much as grafting and inoculating does fruits." Sir Nicholas happily agrees: "'Tis true; I do it; I use it to that end" (2.2.224–30). Not cloning, exactly, so much as a grafting or the conversion of "human" to ovine flesh, these transfusions resplice sheep and not-sheep flesh so as to produce a series of hybrid, transgenic figures. The story of enclosure, the conversion of arable land to pasturage, dovetails with the prospect of endo-colonization, the genetic micromanipulation of flesh.

Shadwell's satire cashes in on a series of actual experiments conducted by members of the Royal Society almost a decade earlier—experiments that the play loosely summarizes as "a rare experiment of transfusing the blood of a sheep into a madman" (2.2.180–81). Scrambling the jargon of the day, the play's report tells us that "the emittent sheep died under the operation, but the recipient madman is still alive. He suffer'd some disorder at first, the sheep's blood being heterogeneous, but in a short time it became homogeneous with his own" (2.2.184–87). The play backs up its claims by referring to "a letter from the patient who calls himself the meanest of my flock" (2.2.207–8) and lauds this patient for his almost ovine compliance. The idea of blood transfusion was hardly new—the story of Medea and Aeon in Ovid's *Metamorphoses* serves as its *locus classicus,* and early scientists in England and on the Continent had conducted all manner of experiments in their research on blood. Signally important in the prospect of successful transfusion had been Sir Christopher Wren's development of the syringe, which enabled experiments "infusing drugs and poison into the veins of animals," and then, between 1667 and 1668, "so many transfusions between animals, of the same and different kinds," were conducted "that to collect them," quips Marjorie Hope Nicolson, "would result only in a dull catalogue of common things."[26]

As Thomas Birch, first historian of the Royal Society, reports, three transfusions from sheep to a clergyman were made between November 1667 and January 1668. They stopped abruptly upon news that a similar experiment conducted in Paris between a calf and a man had resulted in the death of the human recipient.[27] Writing two years later, Dr. Henry Stubbe, not exactly a friend to the society, "discusse[d] in detail what we would call 'incompatibility of bloods,'" offering that "animal blood is not homogeneous with that of human beings, and the blood of some animals is heterogeneous with that of others."[28] Birch tells us that the "madman" in question, who had agreed to the transfusions, was "one Arthur Coga" and that he "was willing to suffer the experiment . . . for a guinea."[29] Represented as variously unstable by members of the society, Dr. Edmund King, who, along with Robert Boyle, oversaw the experiments, described him as "speak[ing] Latin well, when he is in company he likes," but found "his brain . . . sometimes a little too warm."[30] Coga's instability seems, to some extent, an artifact of the experimental reports themselves, which worry his reliability as a witness, especially so given that he does not seem quite a gentleman and was a little too eager to please.[31] Whatever the risks, Coga seems to have prospered. Apparently he felt much improved by the transfusions, and he delivered a series of reports to the experimenters, privately, and more publicly in Latin before the society, though Stubbe finds all this very "doubtful" and records that he "was told [that Coga's] arm was strangely ill after" the transfusion.[32]

Coga had his own theories as to why the transfusions proved successful, theories not entirely compatible with those of the society, though they would tend to explain why their experiments worked while those on the Continent did not. Dr. King tells us that when "one [of the spectators at the first transfusion] asked him "why he had not been given some other creature's blood," Coga quoted John 1:29 and "told them, that *sanguis ovis symbolicam quondam facultatem habet cum sanguine Christi; quia Christus est agnus Dei* [the blood of a sheep has the symbolic faculty of Christ's blood; Christ is the Lamb of God]."[33] "You would have smiled, if you had seen him," offers King, but here Coga discloses his own quasi-Eucharistic theory of transfusion. Switching registers, if not languages (many of the society's records were written in Latin as well as English), Coga's unreliability becomes legible as a rival mode of explanation and also a rival use of Latinity. Call him mad, a bit "off," a clergyman, Coga translates or trans*signifies* his transfusion with the blood of a literal lamb into an actual Communion.[34] Eucharistic poetics trumps the rhetoric of experimental protocol and empirical observation, rendering the society's theatrical demonstrations accidental vehicles for another, divine essence. Smile we may, but Coga's rival theory renders sheep blood "homogeneous" or "compatible" with his own by way of the symbolic efficacy of the Agnus Dei. The blood of the lamb, the blood of Christ, serves as the point of material-semiotic transfer, an absent-present substrate that, whether "Real" or "Commemorative," renders the blood of a sheep and a Christian miscible. The three sheep prove "accidental" (they died) but so also, for that matter, shall the Christian, no matter how seemingly central the role he plays. Molded from clay and inspired with divine breath, Coga's body shall also expire, whatever the fate of his soul. The governing essence lies elsewhere—no matter the exact quantity of blood transferred.[35]

Come the end of the story, Coga disappears into Restoration London, reappearing courtesy of *The Virtuoso* and its parasitic satire and also in the form of a letter that most probably provided Shadwell with his inspiration. The letter's origins remain murky. Published by Stubbe, Nicolson conjectures that it emerged from London's coffeehouse tables. As you may imagine, the letter comes richly laden with rhetorical wares. "To the Royal Society of Virtuosi, and all the honourable members of it," it begins, offering itself as "the humble Address" of the metaplasmically altered "Agnus Coga," who writes to complain of financial difficulties now that he is no longer "his own man" but their "creature," because it was their "Experiment [that] transform'd . . . [him] into another species." Fallen on hard times, he finds it necessary now "to pawn . . . his cloaths" and "dearly purchases your sheeps blood with the loss of his own wooll." Forced into a cycle of self-consumption, the letter switches metaphors, transforming Coga's sheepy body into "this ship-wrack't vessel of his," so that like the crew of the Argos, "he addresses himself to you for the Golden Fleece.

For he thinks it requisite to your Honours, perfect Metaplasts, to transform him without as well as within."[36] Some sheepy Jason in search of the gold the society deprived him of by fleecing him, Coga sues for literal and figurative damages. Reduced to pawning his own "wool," he requests a bit of alchemical liquidity in the form of the capital that derives from the society's keeping livestock such as himself and the even less fortunate sheep—golden coins that jingle in the pockets of the society's members.

Shadwell's *Virtuoso* ends with a similar mooting of a labor dispute that writes the blood transfusions back into a longer history of shepherding and sheep rearing in England. As news of Sir Nicholas's new generation of "sheep" spreads, he finds his house besieged by a "rabble" (5.2.106) of angry ribbon weavers. These men and women have put two and two together and realized that his invention of an "engine loom" (5.2.110), along with his newfound sufficiency in wool, threatens their livelihood. Like Jack and Dick, like Coga, no longer their "own men," they are here to "pull [Sir Nicholas] . . . out [of his house] and tear [him] to pieces" (5.2.106–7). Sir Formal attempts to calm them with his high style and public speech, but he forgets to mention "Henry the Fifth," and so the demonstration is routed by gunfire, though no one is actually hurt (5.2.81). So ends this story of metaplasm amid contested witnesses, labor disputes, and the prospect of strike action.

Between Coga and Dolly, we are able to make out a long trajectory to a Christian pastoral power, exercised on a multiplicity, that territorializes its mechanisms in the flesh it cuts up and manages: by way of the Eucharistic feast that it portions out to its flock, a technique whose material–semiotic function is conserved in the biomedical technologies that "write" successive orders of flesh today; by way of the selective breeding and then industrial production that, as Longvil proposes, might one day "alter the flesh of creatures that we eat" as well as our own. Theology cohabits with technoscience; the Eucharist renders blood transfusion a sacrament; and biopower deterritorializes Christian pastoral power from its sheepy origins to write across the lines of kingdom and kind. Our discourses are marked by the semisufficient transfers and elisions of these tropes and littered with their remainders, such as Coga's grafting of sheep and not-sheep by way of a divine substrate. The issue here does not lie in outrunning this Eucharistic poetics and producing a purified set of discourses or procedures but instead in owning what remains of their theological basis as it haunts us and funds our discourses, just as we need to understand the ways in which theology primes its own technics.

Coga's proposition proves productive, then, for it discloses the way the Eucharist figures already as a biopolitical operation that convokes a collective by and through the conversion of "flesh," a conversion to which Donna Haraway alludes through the language of mess-mates (if not Mass mates) in *When Species Meet* and the convoking function to the companion species relation. The word *companion,* as she reminds, "comes from the Latin *cum panis,* 'with bread.' Mess-mates at table are companions."[37] At issue, here, is not some simple identity between terms so much as the longevity of a constitutive anthropo-zoo-genetic figure in constituting our worlds. Material conditions change. The division of labor between sheep and not-sheep humans alters, for example, with the arrival of industrialized farming, disappearing the historical facticity of sheep from the equation almost entirely, or, more exactly, localizing and encrypting their presence as they are reconstituted as fungible beings or sentient commodities. If Coga recovers the way our co-relations with sheep come filtered through a relation to the divine, are underwritten by incarnational poetics, then Haraway invites us to ask by what ratio may we render the theological a technical input merely while retaining the convoking function it puts to use. The figure of the multispecies and the mess-mate ask whether it is possible to, and by which techniques we might, render our convocations less lethal and so open to asocial, merely serial polities that alter what it means to write the word "us." How do we incline a mode of being that disowns or sloughs off its sheepy correlations into modes that recognize and proceed on the basis of a being and becoming with?[38]

Today, the taxidermied Dolly remains on display in a museum. The fate of the society's three sheep I know no better than that of Coga, who I assume died and was buried. The sheep died also, of course, but they were probably sold on, their bodies cut up according to a series of routinized use-values, all used up, and what was left boiled down for glue. But given that, in both the case of the society's experiments and in Shadwell's play, the stories end by staging (however perversely) a set of labor disputes, it is tempting now to hallucinate the arrival of these three sheep along with Dolly, Coga, Shadwell's ribbon weavers, Jack, Dick, and friends, en masse at the doors of the Royal Society, a flock of angry, striking sheep and their allies. Or, perhaps, in a very different, more sobering register, we should imagine something on the order of Sue Coe's *Sheep of Fools,* whose arresting images seek to render the "historical development of the present-day phenomenon of the live export of sheep" as part of an industrial food chain. Indeed, such a return of a sheepy undertow or substrate manifests in all-out horror mode in the ludicrously and comically pastoral, New Zealand mise-en-scène of the film *Black Sheep* (2007), whose cloned sheep grow incisors and quite literally, and very graphically, revenge themselves on successive groups of lab technicians and other inheritors of Sir Nicholas and company.[39]

Sometimes the sheep and their not-sheep human allies, whose labor does not count, arrive, graphically, upsettingly, or movingly. But they also sometimes come in downright cuter, cuddlier, charismatic versions. Some of these, like Dolly, are stuffed, but they tend to prove more numerous, smaller, plusher.

TROPE 4. SHEEP-LESS NIGHTS: TUNING IN, CIRCA 2000

There was a moment, before they became box-office material and spun off the children's shows *Shaun the Sheep* and *Timmy Time,* that if you happened to turn on your TV or open your e-mail, you might find them, the metaphorical sheep of sleepless nights, let go from their usual mode of employment and left to look for other work. In 2001, the Serta mattress company commissioned a series of commercials made by the Bristol-based Aardman Animations, whose first Plasticine, animated sheep appeared in the Wallace and Gromit film *A Close Shave* (1995). In the first of these ads, now anthologized along with their dramatis personae on Serta's corporate website, a sleeping man finds himself awoken by a strange noise.[40] He looks out his bedroom window and sees a flock of sheep. They gaze at him, and a spokes-sheep for the group shouts, "Hey! It's ten o'clock . . . ten thirty-eight. Want us to put you to sleep?" "It's the counting sheep," the man says to his wife, but she's unimpressed. "Didn't you tell them we got a Serta?" she asks. "Did she say Serta?" asks the disappointed sheep. "Yeah," replies the husband, "it's so comfortable we don't need you anymore." The sheep look dejectedly away but perk up when a neighbor opens the window and shouts, "Hey! Keep it down over there. I can't sleep a wink!" Subsequent ads feature similarly uncounted flocks frustrated in their attempt to relocate to a clinic for insomniacs by the arrival of a load of Serta mattresses; jailed for removing the "do-not-remove-under-penalty-of-law" label from mattresses and pretending to have "[torn] a man to pieces"; crowding into a man's bedroom and eying him menacingly; pleading to keep their jobs; or simply let go and homeless, wandering away from the flock.

In 2002, Serta was graced with a Gold Effie Award for its ads, and the campaign won a strong fan base and a lucrative trade in beanie-baby sheep, which each wear a medal that reads "*Serta*. We Make The World's Best Mattresses" on one side and "Out Of Work, Thanks To *Serta*" on the other. Sheepily named John Wooley, Aardman's producer, claims that the aim of these ads was to "convey a strong, simple message to which everyone can relate." Whereas Shaun the Sheep of Wallace and Gromit fame was a diminutive and unlikely hero, "counting sheep," Wooley goes on, "required more adult characters . . . working men's sheep with voices like union negotiators."[41] It's a witty and pleasant conceit! Serta's mattresses are so comfortable that anyone who owns one sleeps soundly. No more sleepless nights. No more tossing and turning. Enter the sheep. Idiomatic sheep. Displaced sheep. Or, more correctly, sheep reassigned from the imaginative realm of a sleepless bedroom to a figurative dole queue. Here, in the land of perfect sleep, the monotonous, self-identical sheep we are told to count go uncounted and so they begin to speak. But when they do, they follow the now familiar and frightening script of downsizing, of "human resources," the loss of jobs and communities by the migration of manufacturing overseas, and so they plot all manner of minor rebellions—pulling the labels off of mattresses (attacking their object-foe) or keeping the neighbors up by talking too loudly outside (a consumer-side initiative that creates, so they hope, demand for them).

Subtext: stage your revolt *here*; deterritorialize *here*; it's OK to become s*h*eep as you s*l*eep. So, shush! There, there. Go to sleep. Get your rest. For, tomorrow you must go to work. For now, though, luxuriate in the fact that you have earned your bed and so your rest and count no sheep. Serta has done our shepherding duties for us. And so, like the figural sheep of sleepless nights, we have been let go as shepherds. Plantlike, we can simply exhale and relax into our "beds." Yes, you might say that, with Serta's sheep, Jack, Dick, Arthur, or is it "Agnus," Coga, and Shadwell's weavers, along with the three sheep and the Little Tramp, all show up in figural Plasticine form to complain that they go uncounted and so no longer count. But, if they do, the script is no longer quite their own (and perhaps it never was). For, quite soon, if everyone gets a Serta mattress, they shall literally cease to "matter" at all. If you look closely, they are already beginning to lose their form and disappear. Serta's creation of a perfect slumber, a sleep into which, plantlike, we slip, shall cause us not merely to forget them but to forget that we ever knew of them at all. Serta's mattresses are just so comfortable and so effective. They unmake these sheep's world by remaking our beds.

Back in 2000, when I first saw these ads, I couldn't help feeling that these sheep seemed familiar. Eventually, I realized that I had seen them before (as I had also, in a different way, Serta's mattresses), or, if not exactly *these* sheep and *these* mattresses, then the zoographic and the text of leisure they embody. Not on TV but in Thomas More's optative, self-help, or *Truly Golden Handbook for the Best State of a Commonwealth Called Utopia* (*De Optimo Reipublicae Statu Deque nova insula Vtopia libellus uere aureus*; 1516).[42]

TROPE 5. SHEEP BITES MAN: UTOPIA, 1516

Book 1. In the course of an argument over dinner at Cardinal Morton's with an objectionable English lawyer, who maintains that the poor "might maintain themselves" by "farming" or other "manual crafts, if they did not voluntarily prefer to be rascals"(60–61), Raphael Hythlodaeus, who speaks nonsense but figures true representation, summons a flock of sheep to the table:

> "Your sheep," I answered, "which are usually so tame and so cheaply fed, begin now, according to report, to be so greedy and wild that they devour human beings themselves and devastate and depopulate fields, houses, and towns. In all those parts of the realm where the finest and therefore costliest wool is produced, there are noblemen, gentlemen, and even some abbots, though otherwise holy men, who are not satisfied with the annual revenues and profits which their predecessors used to derive from their estates. They are not content, by leading an idle and sumptuous life, to do no good to their country; they must also do it positive harm. They leave no ground to be tilled; they enclose every bit of land for pasture; they pull down houses and destroy towns, leaving only the church to pen the sheep in" (65–67).

Raphael animates the sheep of England as a flock of man-eating beasts that literally consume the rightful human inhabitants of the land out of house and home, feeding on people-grass to fill their owners' stomach-purses. In a perverse reversal of pastoral care by which metaphorical shepherds have literalized their flocks and have been transformed into figural wolves, these sheep enable Raphael to make his point with pleasant and devastating wit. As anthropologists of cloth and ecological historians remark, "because cloth production requires exceptional investments of labor and materials, it is always potentially competitive with the agricultural and military exigencies of the polity ["sheep eat men" was the expression for this in sixteenth-century England]."[43]

More specifically, Raphael takes aim not at the evils of land enclosure per se but, as Joan Thirsk observes, at the conversion of land from corn/cow/sheep farming with small flocks of twenty to thirty sheep and cows to pasturages for intensive sheep farming geared to the production of wool for export abroad.[44] Throughout the first half of the sixteenth century, land enclosures were common, and the reasons for them varied. Not all enclosure was undesirable. Indeed, its origins may lie in the reclamation of land depopulated as a result of the depredations of plague and so repopulated with sheep—or, in multispecies terms, an attempt to mitigate the disaster of a parasitic overcoding of England's infrastructure by *Yersinia pestis.* Raphael's critique remains local, then, allied to the common field system to which Hugh Latimer, Archbishop of Worcester, would later appeal in his iconic "Sermon of the Plough," delivered in 1549 as a stabilizing social force at the height of the "dearth."[45] Accordingly, Raphael animates the sheep of Tudor England as the wolfish prosthesis of profiteering landlords with no regard for the Commons. He seeks to make visible what he regards as a corrosive iteration of a companion species (graziers or "wool growers" and their big flocks) as one violent intragroup within an unreduced Commonwealth of Christian not-sheep "sheep." He does so by momentarily promoting sheep up the food chain so as to disclose the truth to the wolfish adage that "man is a wolf to men," though here the jaws and the teeth belong to sheep and the flesh eaten is the lost livelihood of crops not cultivated and the grass grown and eaten instead that enables the now uncannily plantlike sheep to "grow" wool and thence capital. England eats its own, depopulating its lands while its numbers of sheep rise, as does the price of their wool, sent abroad and then reimported as cloth to be dyed at home. If this system of sheep farming or "wool growing" leads to depopulation and dislocation, it does so because it grafts sheep onto a land cleared of competing human or Christian presences in a way that anticipates Longvil's excitement at the prospect of creating wool-growing sheep-men and women by way of blood transfusions. For Raphael, such transfusions occur on a more elaborated and so less knowable scale between and across plant and animal kingdoms, between and among competing groups of Christians, in a way that disanimates sheep entirely, disappearing their status and their presence so that England simply grows its staple commodity.[46]

Raphael goes on to point out as many of the negatives to this economic structure as possible, focusing on the way it recodes the land by driving people from it, concentrates labor in the hands of a small group of shepherds, and leaves the hands of formerly agricultural laborers idle. For More, enclosure represents the elimination of many of the ties of reciprocal care between flocks of human subjects and their not-sheep shepherds, a cancellation of Christian pastoral power. Sheep farming, "growing" wool, rewrites the biopolitical settlement that More takes to be the governing function of the state and the sovereign, violating that "mutuality" and those "obligations" that obtain, or ought to, between fellow Christians.[47] Raphael's description presciently posits a realm without institutions of pastoral care—a realm whose monasteries are already in self-dissolution.

At one moment, he wryly notes that those churches that remain on land devoted to pasturage are used as "sheepfolds," their congregations having morphed from figural into literal "sheep."[48]

Raphael's evolved figure of this cannibal sheep-grass-landlord-wolf proved so compelling that it calves a predatory oveme (minimal unit of sheepy antagonism) that travels through the period and beyond. His predatory sheep speak the truth about one order of human labor relations—a function that Serta's uncounted sheep both replay and, perhaps, pleasurably encrypt. Figures of predatory or brute, literal sheep as a rival multiplicity predominate throughout Tudor England. In *A Discourse of the Commonweal of This Realm of England* (1549), when the long-suffering knight complains that he can only make ends meet by keeping sheep, the husbandman moans, "Sheep, sheep, sheep."[49] Aphorisms such as "the more sheep, the fewer eggs a penny" and "we want foxes to consume our shepe" circulated throughout the 1540s. And Thomas Becon's dialogue *The Jewell of Joy* (circa 1550) discloses the real scandal to the figure, explaining that "those beastes which were created of god for the nourishment of man doe nowe devoure man."[50] By the end of the sixteenth century, however, the phrase had lost much of its bite and become a necessary item of ovine lore to be replayed, as it is by Edward Topsell in his *Historie of Four-Footed Beastes* (1607), a curiosity, referring to depredations that now no longer mattered quite so much as they once did, given the collapse of wool exports mid-century and the downturn in cloth prices.[51]

But, in March 1549, during Edward VI's minority, when aggressive enclosures were at their height, Raphael's figure must have seemed almost exactly right. And More's inheritors, Lord Protector Edward Seymour, Duke of Somerset, and a group of like-minded second- and third-generation humanists known as the "Commonwealth Men," good shepherds that they attempted to be, set about actually counting England's sheep. They promulgated legislation "grant[ing] Edward VI the proceeds of a tax on sheep coupled with a purchase tax on cloth."[52] As M. W. Beresford remarks, historians had long regarded the tax as never having been implemented, making it "the shortest lived tax in English fiscal history," but, in fact, the measures were carried out, if spottily, and not particularly successfully. Somerset fell from power in autumn 1549, and the measures were repealed immediately under the Duke of Northumberland.[53] The taxes sought to install a measure of commutative justice in the tax system by discouraging pasturage and so stemming the conversion of arable land to pasture. The sheep tax required a census of flocks to take place on June 25, 1549, effectively maximizing the number of sheep by focusing attention on sheep-shearing season. Preparatory materials for the legislation reveal that in 1546, estimates of the number of sheep in England ranged between 8,407,819 and 11,089,149 but were revised downward by John Hales to 3,000,000 in a memorandum titled *The Causes of Dearth,* produced probably as part of the 1548 Commission on Enclosure.[54] The proposed taxes were progressive, exempting commoners with flocks of less than twenty sheep and charging them a halfpenny a head for fewer than ten and a penny a head for eleven to twenty (encouraging mixed sheep/cow/corn farming) and taxing flocks grazing on common land at a lower rate. The tax differentiated between three categories of sheep, each of which was to be taxed at a different rate:

1. three pence a head on ewes kept on enclosed ground for the greater part of any year, whether the enclosed ground were marsh or pasture, "that is to saye, groundes not comen nor comenlie used to be tilled"
2. a tax of two pence a head on wethers and other shear-sheep on these same enclosed grounds
3. a lower rate of three halfpence on all sheep on the commons or on enclosed tillage lands[55]

Taxes calculated on sheep should only be paid, however, if they exceeded the usual taxes that would be paid on goods, effectively exempting small farmers, and "if [the] sheep tax did exceed . . . property tax [a farmer] was only to pay the difference." Estimates on the revenue that might be generated varied greatly.

Implementation of the tax, as with any undesirable, centrally planned initiative of the Tudor regimes, was erratic at best, and the local priests and "honest men of the village" named as census takers failed in many cases to record the existence of any sheep at all—or were seemingly unable to identify one. In North Riding, only four or five villages were represented at all, and in these the flocks consisted of two hundred sheep or fewer.

Likewise, the reduction in tax for sheep grazing on common land meant, predictably, that most of the sheep reported tended to be of this type. And "nowhere did the commissioners report flocks near the prohibited size of 2400," as mandated by an earlier act of 1533.[56] The act met with widespread resistance and was repealed at the end of the year when Parliament returned. Somerset was already under arrest and in the Tower of London, his power having fast diminished as a result, in part, of the mass riots that summer, culminating in Kett's Rebellion in July 1549. Still the tax represented a strategic attempt to intervene in the dominant writing machine of the period, to rewire its relays and so redraw the relations between human persons and sheep at a moment when rural distress fueled popular protest. Implicit to its terms and its mode of implementation is the language of Christian pastoral devotion, and the preamble to the act reminds landlords that they are but "tenants." It calls on God to protect "this lytle Realme and us His poore Servants and little flock, taking to his charge and defence our little Sheparde."[57] Here the universalizing potential of the terms of pastoral care (however feeble their reality) sought to remind members of Parliament that they were on earth only for a "short staye." Enter the Big Other or divine shepherd as we are counted, losing our differences to become his interchangeable sheep.

Historically particular sheep presence in the act, but they do so only to the extent that they serve as relays (positive and negative) between differently abled human persons and refer to differing ecological niches. The different rates of taxation seek to prescribe desired modes of land use by polities of sheep and not-sheep "human" persons, while discouraging others. Thus shall risk or "dearth" be managed by a zoo/biopolitical fix. That said, still more immediate and violent particularization of sheep as historical beings contributed to the failure of the tax. Among other reasons, Kett's Rebellion was vilified for its scenes of mass ovicide, as the rebels reversed Raphael's figure by slaughtering an estimated twenty thousand sheep to feed themselves.[58] Thus were sheep put to use as a food item and deployed as a form of butchered blood writing, the rebels' knives, like that of Dick the Butcher, razing the writing of the land that the Commonwealth Men had sought to legislate against. Once beaten, of course, like Jack Cade, Kett's rebels became mere "sheep" again, "when they ran confusedly away," as the same chronicle opines—their quasi-utopian interruption of history merely a momentary fitting of the general text.[59]

Sheep, we know, "write" successive landscapes. Domesticated as early as 10,500 BCE, the wear and tear of a multitude of hooves obliterates individual traces and carves their collective presence into the land, for good and ill.[60] As Karen Raber reminds, the ability of sheep to alter their biome derives from the way their numbers are indexed to their ecological niche. Overgrazing, for example, can set off a cascade of factors that result in what is called "ungulate interruption," the ballooning of numbers and then their sudden collapse.[61] Beyond the exportation of such flocks as tools for colonial expansion, rewriting the land and displacing or overwriting indigenous multispecies, as in Scotland, Australia, Mexico, and elsewhere, such megaflocks were and remain subject today to successive pandemics, whose management by the industrial use of antibiotics creates its own negative feedback elsewhere in the flesh of the collective.[62]

Raphael's ovine coinage offers merely one historically and culturally specific instance of this wider lexicon and repertoire of tropes as variously sheepy figures are generated to make manifest the infrastructure that their presence (un)makes. Similar tropic instances generate in other times and places keyed to the way in which sheep figure within the multispecies relations of that locale.[63] Sometimes these summonings speak directly, as in the case of More's *Utopia* or Serta's mattress ads. At other times their forms may prove subtler, less immediately recognizable, even as the claims they make, upon consideration, may prove more visceral still.

TROPE 6. "AN ABSENCE IN THE LANDSCAPE": SCOTLAND, 2001

In his native Scotland, among other time- and space-bound art, Andy Goldsworthy cards wool, trailing it atop stone walls (Figure 3), encasing stones in a woolly coat (Figure 4), all to do away with or to divest sheep, so he says, of their wooliness and so to deliver up what he terms their "power" to make the land take their impression. Goldsworthy's installations aim to make present "the absence in the landscape" (depopulation) occasioned by the land clearances orchestrated by English landlords in

Figure 3. Andy Goldsworthy's carding of wool along the tops of stone walls. Image capture from *Rivers and Tides: Working with Time* (2001).

the eighteenth and nineteenth centuries. "First the labourers are driven from the land, then the sheep arrive," writes Karl Marx, summarizing the process in the first volume of *Capital.*[64]

He does so by "writing" with the wooliness of living sheep. He generates all manner of uncanny, hairy stones, stones whose inorganic bulk now evinces some order of sympathy with the living, even as they stand there like some set of ovine tombstones or standing stones, commemorating the erasures (no more trees; no more people; long gone or diminished flocks).

The sheep that once grazed here have passed on, but their traces remain. Goldsworthy uncovers or re-covers these sheep's now covered tracks, creating faux-hybrid-stone-sheep and enlisting the labor of wool making to delineate or rubricate the land—the stone walls that crisscross the land an organizing syntax indexed to the methods of sheep farming. Goldsworthy's metaplasmic refolding or reknitting of surfaces produces an aesthetic recoding that registers historical elisions by causing a sheepy archive to presence in and as this woolly supplement. It would be only too easy to imagine other such installations around the world—in England, Mexico, Australia, New Zealand, or the contemporary Middle East, where overgrazing by sheep is regarded as a significant vector for ecological collapse.[65]

Figure 4. Andy Goldsworthy's wool-encased standing stone. Image capture from *Rivers and Tides: Working with Time* (2001).

Such figural presences, reanimating less the now absent sheep than the efficacy of those sheep, correspond to the likes of Marx's famous table in *Capital* that "not only stands with its feet on the ground, but in relation to all other commodities, . . . stands on its head, and evolves out of its wooden brain, grotesque ideas far more wonderful than if it were to begin dancing of its own free will." This plastic, kingdom-crossing table performs not the agency of wood so much as the involutions of the commodity form and its disanimating relocation of the labor of making as the commodity form takes hold.[66] Raphael's and Goldsworthy's figures do something similar. They recognize the historically particular presence and effects of certain flocks of sheep—sheep they cause to presence and whose agency they may be said to account for—but they remain hybrid figures that, even as they point to actual beings, use them to diagnose antagonisms between rival multispecies, rival subgroups of animal, plant, and human actors.

The expression or "content" of these figures remains coded by their points of enunciation: the uncertain status of ephemeral, time-bound, not quite outsider but maybe "cute" art within a semiautonomous but still post-Union Scotland; or, at Cardinal Morton's table, a scene recognizable to More's humanist readers as a disputation of sorts and legible also as a pedagogical lesson for readers in how to make effective arguments.

In More's case, the scene plays as a rehearsal of knowledge and rhetorical skill in service of rational analysis and so as a performance of the role of a humanist, who speaks back the truth or a truth to power. Goldsworthy's installations raise a not-unrelated dilemma: what exactly is the order of labor his art objects deploy? On one hand, the objects call into question the difference between artisanal labor and art making; but they also, by the same gesture, seem to offer some kind of reparation that might be reckoned only in terms of the affective response or captivation of the viewer (or filmgoer). The concern, then, lies, for both Goldsworthy and More, with gaining or keeping an audience and with what their ovine figures make visible. The metonymic tit for tat that produces a flock of wolfish sheep and these woolly stones offers a mode of reading that pays attention to lost or missing causations, discovering the occluded agents via the metonymic linkages from which the final result—so many poor people in England or a deserted Scottish Highlands—appears detached.

In Goldsworthy's case, the stones manifest as some order of compensatory signage that recovers what was erased by and through the writing of land by flocks of sheep. In Raphael's, the product of the encounter proves even subtler. His rhetorical display reveals the workings of enclosure as an economic force, but more importantly, its rhetorical flair compels Cardinal Morton to allow him to continue speaking over the objections of said English lawyer. "Even while I was giving this harangue," Raphael tells us, "the lawyer had been busily preparing himself to reply," but just as he starts, the Cardinal interrupts him and tells him to "hold your peace" (70, 71). The true "product" of this scene is the sound of Raphael's voice at a table, a voice that manages to transform a disputation into a monologue, a monologue that shall quite quickly take on the flavor of a "white discourse" or that serves, in Frederic Jameson's words, as the "very prototype of a narrative without a narrative subject and without characters."[67] After the break following book 1, Raphael's description of the island of Utopia in book 2 unfolds as an apparently neutral rendering. The text simply becomes what it describes, an order of humanist habitat or optimized state in which all rational endeavor immediately finds a use, seeing off at once the threat that thinking is idle and so vain and also the debates that underwrite More's thought experiment: disagreements over whether to tune in to Ciceronian arguments in favor of the tasking of *otium*/empty time and the acquisition of knowledge by the worlds of use and business (the Commonwealth) or the opting out of Platonic withdrawal.

Marked generically as a *libellus,* or handbook, *Utopia* announces its genre as a guide for the optimization of persons and collectives. It encourages its humanist readers to esteem it in these terms, as a self-conscious rhetorical exercise but also as a set of routines by which, despite their canniness, readers shall find themselves refashioned, much as Sir Edward Seymour and the Commonwealth Men hoped their legislation might refashion the Commons. You too can be a humanist, the text offers, recasting the age-old question of philosophy, "how to live well," how to manage your *soma* and *psyche,* as a question of the management of a Commonwealth or collective. "*Omnes et singulatim.*" The care offered to each and every utopian as she is understood to "live well" collectively remakes the Commonwealth. Like the Royal Society, the Roslin Institute, like Jack and Dick, and like Serta's mattresses, More seeks to rewrite or retie the bonds that form between different articulations of flesh, here sheep and not-sheep "humans"—how we labor, how we rest, how we make love, how we sleep. Book 1 announces the project and book 2 imagines one order of biopolitical recalibration of our flesh, a calibration that has given as many readers pause as it has excited. Louis Marin, for example, reads Raphael's flock as so much "raw material," presaging the way for what, in book 2, constitutes a conservative response to the disclosures of book 1, turning back the clock to produce a feudal model of the manor.[68] Richard Halpern remarks on how these metonymic sheep serve as "dummy subjects," perfectly "parodying the corresponding metonymy that allows the lawyer to blame [enclosure's] human victims."[69] But he too cautions us that in book 2, we shall find ourselves introduced to a model of the bourgeois citizen-consumer who wishes to achieve a state of equilibrium in his exchanges with the world (input equals output).

Utopia, the space of the good, the well, the true, that is also the no-place, may cancel out the differentiation of ostensibly identical Christian subjects into differently articulated groups based on predatory economic practices that create rival multispecies alliances of animal and plant actors.

It may offer news of a way forward by and through the way its removal of private property, the bad fetish of the commodity that articulates us all, at base, as some order of living "stock," but it seems difficult to read it as necessarily producing a compensatory or more hospitable fetish relation that we may positivize and put to use.[70] The key, perhaps, lies less in its apparent contents than in the manner of its relation, the series of antimimetic effects it generates, on the order of Raphael's uncanny sheep. What we have been reading, so it turns out, proves to be a tutorial as much in the pragmatics of effective rhetorical performance as in proto-ecological critique, though the difference between the two may prove moot.

What, then, of Raphael's and England's plant-animal-sheep? What of their congealed labor-time—the labor of living and dying, different as that might be (or not) from their shepherds? What would it mean to consider their historically particular existence in More's and Marx's terms, if not also within the aesthetic register that Goldsworthy offers? What figures do we need to generate to accomplish that task, a possibility preserved by Marx, as he winks at us and offers the prospect of something "wonderful" but also, he implies, less "wonderful" and so more credible than the fetish of the commodity, that his strange, metaphorical table might, of its own "free will," decide that it might like to dance? Tempting as it is to doze off and allow Raphael to do all the talking, I shall not take my rest just yet but will keep on counting and concede that, for More (and perhaps Goldsworthy), *Utopia,* the space of the good, the well, and the true, might lie in what he considers a well-used "sheep" and, for that matter, a well-used Christian not-sheep and shepherd.

TROPE 7. "COME I BID THEE": SUFFOLK, CIRCA 1592, AND WYOMING, 2010

Eighty years or so after More wrote his *Utopia,* writer and translator of husbandry manuals and how-to books Leonard Mascall wrote a short poem titled "A Praise of Sheepe" in his *First Booke of Cattell* (1591). According to the rules for such encomia, he blazons his subject, a ram, and inventories his exhaustive use-values. Extolled throughout the period for their supreme usefulness, *utilitas,* or, in humanist terms, their "profit," sheep were regarded as that most ideal of creatures, proof of the bounty and beneficence of the divine shepherd. Edward Topsell writes of the superlative "oeconomical profit" of English sheep.[71] The entry for sheep in his *Historie of Four-Footed Beastes* (1607) is the longest devoted to any animal, and he processes their multiplicity of breeds or kinds as further evidence of their usefulness. Topsell assembles a national guide or gazetteer to different breeds of sheep that corresponds to many lands of the world as he knew it. Each breed comes suited to its ecological niche or biome. In today's sheepy lexicon, we would recognize this sheep to be Dall and so not Romney, Big Horn, Texel, or Turki, to name just a few of what compose nearly a thousand distinct breeds or kinds—it would take too long to name them all. We would note also how far this sheep might have traveled from where it began and begin to process the costs of its translation—costs Goldsworthy's installations, for example, seek to reckon by transposing them into an aesthetic register.

But for Topsell, this interest in historically particular and localizable breeds proves more immediately interesting, more attuned to a national agenda. For England's sheep, so it turns out, are especially happy, given the absence of wolves in England, which makes their fleeces the softest in all of Europe. They are therefore especially profitable and economical. They manifest as nothing short of a national treasure, proof of the special status of England's pleasant climate, a naturalized affordance.[72] Indeed, there seems almost no use that a sheep cannot be made to serve:

> These Cattel (Sheepe) among the rest,
> Is counted for man one of the best.
> No harmfull beast nor hurt at all,
> His fleece of wooll doth cloath vs all:
> Which keepes vs from the extreame colde:
> His flesh doth feed both yonge and olde.
> His tallow makes the candles white,
> To burne and serue vs day and night.
> His skinne doth pleasure diuers wayes,
> To write, to weare at all assayes.
> His guts, therof we make wheele strings,
> They vse his bones to other things.
> His hornes some shepeheardes wil not loose,

> Because therewith they patch their shooes.
> His dung is chiefe I vnderstand,
> To helpe and dung the plowmans land.
> Therefore the sheep among the rest,
> He is for man a worthy beast.[73]

In Mascall's verse, every single part of the sheep appears destined or "written" for humans. Hence the sheep serves as the emblem both for the generosity of the divine and also for Christian meekness and subordination. Yes, sheep were work. Yes, they required care and concern. And, yes, that caregiving could prove exhausting. But so what? Good shepherds take on an exhausting set of labors that aim to exhaust the needs of their sheep. Such attention begins, as Mascall's contemporary Gervase Markham commands, with the earth itself: "know then, that whosoever will stocke himself with good Sheepe, must looke to the nature of the Soyle, in which he liveth: for Sheepe according to the Earth and Ayre in which they live, doe alter their properties." Attend to the biome you share with them. Doing so may mean that the "barraine sheepe" become "good."[74]

This risk management of the health of the flock confuses or combines registers that we may think of now as separate. "Many times it falleth that the Ewe dyeth in the yearning of a Lambe," writes Topsell. "When she is ready to be delivered," he continues, "she travaileth and laboureth like a woman, and therefore if the Shepheard have not in him some mid-wives skill, that in case of extremitie, he may drawe out the lambe when the members stick in the matrix," then all might be lost.[75] So it is that the Greeks, Topsell claims, call the shepherd an "*Embrucoulos.*" The shepherd (or shepherdess) turns midwife, turns sheepdog or recruits one. So much is required of him or her, then as now. Such minute care and concern become an explicit question also of communication, of responsiveness. Certain plants may induce a miscarriage, so a shepherd must have a "voice or whisell intelligible to the sheepe, whereby to call them together if they be scattered abroad feeding." You must be able to convoke the flock, now, then, forever, wherever, be it in Suffolk, Devon, Qatar, or Wyoming.

The film *Sweetgrass* (2010), which follows a sheep drive across Wyoming, opens with the sound of the wind and, above it, the light tinkle of a bell set in motion not by a human hand or by the wind but by the rhythms

a

b

Figure 5. Calling together of the flock by a shepherd calling "Coombidee." Image capture from *Sweetgrass* (2010).

of a sheep chewing. The shot dwells with this sheep, who catches sight of the camera or cameraman and freezes, freezing us (reaction or response?). The screen dissolves to show the title and our location as part of the in-title sequence. The sound of the wind dies to be replaced by a new sound—that we quickly person with a voice as we detect movements on the edge of vision that pass as human (Figure 5a).

The sound turns syllabic: "coombidee, coombidee, coombidee," a phrase that repeats over and over. We watch as the sheep gather in response; slowly at first, and then as a mass of moving whiteness across the land (Figure 5b). "Coombidee, coombidee, coombidee . . ." The sound turns out to be a contracted or metaplasmically reduced form of a divine "Come. I bid thee" that calls the faithful to join in a psalm. In an interview with *Cinema Scope* magazine, sound recordist Lucien Castaing-Taylor comments that these words derive from "the Tudor/Stuart practice of calling in your ducks to protect them from foxes. Come. I bid thee."[76] Deterritorialized from the prayer meeting to the farmyard or the fields, the phrase contracts. By its repetition, by the loss of sense that comes with syntactical spacing, the words lose their separation to become a noise or paralingual form that gains in clarity to ovine ears.

Satisfying to both human and ovine ears, but belonging to neither, "coombidee" both retains and forgets its origins. The "I" that bids, calling the sheep of the flock together, finds itself reduced to a vowel amid consonants, the shepherd become a convoking function. Speaking across the lines of species or kind, this "coombidee" becomes a translational node, a transspecies pidgin that serves as a minimal anthropo-zoo-genetic archive of our co-making. The threat of predation; of loss from an errant, poisonous plant taken as food; from the chancy, contingent cascade that is birth—all these factors and more produce an ethical discourse that cohabits with the extraction of "profit" or use-values from the flesh of the flock or multiplicity that the shepherd divides, treating the flock, managing it, one sheep at a time. The coarticulation of shepherding as a practice of care and concern tied to an ideal of sacrifice and "profit" produces a double coding: the recognition of sheep as agentive creatures, with ideas, desires, weaknesses, frailties, strengths, faces, sometimes names; and of shepherding less as a human endeavor or preserve than as an orientation, a position to be variously occupied by another being than your sheepy self, be it human or divine.[77]

This doubleness is captured quite beautifully in an illustration for the month of June from epigrammist Thomas Fella's 1585–88 manuscript *Commonplace Book* (Figure 6), with its scenes of milking cows, of washing and shearing sheep, that stand as positive exempla of human labor against the uncertainly coded wine making that appears on the facing page. There is industry and care in the winehouse, even as that figure comes coded by an uncertain biblical trajectory in which the cultivation of grapes in Genesis leads to the undoing of Noah by the agency of a plant. In the image, one figure attentively decants wine or grape juice from one barrel. But another guzzles down the fruits of their labor. The figures' absorption in their tasks and pleasures figures a negative version of the ethic of care and absorption represented by the labor of sheep farming on the facing page. Up in the rafters, an uncertainly sexed person straddles the troughs below, a stream of urine and feces descending, either into the container holding the wine or behind it. "The very image of idelnese" is how the motto captures the scene, coding its uselessness and expenditure of resources.[78]

Meanwhile, the virtuously sheepy labors outdoors (milking cows, washing the fleece of the sheep, preparing for shearing) are described as follows:

> This hurtlesse beast with meeke moods yelds wool
> And skin to cloth our naked clotte of claye.
> He gives his flesh to feede our bellies full.
> Nought for him selfe he brings but for our staye.[79]

As in Mascall's poem, this short verse paints the world as one great sheepy buffet. And the sheep do not mind. Or, better yet, they have accepted that this is their purpose—to function as a series of animated use-values for the burgeoning numbers of "naked clotte[s] of claye" who put them to use. "Coombidee." Come. I bid thee.

It's tempting to imagine Jack Cade reading Mascall's encomium and Fella's visual epigram, struck by the omnipresence of sheep and sheep products in his world. Replay Jack's logic here, from an "ironic standpoint," and we might find ourselves feeling that the Christian "clotte of claye" now resembles precisely some parasitic, predatory, mimetic sheep-being or human-sheep hybrid—clothed in sheep, full of sheep, reading by sheep-light books bound in sheep, the words written on sheep or on paper "sized," enabled to bear ink from a pen such that we can take notes in the margin of a printed book, by being steeped in boiled sheep bones.[80] Beneficiary of natural law that he is, this warmly dressed, well-shod, well-fed Englishman has the means to provide for the pen, paper, and books

Figure 6. "June." Manuscript illustration from Thomas Fella's Booke of Divers Devises (Folger Shakespeare Library, V. a. 311), showing the labors of the season. By permission of the Folger Shakespeare Library, Washington, D.C.

The Image of Idelnes
T·F
The house and presse wherin the grape is converted to wyne
good wyn

in his closet as he is reading or perhaps writing a letter by candlelight. As he does so, he reflects gratefully on the kindly nature of English sheep by celebrating their, here "his" (this is a ram), usefulness, thanking him, or his own divine shepherd, as he must, for the virtuous, self-sacrificing (or, more properly, always already self-sacrificed) sheep become naturalized antinarcissism.[81] It's this antinarcissism or constructed passivity that renders sheep apparently plant-like, ambulatory growths or effusions of human destined use-values (wool, milk, skin, bone, flesh). It's this same structure also that renders the Christian viewer at once as beneficiary of this ovine bounty and yet strangely also the sheep's inferior. For who, except a sheep, could rise to such an absolute subordination of self?[82]

While it may seem that the coextensive zoomorphism/anthropomorphism to these *encomia* goes missing, it would be more accurate to say that it serves as their terrestrial grounding, a grounding that Jack derealizes in *Henry VI, Part 2* and whose abuse Raphael diagnoses in book 1 of *Utopia*. Within the likes of Fella's illustration, the logic of pastoral care manifests as a hierarchical series of metaphors that stabilize the contradictions that might arise from the crosscutting of the lines between *bios* and *zoë* that the motto sets in motion. With Fella, in other words, we watch and read the set of operations that cowrite or co-make sheep as "sheep" and "clottes of claye" as Christian persons, renders the sheep virtuously "sheepy" and the "clottes of claye" gratefully "sheepish," forced, that is, to read the sheep's essence as the standard that his own Christian devotion should emulate. For these sheep do not sacrifice themselves for "us" exactly but for our "staye," which resonates with a sense of both security and precariousness for these "clottes of claye," who, like the sheep, are finite, fragile beings, pressed together out of the earth to which they shall return when their lease is cancelled. The "clotte of claye" shuttles between gratitude for her nonsheepy privilege that enables her use of the world and a sheepy insufficiency or sheep-envy as she addresses the divine shepherd. The post facto attribution of a "selfe" to the sheep, a "selfe" that is constantly sacrificed, available and attributable only by the sheep's "use," represents the logic of deprivation that funds the apparent universalism of the sheep as emblem of a perfect subordination to divine will and divine planning. Thus, as the mottos to Fella's illustration imply, when you contemplate your own creaturely life, "clotte of claye" that you are, know that you must be or become a sheep, metaphorically, at least, in your conduct with regard to the divine.

Use the sheep. Eat the sheep. Shear the sheep. Clone the sheep. Boil the sheep down to make glue. But become the sheep in how you orient yourself toward your shepherd. We have been bidden, and so we come together to be counted. "*Omnes et singulatim.*" "Individual" "clottes of claye" emerge out of and disappear back into this undergirding *oves,* at once singular and undifferentiated, that stands for the many. Every day, everywhere they went, Jack, and his betters, such as might afford and read Mascall's or Topsell's book, lived out a series of sheepy metaphors, oscillating between the position of sheep and not-sheep. Sometimes it was necessary to own up to one's sheepiness, while at other times it had to be roundly denied. No apparent contradiction resulted. Sir Walter Raleigh's worldly antipastoral self wittily refuses an office he deems beneath him with the words "I would disdayne it as miche to keap sheepe," but he finds no difficulty declaring his pastoral poetic willingness when it comes to Queen Elizabeth I to cast himself as both a shepherd and a "gentill Lamm."[83] Yet, as Raleigh also knew, there were times when it was important for a not-sheep shepherd of men to insist that he really was, in fact, a sheep or a lamb, such as when he found himself, along with everyone else, singing Psalm 23, "lying down in green pastures," "the Lord [his] Shepherd" and "he his Sheep." Or, perhaps, one day, signing letters addressed to Queen Elizabeth I, with the epithet "your majesty's 'sheep,' and most bound vassal," as did Lord Chancellor Sir Christopher Hatton in the early 1590s—Elizabeth coming to refer to him affectionately, as Robert Cecil would opine later in a letter, as "her mutton," fantasizing Hatton's presence with her as she recreated out on the "Downes covered with sheep." Sometimes it is good for the human not-sheep to be a sheep—reassuring, or, maybe, even, given the right shepherd, a little sexy.[84]

Such are the benefits of sitting at the end of a parasitic chain of the relations of use that obtain between persons and the sheep, from which you extracted so very much "profit." Though, on occasion, when things turn nasty, you might find yourself treated as if a sheep, running confusedly away from the battlefield that your protest march has become, or, as in the case of Thomas Granger, a servant of some sixteen or seventeen years of age, in the Plymouth Plantation, sentenced to death for "buggery, and indicted for the same, with a mare, a cow, two goats, five sheep, two calves and a turkey."[85] At his deposition, he was forced to name these creatures, which were rounded up and killed along with him, for the "health" of all. But when "some of the sheep could not so well be known by his description," they "were brought before him and he declared which were they and which were not." He and they were executed "about the 8th of September, 1642. A very sad spectacle it was," the account offers:

> For first the mare and then the cow and the rest of the lesser cattle were killed before his face, according to the law, Leviticus xx.15; and then he himself was executed. The cattle were all cast into a great and large pit that was digged of purpose for them, and no use made of any part of them.[86]

Before he died, and to complete the sentence, Granger was forced to count the sheep one last time, counting them now as part of the distributive process that shall cut them from the flock. So it is that the contaminated of both flocks are purged regardless. Here counting becomes some perverse, distributive lineup in which the dissident "sheep" who has caused scandal names his fellow animal accomplices. The spectacle unfolds as a parody of recognition as these creatures are convoked as a blaspheming polity and condemned to death in a mass grave. Thus bidden, so we come, whatever our fate.

How to alter this settlement? To begin to turn these tropes differently, to own up to our sheepy co-relations, I need to follow one spooky spectralization of Dolly the sheep into the future, or to one, now dated, version of it, and reintroduce you to Rick Deckard and his wife, Iran, the central characters of Philip K. Dick's *Do Androids Dream of Electric Sheep?*

TROPE 8. "MY SHEEP'S ELECTRIC": SAN FRANCISCO, 1968/2021

Rick Deckard you may have met before in Ridley Scott's filmic adaptation *Blade Runner* (1982), but Iran you probably have not, unless you've read the novel—she's a casualty to the adaptation process, as the underachieving, slightly seedy, bounty hunter cum Cartesian knock-off of the novel becomes the dark, brooding, species boundary–crossing Harrison Ford of the film. In the novel, Deckard's dreams are a bit more mundane. After agreeing to "retire" the four remaining "andys" on earth, Rick tells his superior, Inspector Bryant, "If I get them, I'm going to buy a sheep." "You have a sheep. You've had one as long as I've known you," replies Bryant, referring to Graucho, the now defunct electric sheep of Dick's title. "It's electric," admits Deckard, hanging up on Bryant. "A real sheep this time," he continues to himself. "I have to get one. In compensation."[87] It's a little hard initially to understand the exact compensation Deckard imagines, or the exact nature of the exchange. The act of killing four more "andys" (electric or differently neuroelectric "humans") somehow obligates one sheep, one "real" sheep. And this difficulty serves as an index to the changed status of animals in the postapocalyptic San Francisco that Deckard inhabits.

In this San Francisco, so-called natural animals are at a premium. The social cachet of ownership that results from their scarcity value is most certainly at issue. These animals remain far more profitable whole than carved up in the manner of Mascall's blazon. More relevant still is the altered affective register in which animals appear, and so the sense in which the pastoral care (welfare) of persons is now indexed to the care they take of animals, to the "affective labor" they perform.[88] In the wake of the nuclear tragedy that serves as the novel's occluded genesis, the animal–human threshold has shifted. "You know how people are about not taking care of an animal," Deckard explains to his neighbor. "They consider it immoral and anti-empathic. I mean technically it's not a crime like it was right after W[orld]. W[ar]. T[hree]., but the feeling's still there" (13). What holds the remaining humans of the world together is a law of something on the order of the conservation of empathy (or Being) in a world in which everything else is turning into kipple, "living kipple." Empathy has become communal property, a collective duty to be performed publicly. Traumatized by the imminent sterility of the species, the fractured human *Dasein* finds itself even more intricately bound up with its animal-others, miming care and goodwill in the collective responsibility of caring for an animal—although it has become difficult now to adjudicate exactly who or what does the shepherding or what exactly is to be shepherded. It's this subtle rewiring of affect that now stands as the guarantee of human status, so much so, in fact, that those humans who cannot afford a "natural" animal are encouraged to buy an electric one to practice on and with, to make both they and their neighbors feel not better or "well" so much as anything at all.

At the same time, the state refines the distributive side of its management of the flock by performing tests on its citizens to ensure their genetic purity (within certain limits) and also their cognitive and emotional responsiveness. Fail these tests and you shall find yourself declared a "special" or "chickenhead," as in the case of J. R. Isidore, through whose eyes we observe the remaining andys' moments prior to their "retirement." The Voight-Kampff test that bounty hunters administer to suspected androids before "retiring" them performs an equivalent function: physically identical to humans in all but their bone marrow, the test measures the emotional response time of its subject to questions that depict animals fragmented into parts, transformed, I am inclined to say blazoned, as they are put to use. Mascall's "In Praise of Sheepe" might almost provide the script for one of these lethal quizzes. Deckard gives the test to Rachael Rosen, presented to him as the niece of Eldon Rosen, inventor of the escaped Nexus 6 androids, whom we discover shortly thereafter to be an android herself: "You are given a calf-skin wallet on your birthday" (48), proposes Deckard. "I wouldn't accept it," Rachael says. "Also I'd report the person who gave it to me to the police." Rachael is passing, performing her humanness adequately. "'In a magazine you come across a full-page color picture of a nude girl.' . . . The gauges did not register. . . . 'Your husband likes the picture,'" Deckard continues. "Still the gauges failed to indicate a reaction. 'The girl,' he added, 'is lying facedown on a large and beautiful bearskin rug.' The gauges remained inert, and he said to himself, 'an android response,'" by which he means, in Cartesian terms, an automatic or automaton reaction.[89] Failing to detect the major element, the dead animal pelt, means you fail the test. Rachael is insufficiently empathic; or, rather, her mimetic faculty is only so good.

As we learn, the Voight-Kampff test depends upon an ontology that insists on a fundamental disconnect between solitary predators (spiders, cats, sociopath humans, androids) and herd animals: "empathy, [Deckard] once had decided, must be limited to herbivores or anyhow omnivores who could depart from a meat diet" (31). Of course, the novel complicates this distinction even as it may reinforce it. We learn that there has been a succession of ever more subtle tests, each of which must delimit with ever greater or ever slimmer certainty the diminishing difference between android and human response times. Likewise, certain "specials" may not be able to pass the test, but, still, they remain "human." Then, of course, comes the irony that many of the novel's readers from 1968 to the present may also not fully pass the test, were they to take it, failing to pick out the archival remains of the animal that serves as backing or material for this or that object in the story or scene shown to them. In the end, Deckard's homegrown ethological theory that "andys" are merely "solitary predators" (31) enables him to kill them with little to no remorse, and he rewards himself with a real, live goat. And courtesy of J. R.'s horror at watching the "andys" dismember a spider, the novel seems to encourage us at least to understand or even accept the slim difference not between what we may call a supposedly organic "response" and a supposedly machinic "reaction" so much as the differing capacities and orientations of herbivores and predators or meat eaters.[90]

Further problems arise when Deckard realizes that he has begun to feel empathy toward female androids. He no longer wants to do his job. Rachael takes him to a hotel. They drink real coffee. They have sex. Afterward, Deckard looks sad. "'You're not going to be able to hunt for androids any longer,' Rachael says calmly. 'So, don't look so sad. Please.'"—"No bounty hunter has ever gone on" (198). Sex with Deckard constitutes a survival strategy of sorts, a ruse the androids have developed to create ties of affinity with the very humans who are sent to kill them. The question of real ties is not voided but merely set to one side. Deckard determines to kill Rachael—"if I can kill you," he reasons, "I can kill them." In the event, he cannot kill her, but he does kill the rest of them. It costs him dearly, though. He calls Iran to tell her it's all done, that he's done, and learns that his goat is dead. "I'm sorry," she says, "the goat is dead" (226). "I saw her very clearly," Iran continues. "A small young-looking girl with dark hair and large black eyes, very thin. Wearing a long fish-scale coat. She had a mail-pouch purse. And she made no effort to keep us from seeing her.

As if she didn't care" (226–27). "It's so awful," she says, "so needless" (227). "Not needless," replies Deckard. "'She had what seemed to her a reason.' An android reason, he thought." Deckard heads north into the wilderness, where he finds, as he thinks, a real, live toad. He takes the toad home to Iran. And they start their lives over—only to find, of course, that their toad's electric (242–43).

The novel ends, as Ursula K. Heise notes, with an acceptance that "electric animals" have some sort of life and by extending rights of semi-citizenship to artificial animals as surrogates for the real thing, but systematically seeing off the androids.[91] And like all good domestic fiction, at the end of the novel, Deckard and Iran reconstitute their relationship as caregivers, and he enjoys the patriarchal privilege, not of raising children, but of caring for an electric toad.[92] Iran orders up a "pound of artificial flies that really fly around and buzz." Deckard quits the bounty hunting business and gets some rest. In the end, he accepts "electric animals" because he's ceased to worry about fetishes or denunciations or being exposed as an owner of a "fake" animal. He understands that precisely because he cares for this toad, precisely because he treats the toad as he would a theoretically more "natural" animal, it becomes a full as opposed to a dummy subject. And for Deckard, this is what it means to become "well." It means also that, while "andys" remain other to him, that is merely to say that they are different, different in kind, but still kin, predators as opposed to herd animals, as he posits some constitutive mimetic commonality among ungulates and their not-sheep humans. This does not mean that "we" are all one big, happy family. On the contrary, collectives are now constituted based on ties of affinity and practices of care. The mimetic basis of the human, mimicry, becomes less a point of demarcation or division than a zone of appearance for signals that may be processed as both a reaction and a response. Mimesis becomes, on this basis, the zone in which differing, finite beings encounter the rhetorical process of others whose gestures, signals, movements they absorb, alter, and reenact. Repetition, iteration, produces successive patterns that comfort (or do not), that build something that may tie beings together. To own this insight and to choose to live it, for Deckard, is to become "well," or at least to feel genuinely tired out by real, affective, if immaterial labor.

At the end of the novel, he falls asleep without the mood-altering organ with which the novel begins. He forgets about Graucho. He counts no sheep, electric or otherwise.

"Does the animal dream," asks Derrida. "Does the animal think?" "Does the animal produce representations . . . a self, imagination, a relation to the future as such? Does the animal have not only signs but a language, and what language? Does the animal die? Does it laugh? Does it cry? Does it grieve? Does it get bored? Does it lie?" If the novel answers affirmatively for the "andys" and for Deckard, even as it may decide that those "dreams" and "representations" prove incompatible, it opens the possibility of inquiring into the ways in which the dream-life or representational worlds of our animate others might begin to "count," in the second sense of the word, and so begin to matter, to phenomenalize as a matter of care and concern in our various presents.[93] As Deckard tells Iran, Rachael's killing of their goat was not necessarily a senseless, irrational behavior; she had an android reason or rationality for doing so. It is not, as Derrida elaborates, that hard to conceive of forms of "non-bare *zoë*," which is to say, merely, differently "backed" forms of life. To do so requires us only to understand that "at bottom everything we have spoken about came down to problems of translation. Translation in a sense at once fundamental and diverse."[94]

Making good on my sheepy reading, passing beyond the logistical counting or inventorying of the ovine figures that I have performed in this chapter, requires knowing what Deckard knows at the end of the novel, that the connections or ligatures between persons and things that produce beings on the order of Dolly, Serta's uncounted sheep, Raphael's wolfish flock, Mascall's sheep–humanist hybrid, or an electric sheep named Graucho, and categories of being (an ontology, the discourse of species), manage to do so only if the final instance, once the end point in a chain of making, is severed from the whole and offered up as a description and rationale for the whole. Our actions, our mimesis, prove constitutive. They (re)make worlds, establishing certain routines by whose repetition those worlds remain the same. The "sheepy" reading on which I have embarked in this chapter, counting in order to stay awake, counting so as to resist the sleep that seems to come when we reach an end or a discourse of "ends," means regarding the whole network of associations that constitute animals, plants, and world, all of this immense "us," all of us, considered now as partial, hybrid, folded forms of being. The difference between entities lies in their distribution or stacking throughout this web—at the center of which, throughout this chapter, has sat that most elaborately networked subject of all, the English *anthropos* speaking Mascall's poem, the arch-parasite, who is king because he sits still while the rest of the world is in movement.

Is it possible to reverse this settlement, to place sheep at the center of this web? How then does their world look? What happens if we take Derrida at his word and regard questions of ontological difference and category breaks as merely a matter of translation and so of the media ecologies necessary to enable the transfer between domains? What happens if we allow sheep to dream and so to author their own multiplied, altering sense of a "we" or an "us"?

TROPE 9. THE BIG SHEEP: DEVON, 2015

I am happy to say that I have found Utopia. It's not the United States (sorry) but is in fact a tiny *locus amoenus* in North Devon. There is a map, and I am happy to provide directions. Get on the M5 outside London and take that to the A361, aiming for Bideford. Then just keep your eyes peeled for the "BIG flag!" There you shall find "Ewe-topia."[95] The price of admission is much less than the cost of fitting a ship for a sailing expedition, such as Raphael undertook: twelve pounds ninety-five per adult, six pounds for toddlers and older, but under threes are FREE. There is a family ticket. And there are discounts. What delights await you? Indoor and outdoor play zones, such as Ewe-topia, the chance to stroke sheep, a sheep show, duck trialing, sheep racing (including such obstacles as Shepherd's Brook and Ewe-turn), sheep milking, sheepdog trials and training, lamb feeding shows, and Nature Trail. There's also the Sheep Shop and the Shepherds restaurant. And it turns out, by the way, that Utopia, as I suppose it always had to be, is "An All Weather Attraction," proof against the vagaries of time and place.

Perhaps I should stop here—except to point out that *Utopia* today consists of adequate and affordable day care that enables parents of a certain means to enter the paradise of a little conversation and a bit of time alone while the kids run around whatever sheep-themed maze is provided. Ewe-topia manifests as but one instance of the historical development of capital, the naturalization and extension of *otium* (leisure or idleness) to all who "work" or who work (consume) as they "play." If you have a car—but I am sure there are buses—you can get there. And it is probably a good day out at a fair price. They have strict ethical standards on how they treat the sheep. Ewe-topia speaks directly, then, to the status of British farming in a landscape refigured by EEC legislation. It participates also in the increasing struggle between urban (London) England and the declining rural constituency, recirculating the landscape and the labors tied to it as fungible assets. It is then a rather canny, though by now routine, postindustrial conversion from production to leisure services. And, in Ewe-topia, sheep—O, happy sheep!—prove more valuable alive than dead.

But this is to read Ewe-topia merely as a contemporary site. Its name, as the intralingual pun implies, places it securely within a longer tradition, which it parodies, or rather recirculates, cites, and situates, reanimating Utopia as itself a kind of capital, worth a giggle, worth a visit by phenomenalizing the silent female sheep that lurks within its title. In this sense, it performs the same kind of cultural work as More's own text, albeit in redacted and reduced form. Ewe-topia completes the rhetorical project, cashes out the intralingual pun to the title and, in the process, brings Raphael's rhetorical sheep into being as they take their place alongside More and his fellow humanists (then and now) on the service side of the text. Sheep are momentarily promoted from the food chain to the service industry. They remain "stock." They are still consumed, though now mostly in metaphorical terms. This shift in status is incidental, of course, but the differences it might generate remain unknown. It has yet to be decided. Does Ewe-topia, in other words, serve up, in touristic mode, affective ties long since past that may have formed on the small family farms such as Thomas More's *Utopia* and his inheritors, the Commonwealth Men, sought to protect from the depredations of enclosure? On such farms, the sheep, cows, ducks, and chickens that were part of the laboring circuit might have had names, and might also, on occasion, have found themselves named in wills and testaments.[96] It is tempting, then, to say, yes, that it is so, and that the ties that form between the children and other human animals that visit the sheep of Ewe-topia sustain the possibility of something more even as or perhaps because they come routed through the figural sheep that stream into our living rooms daily—the descendants of Serta's uncounted flocks.[97]

In "Bitzer Puts His Foot in It," for example, an episode of *Shaun the Sheep,* the series that individualizes the let-go sheep of sleepless nights from Serta's advertising campaigns, the animators imagine a scene in which the show's sheep with their sheepdog turned co-conspirator run amok with a patch of wet cement, creating all manner of sheepy impressions that present on or by the hoof. In one frame, Shaun the Sheep and friends have drawn a Hollywood star in the cement (Figure 7). They register their achievement with an instant Polaroid moment. The episode presents as a succession of archival or autoarchiving moments in which the emphatically singularized "Shaun" finds himself ennobled by the success of the show and so written into the text of stardom as hosted by a hastily constructed stretch of Hollywood Boulevard down on the farm. Shaun—even if he represents merely the show itself—makes his mark, takes up a relation to the trace, and becomes, if not a subject, at least a star.

When their farmer/owner/"father" turns his back, the otherwise fairly sheepish sheep get up to all kinds of crazy writing games, of which Shaun's stardom is but one instance. Does the name "Shaun," which emphatically presents a denuded self, a naming that enacts or registers as deprivation, speak to a desire to write "sheep" differently, a desire to encounter traces of their own making, a desire for differently articulated tracks? He is the only sheep on the show that has a name—until the spin-off aimed at even younger audiences, "Timmy Time." Of course, the iterative scheme of the show requires that not much can change. Each episode unravels itself by shoring up, however implausibly, however comically, the Plasticine farmer as head of this multispecies household, and all of the sheeps' inscriptions are erased. But it is "Shaun" who receives top billing and whose star power this episode celebrates. What would it

Figure 7. Shaun proud of his Hollywood star and hoof print captured in cement. Image capture from *Shaun the Sheep: One Giant Leap for Lambkind*, episode 4, "Bitzer Puts His Foot in It" (2007).

mean to take our cue from Ewe-topia, from Shaun and Timmy, and extend More's humanist table to sheep, to invite them to participate in the nexus between work and leisure that his text explores? What would happen if we tried to offer them a place at such a table or Commonwealth, however they might have to be reconfigured to accommodate creatures with hooves as opposed to hands and feet?

Such a project would entail, as a start, hearing the word "count" differently, according to its second sense. We shall have to ask what "counts" for sheep, with us and without. It's time, in other words, to play the shepherd again, to count the sheep, all over again, as Foucault knew we would have to. *Omnes et singulatim.* But this time, let's change the terms of the counting to see what it may yield. No more counting to reassure myself that the risks to the flock have been managed, that one of our lively, animate stock hasn't gotten it into his head to wander off or be stolen. No more counting so that we may go to sleep. I count to stay awake. One, two, three, four, each number, each iteration, exists not as the disanimating marking off of a series of interchangeable units (sheep) but instead as a measure of time that invites our tropes to dance, offering the counting up as the potentially enlivening scene of an encounter between beings whose contours shift and have yet to settle into a particular shape.[98] For I am no longer exactly sure what a sheep, let alone a singular, historical sheep, or a single, historical flock *is* exactly (to say nothing of ourselves). I am not even sure whether it makes sense to speak of a singular sheep, a sheep plucked from the flock or some other yet-to-be-named intragroup. And this static or interference proves productive.

As primatologist turned sheep observer Thelma Rowell cautions me, I have to proceed on the basis that much of what we know of sheep derives from the way they have been rewritten by the modeling of sheep and humans as coeval multiplicities. Rowell argues that the selective breeding of sheep, their manipulation as livestock or living capital, has essentially rendered them "sheepish." The traditions of animal behavior studies have dictated that those animals who lead interesting lives (that is, lives deemed interesting to us) have tended to serve as privileged experimental subjects—especially if they may be grouped as among the relatives of a certain *Homo sapiens.*[99] Animals, "most animals," she observes, who "spend the majority of their time doing nothing," tend to be neglected or asked only the most boring of questions. "Sheep behavior studies," she elaborates, "are mostly to do with what they eat, and sheep are not, generally, permitted to organize themselves."[100] What would it mean to reverse this settlement and allow sheep, historically particular sheep, to presence as if a subject and allow them to constitute whatever mode of collectivity corresponds to a "we" or an "us" for them? To find out, let's take our cue from Rowell and head off into the countryside where she watches certain historically particular sheep and, in watching them or watching with them, attempts to undo one particular modality of species hegemony that has rendered them (and us) "sheepish." Rowell accomplishes this goal by creating a highly specific multispecies writing machine in which she "decide[s] to watch sheep in the same way . . . [she has] been watching monkeys," promoting sheep up a deleterious species hierarchy. The effect, I shall argue, is to rewire the relations between *otium* (leisure/idleness) and work *(negotium)* that were programmed for us in More's *Utopia,* across and the lines of species and kind. Thelma Rowell writes pastoral for sheep. Or, better still, she invites them to write their own and so opens our discourses to still other genres or species of writing.[101]

There. Perhaps I have finally drifted off like all the West's shepherds, tranquilized by my counting. Or, perhaps, for the first time, I am truly awake? Although, if I am, I find myself left with a question, a question, unlikely as it may seem, that I hope might enable us to prorogue the disappearance of the ovine substrate within which we are made. And so, I beg to be reminded, one more time, and just so that I can be sure: what was pastoral (again)?

2

What Was Pastoral (Again)?

MORE VERSIONS (*OTIUM* FOR SHEEP)

> Today I would like to finish with these histories of the shepherd, the pastor, and the pastoral, which must seem to you a bit long-winded, and return next week to the problem of government, of the art of government, of governmentality from the seventeenth and eighteenth centuries. Let's finish with the pastoral.
>
> —MICHEL FOUCAULT, *Security, Territory, Population*

> "Does not even a consideration of the adaptation of man's limbs to their functions convince you that the gods do not require human limbs? What need is there for feet without walking, or for hands if nothing has to be grasped," or for other "parts of the body, in which nothing is useless . . ." What need, we might equally ask, has *Otium* for hands?
>
> —BRIAN VICKERS, adapting Cicero in *De Natura*, "Leisure and Idleness in the Renaissance, Part II"

At the end of the last chapter I may or may not have fallen asleep—along with all the sometimes sheep, sometimes shepherds I was in the middle of counting. Some of these sheepy shepherds ended "well" or well enough, *soma* and *psyche* recalibrated, like Rick Deckard in *Do Androids Dream of Electric Sheep?* Others, all those lucky owners of a Serta mattress who count no sheep, simply conked out to wake the next day bright eyed and bushy tailed. Still others arrived at a literal, graphic, horrid ending, violently cut from the flock, as in the likes of Jack Cade, Thomas Granger, Antonio Gramsci, and countless more now deprived of name. It seemed difficult, as I counted, to keep the node or switch between coeval multiplicities (sheep and not-sheep humans) in focus. The tropes turned. Our sheepy skins appeared only to disappear, flickering in and out of view, as did the variously animate presences of the sometimes literal, sometimes figural sheep that we come into being with. How, then, to maintain this focus? How to render the matter of this shared metaphor available for shearing? How do I incline the logistics of this counting toward a scene of translation in which the serial rationality of counting, counting as calculus, as *logos*, morphs into a question as to what "counts" for sheep and so also for "us"?[1]

This chapter is eco-friendly but trades in echoes. Its title reprises Paul Alpers's synoptic *What Is Pastoral?* (1996) by way of a parenthetical repetition that demands to be reminded, one more time, and just so that I can keep things straight, "What Was Pastoral (Again)?" My answer, or simply what follows, "More Versions," may be understood as an addendum to William Empson's *Some Versions of Pastoral* (1935), in which the pastoral mode proliferates, becomes a way of describing the structure of being, a structure that the further proliferation of *otium* (leisure, idleness, boredom) to sheep, *otium* "for" and "by" sheep, might be said to alter.[2]

My aim is not to recover some idealist definition of pastoral that will see off competing models or settle the matter once and for all. Nor shall I attempt to revalorize pastoral as a mode of nature writing or environmental critique (as on occasion it may prove to be).[3] On the contrary, by troping Alpers's and Empson's titles, I seek to reinhabit critical skins too quickly sloughed and to fill them out in a way that enables us to understand the full biopolitical significance of primatologist Thelma Rowell's decision to watch sheep as if they were chimpanzees. For by so doing, she invites them to cowrite something like pastoral, but pastoral that now plays out across the discourse of species, folding together humans and sheep in unexpected ways. If outrunning our sheepy metaphors and sheepy anthropo-zoo-genesis proves difficult, then perhaps there remains something (however small) still to be had from pastoral, something into which Rowell taps, even as it may otherwise manifest under the sign of boredom, idleness, or fatigue. It may prove helpful, then, to consider this chapter a foray into a comparative ethology or an ecology of practices that inquires into the way a field science such as animal behavior studies, and softer endeavors such as literary history and pastoral load different entities (creatures, texts, temporal effects) into discourse. With Rowell, I argue, pastoral (or something like it) decamps from literary historical discourse to ethology: its terms and tropes turn in the process even as they continue to provide the scenography for her encounters with sheep. Is the result, shall the result, "must" the result, be pastoral? would be Empson's question. It shall also be mine.[4]

I begin by offering a partial, necessarily impressionistic redefinition of *pastoral* in relation to a long history of *otium* as a technique or calculation and so to a script that codes labor and leisure via a constitutive but incomplete anthropogenesis. There remains something potentially scandalous to *otium,* this neutrality, idleness, or merely empty time. The word gestures toward a basic, grounding, plantlike *phusis* that manifests not in the form of sleep but as an ambient disengagement from our routines or ways of being.[5] *Otium* names and exceeds the necessary passage that must be crossed and recrossed each and every day as the motive force of bodies and minds that labor is managed. It names a state of rest, a zero degree, through which we pass, a vegetal being allied to pure growth to which we momentarily return, conscious but occupied, our agency and awareness turned inside out or shifted to one side. We find ourselves configured as if a plant, as if a being of pure growth, become neighbors to all the plantlike sheep of England's enclosure movement whose animate, living "stock" were used to "grow" wool and flesh. Such a wakeful but disassociated state proves productive. It comes allied with the process of thought or knowledge building in humanist vocabularies, with a mode of counterfactual thinking, that we name utopian. But get stuck there, refuse to make the calculation—worse still, choose to linger there—and you may precipitate some deevolution or deformation of the "human" via a loss of handedness that paradoxically causes us to cohabit with the divinities for whom hands are superfluous and the beasts who are said to lack them. While the classical tradition might laud this state of intellectual repose, its humanist readers also inherited a commentary tradition from Christian homilists that cautioned against this deactivation or lethargy, this sloth or idleness *(acedia).* For by what ratio, by what measure, could you know exactly the difference between virtuous leisure and mere idleness or sloth?

Brian Vickers captures this ambivalence quite precisely when he revises Cicero in *De Natura* and asks "what need . . . has *Otium* for hands?" None is the answer, none at all.[6] For with *otium* we enter into a state of being in which a relation to *technē,* and to the trace, turns on itself to produce an overwhelming, intoxicating high (or low)—something Vickers illustrates by way of Andreas Mantegna's *Pallas Expelling the Vices from the Garden of Virtue* (before 1503) (Figure 8), in which *Otium* is personified in the foreground as a man without arms, led forth or away by *Inertia.* To the far left of the image, wound with allegorically inclined mottos, Mantegna places a strangely evolved tree or deevolving human figure, part person, part plant, that both watches and frames this scene of enclosure.

The painting serves as an iconographic and allegorical superlative that encapsulates both the negative valuation of *otium* as idleness and the captivating allure or strangeness of the bodies it produces. Ovid's *Remedia Amoris* may provide the orienting commentary, "Otium si tollas periere cupidanis arcus" (Eliminate idle pleasure and Cupid's bow is broken); Justice, Temperance, and Fortitude may look down upon the scene, but even as our eyes are drawn right by Pallas's lines of force, the depth of greenness (the natural–cultural topiary) to the left draws us back to the sentient immobility of the tree-woman or woman-tree as a figure of the surplus or reversibility to *otium* and its attendant vices and virtues.[7] Pallas moves through the picture space with the intent of emptying this garden, of restoring movement to its languor—of shoring up categories—but her movement does not rewrite the scene; the mixed animal–plant bodies remain, an iconography of creaturely indistinction that registers the interpenetration of animal and plant as such. The painting captures thereby the way the condition or passage, the arrested production or productive stilling we name *otium,* crosses, folds, or forgets this vegetal being, revisiting it or titrating it in small, many dosed pleasures that salve or assuage our pains.[8]

Thelma Rowell's protocols of observation interfere with this process. By giving time to sheep, by waiting upon them and what they do, she voids this calculus that renders *otium* both a virtuous, rationalized technique and a desirous, sleepy remainder. If sheep have traditionally served as the abject mode of such a plantlike state, their animation, their very creatureliness, just so much backing to their flesh, Rowell offers them some measure of practical sovereignty, a chance to decide, so it appears, on what counts for them. In doing so, she rezones *otium,* siphoning off

Figure 8. Andreas Mantegna, *Pallas Expelling the Vices from the Garden of Virtue,* before 1503. Tempura on canvas. Musée du Louvre, Paris. Photograph by Erich Lessing/Art Resource.

its metaphysical aura so that it comes to serve as a figure of empty time merely, a waiting, or pause, as the drama of subjectivization plays out across and between the so-called divide of species. In her hands, or rather somewhere between her sheep's mouths and hooves and her eyes and ears, *otium* comes to reside in the media ecology, translational, or anthropo-zoo-genetic practice crafted to enable sheep to "count." *Otium* becomes a technically necessary input merely, a period of observation during which we do not act and so are forced to wait upon the actions of historically particular sheep. Paradoxically, as Rowell does so, pastoral returns to its origins as a form of modeling, a media ecology or tool of translation, that creates the appearance of depth, of an inside as well as an outside. Redistributing its terms and effects across and between animals (human and sheep), Rowell invites us to reprogram the anthropo-zoo-morphic calculation that is *otium,* the biopolitical switch or relay that calibrates our sheepy shepherding.

So, let's not finish with the pastoral quite yet—even if in not finishing I risk the appearance of being long-winded and even a certain element of boredom or fatigue. I understand the desire to finish, to be "done" (and so to sleep). I understand also the desire to render the anthropological machine that creates botched instances of the "human" by sloughing off the "animal" inoperable and to assert some order of suspension or deactivation. It would be good to finish. It feels good to be "done" and so at rest.[9] I fear, however, that we are only just at the beginning. For, let's not pretend that our sheepy, shepherd skins shall one day simply fall away. They remain, even as they may come to take other forms. And let's not assume either that we know what pastoral *is* or may become even as we may agree with the Leo Marx on a famously minimal definition: "no shepherds, no pastoral." To which we shall have to add sheep. "No shepherds, [no sheep] no pastoral."[10]

LOOK, NO HANDS!

Like most Westerners, that is to say, historically particular animals "cut" from the general flesh so that they enjoy certain privileged social hieroglyphs of species, race, gender, and generation, my first encounter with pastoral was in front of the television. I do not mean to say that as a child I was deprived or didn't get outside much but rather that my world was punctuated by the times at which certain programs began and ended, by what, in *Television* (1974–75), Raymond Williams names "the mobile concept of 'flow'"—the experience of temporality as mediated or emitted by the technologies of broadcast television.[11] Growing up in the United Kingdom before the advent of the remote control, the VCR, the CD, DVD, DVR, a multiplicity of channels, let alone the binge-worthy remediation of television shows or their production for any number of streaming digital platforms, most evenings I got to luxuriate in the "flow" programmed for us down in London that materialized up North by the turn of a knob on the "set" in the living room. In the evenings, 7:15 P.M. heralded the end of the national news and assorted local news magazines and the airing of this or that rerun of *The Rockford Files, Star Trek,* or any number of newly minted sitcoms and nature programs. Life as mediated by the pastoral care

I received from the "box" seemed rich. This "box" teemed with life. So when, now, I invent my childhood or media infancy all over again, I can pinpoint with seeming clockwork accuracy the arbitrarily decided moment at which, with the *negotium* (business) of schoolwork handily put away, time refolded itself into the collective *otium,* leisure, downtime, or release that the TV afforded.

"It's good to put your mind in neutral, to idle, to vegetate" was the moral my home-grown humanist father deduced from this activity—speaking a democratized version of the script that Quentin Skinner finds rehearsed in book 1 of Thomas More's *Utopia,* where the Platonically inclined Raphael Hythlodaeus holds forth while the more sanguine, Ciceronian Morus listens, contemplating the pros and cons of Platonic withdrawal versus a life of public service.[12] In book 1, Raphael voices the standard Platonic elevation of *otium* as the highest and most moral state of being only to be attacked "point for point . . . not merely from the general perspective of a Ciceronian civic humanist" but by More, who speaks the script Cicero provides in *De Officiis.*[13] Depending on whether you agree with Skinner, the script More lifts from Cicero seems to win out, and book 2 imagines a world not of self-cancellation or human annulment so much as of humanist self-actualization, in which *otium honestum* or *negotiosum,* good, which is to say, useful *otium,* is available to all and mere idleness has become a structural impossibility.[14] If, in other words, *Utopia* imagines a humanist collective or habitat that fosters engagement with the good of the whole, then this habitat is premised on the figure of the neutral, the idle, of leisure, or the space that comes between, winking in and out of being, ratified by its proleptic tasking with the good of the Commonwealth, work, and world. Cradled by the guarantee that "tomorrow" meant a return to the worlds of school and work, we idled in the teletopical pastoral idyll afforded by the programmed flow that issued from the box. Handless, my parents idled as they sat and watched my sister and me play remote control, nipping up and back to the TV as needed to change the channel or the volume.

I have begun with this self-indulgent (idle) snapshot of a fictive (idyllic) childhood to make the point that the story of pastoral is keyed to a long history of *otium* and the humanist calculus that programs a set of relations between home and work, leisure, labor, and media. As Anthony Grafton and Lisa Jardine have argued, scripts, such as those Skinner finds retooled in More's *Utopia,* remain operative today in the educational protocols and prerecorded ideologies of a for-always-embattled humanities.[15] Obviously, by their translation to new media platforms and differing demographics, those scripts change—the technical humanist term *otium,* for example, ceases to be keyed to the production of knowledge in the form of what Timothy Reiss calls a "passage technique" and instead morphs into a notion of *tempus* merely (free time; downtime; the time that comes between; the plantlike "sleep" that Serta's consumers buy when they buy a mattress; the entry fee for a visit to the Big Sheep theme park).[16] Distributed more freely through the collective, *otium* become *tempus* recasts Platonic withdrawal or idleness now as a weekly or daily technique, by whose observation human subjects are enlisted in maintaining their stability and happiness, and so also the stability of a labor force. The allied discourse of "wellness" and wellness programs similarly replays the age-old question of philosophy, "how to live well," as a question of optimization merely, of a labor pool's good somatic and psychic hygiene.

More's *Utopia* programs the calculus, settles on the script. In a letter to Ulrich von Hutten in 1519, Erasmus observes, "More had written the second part [of *Utopia*] because he was at leisure [*per otium*]" in Antwerp, "and the first part he afterwards dashed off as opportunity offered [*ex tempore per occasionem*]."[17] More fosters this image himself in his prefatory letter to Peter Giles apologizing for how long it took him to write up this "little book," which, as he says, merely transcribes Raphael's narration. The problem, he writes, was that his "other tasks left [him] practically no leisure [*temporis*] at all."[18] These "tasks" include the law; courtesy visits to important men; and then, "when [he has] returned home . . . talk[ing] with [his] wife, chat[ting] with [his] children, and confer[ring] with . . . servants." "All of this activity," More concludes, "I count as business [*negocia*] that must be done . . . unless you want to be a stranger in your own home." But "when then can we find time to write?" he asks. "Nor have I spoken a word about sleep, nor even about food, which for many people takes as much time as sleep—and sleep takes up almost half a man's life." *Utopia* was written with whatever time *(tempus)* More "filch[es]" from sleep and food. Writing finds itself subtracted from eating and sleeping, from the maintenance of the body, personal and social.

The letters from fellow and largely adoring humanists that come to preface *Utopia* perform minor variations on the script. And at the end of book 1, just as Raphael is about to describe Utopia proper, the text calls attention to the specific calculus between *negotium* and *otium* that More performs in his own prefatory letter. "My dear Raphael," he says, "I beg . . . you, give us a description of the island. Do not try to be brief, but set forth in order the terrain, the rivers, the cities, the inhabitants, the traditions, the customs, the laws, explain in order their fields, rivers, towns, people, manners, institutions, laws—everything which you think we should like to know. And you must think we wish to know everything of which we are still ignorant" (108/109). In response to this call for a total description—a description, which as Frederic Jameson observes, produces the "very prototype of a narrative without a narrative subject and without characters"—Raphael asks not for time but for *otium*. "There is nothing," he declares, "I shall be more pleased to do . . . but the description will take some [leisure] [*sed res ocium poscit*]."[19] "In that case," replies More, "let us first go in to dine. Afterwards, we shall take up as much time [*tempus*] as we want." More, Raphael and Giles go into dinner and then return emphatically to the same place ("pransi in eundum reversi locum, in eodem sedili consedimus"), and Raphael begins. Book 2 is a lot to take in in a single sitting also, so at the end of Raphael's narration, a light supper is provided: "manu apprehendens intro cenatum duco"—More takes Raphael by the hand and leads him into supper (244/45).[20]

The very conventionality of the meal as framing device is the key to its significance. As Michel Jeanneret observes, the use of meals in humanist texts is part of their language of affiliation with Platonic symposia and dialogue traditions.[21] The Erasmian dictum *eddere et audire* (to eat and to listen) figures the "double pleasure" of "tak[ing] in stories while eating . . . supper."[22] And in the table-talk tradition, the meal itself is usually described, and usually accords with the nature of the conversation, because, famously, digestion (*rumination*—"chewing the cud") serves as the primary metaphor for acquiring humanist learning. Book 2 of *Utopia,* however, unfolds as a different order of narration. The utopian *res,* the total immersion or description More requests, demands something more and less than time. It demands *otium,* freedom from bodily and worldly concerns, a state of suspension. How then to derive enough *tempus*—all the time he needs—from their day?

The three men retire and eat and then return to exactly the same place. Utopia or *Eu-topia* (the place of the good) unfolds in the interval between dinner and supper. Utopia proves post- or interprandial: it may only be spoken of or thought of on a full stomach. Eating literally takes no time. Food serves as a given, a necessary input to this narration that is also an act of making, of knowledge production. During this interval, the men are at rest. Their bodies idle while their minds extend into the fictive space brought into being by Raphael's descriptive imaging technology that builds worlds. Sitting in their garden among their plants—the ambiguously valorized *hortus conclusus* or *locus amoenus* that Hans Holbein depicts in his 1518 engraving (Figure 9) synonymous with both the highest virtue and the worst depravity—Raphael, Giles, and More come to resemble the Utopians in book 2, who, as Raphael tells us, are "very fond of their gardens . . . spend[ing] one hour in recreation" there "in the summer," while, "in winter," they sit indoors and "play music or entertain themselves with conversation" (129).

By their careful social engineering based on communal living and the elimination of private property, the Utopians create a collective that enables them to regulate their bodies, individual and collective. The needs of the body remain continually satisfied, leaving no opposition or tension between the worlds of *otium* and *negotium.* Accordingly, they have "scarcely" any need for medicine, "the knowledge of [which]" they regard "as one of the finest and most useful branches of philosophy," using it to explore "the secrets of nature . . . and so [to] win the highest approbation of the Author and Maker of nature." Invoking the figure of the *Deus Opifex,* they assume a divinity that "hath set forth the visible mechanism of the world as a spectacle for man, whom alone he hath made capable of appreciating it" (183). In doing so, they also attain the fantasy of an Archimedean point of knowing the rules of their system even as they participate within it—a system in which they may intervene even as they make or invent. The Utopians permanently inhabit the hiatus effect that

Figure 9. Hans Holbein, engraving for Thomas More's *Truly Golden Handbook for the Best State of a Commonwealth Called Utopia* (1518), showing Thomas More, Raphael Hythlodaeus, and Peter Giles in conversation in a garden while they are attended by John Clement. By permission of the Folger Shakespeare Library, Washington, D.C.

More's text deploys. As busy as they are, the Utopians are always idling, but they are never idle. Idleness, finally, finds its use-value and *otium* becomes a perfectly regulated technique that enables the humanist project.

Of course, in Holbein's engraving, little John Clement, More's "pupil servant" and young "plant" *(herba)* (40/41), is pictured running to and fro, carrying a flask of wine or ale. He lubricates the scene to keep the conversation flowing. By the translation of the text from the verbal soundscape of a dialogue to the visual regimen of the engraving, Holbein recovers the movement of persons that permits the movement of voices—highlighting the division of labor that produces the luxury of *otium* and also the figure of the humanist as a kind of technique or technology of reason. Remote control Clement helps to produce this disembodiedness; preserves a relation to the hand; permits the creation of More's "free" time. But that's OK, for, as More offers, he "does not allow him to absent himself from any talk which can be somewhat profitable." Clement catches what he can in between his tasks. He listens in on the Latin oration as he nips in and out, attends to the drinks, and learns what he can in the process. More ensures also that while this utopian narration, this scene of *otium,* unfolds, the impressionable Clement remembers his hands.

GOING MOBILE

What was pastoral (again)? If my staging of pastoral *otium* as a calculus of labor and leisure keyed to its role in the production of hyperrational(ized) human persons seems overly utopian and so rather esoteric, it is because I seem to be demoting key elements that are frequently taken to define the activity—shepherds, their "natural" milieu, the "natural world" itself—to the role of scenery. Venture to the mid-seventeenth century, for example, when, as Keith Thomas tells us, "sophisticated city-dwellers, like Queen Henrietta Maria . . . dallied at Wellingborough because she liked the countryside," and pastoral seems to have been instantiated as an orientation to some *thing,* the countryside specifically, for which "sentimental longing" and "idealization" were encouraged.[23] Thomas goes on to cite Samuel Pepys's 1667 encounter with "an authentic country shepherd and his son on the downs near Epsom" as a case in point. Yet to read Pepys's description of his encounter is to follow him through an overlay of time frames and places, each with its own codes that shape the way he takes his Sunday rest by going mobile. His pursuit of *otium* requires just as much careful preparation as Thomas More's elaborate framing devices.

We wake with Pepys on July 14, the "Lord's Day, a little before 4," to feel his "vexation" with his wife, who delays their departure, and get under way a little "past 5 a-clock," as they board the coach whose windows frame the "very fine day," and "the country very fine; only, the way, very dusty."[24] They get to Epsom by "8 a-clock." Pepys drinks "four pints" (338) of water "and ha[s] some very good stools by it." They stop at a tavern; visit friends nearby; take the "ayre" by coach, "there being a fine breeze abroad"; and then take a walk. Pepys gets them lost in the woods, "sprains [his] right foot," manages to walk it off. They go up to the downs where the sheep graze and are greeted by

> the most pleasant and innocent sight that ever I saw in my life; we find a shepheard and his little boy reading, far from any houses or sight of people, the Bible to him. So I made the boy read to me, which he did with the forced tone that children do usually read, that was mighty pretty; and then I did give him something and went to the father and talked with him; and I find he had been a servant in my Cosin Pepys's house. . . . He did content himself mightily in my liking his boy's reading and did bless God for him, the most like one of the old Patriarchs ever I saw in my life, and it brought thoughts of the old age of the world in my mind for two or three days after. (338–39)

Pepys notices the father's "woolen knit stockings" and "shoes shod with iron . . . both at the toe and heels, and with nails in the soules of his feet," which he pronounces "mighty pretty" (339). The father explains that the heavy boots are necessary because "the Downes, you see, are full of stones, and we are fain to shoe ourselfs thus; and these . . . make the stones fly till they sing before me." Pepys gives the "poor man something" and observes that he "values his dog mightily . . . and that there was about 18 scoare sheep in his flock, and that he hath 4 s[hillings] a week the year round for keeping them." The conversation over, we head back to town, stopping for milk from a milkmaid "better than any creame" to be had in tavern or town. As the coach flies, "it being about 7 at night," it frames the "people [he sees] walking with their wifes and children to take the ayre." "The sun by and by going down," Pepys tells Mrs. Turner, one of their companions, "never to keep a country-house, but to keep a coach and with my wife on Sunday and to go sometimes for a day to this place and then quite another place; and there is more variety, and as little charge and no trouble, as there is in a country-house" (339–40). The glowworms appear. Pepys finds them "mighty pretty." But his foot still hurts—and even Mrs. Turner warming it gently does no good. He needs help down the lane to his house and has to spend the following day in bed.

While the countryside presences here, it serves as an effect to be generated and also as a vector or anchoring point for a carefully reconstructed account of an economical day out—saving the expense of a country house via the maximized convenience of a coach and horses. Indeed, the week is punctuated, so Pepys says, by these Sunday jaunts. And framed by the double invocation of the calendar, solar and liturgical, Pepys's reconstruction is mediated by the technical apparatus that programmed the flow of his day: the pocket watch, the coach, and the roads. Sunday remains the Lord's Day, but in small. The shepherd and his son, literal or figurative, their Bible between them, locate in the heart of the downs a scene of innocent instruction that transports Pepys back into the world of the Hebrew Bible as the pastoral scene becomes suffused with the residues of religious experience. The sheep, whose appearance overwhelms, serve as a summary synecdoche of the scene. They constitute its material occasion, for it is by their minding at four shillings a week that father and son share this life and have the leisure that they put to such good or innocent use. But if Pepys pronounces the scene "pretty," then it is in part also because at every point he pays his way—the reckoning at the tavern, the successive "somethings" he gives to the son and his father, whom he learns was once in service to his cousin, who owned the country house that Pepys claims he's better off without.

As removed as the world of the country seems from the city—its water serving as a corrective enema to Pepys's dodgy bowels; its milk better than the cream in the tavern; its air cleaner than the road the coach's passage renders dusty—its "prettiness" is funded by a series of monetary exchanges that record the linkage of places and the interpenetration of country and city and points between as traversed by coach. Of course, as Pepys's rendering of the day makes clear, he does not belong on the downs. He gets everyone lost, sprains his ankle because he wears the wrong shoes, trips on stones that the shepherds send flying—the skipping of the stones at the nails on their toes their song, the biblical text the boy reads their lyric, their walking a kind of archaic writing that only Pepys can recognize, read, and record. It is this lack of belonging (his having the wrong shoes; writing with a quill and not the nails on his shoes) that permits Pepys's sense of having reached an outside that is nevertheless familiar, pictured in miniature, as a series of stations or *topoi,* the coach annihilating the space that comes between, its window framing the scenes, and everything bought and paid for, funded and maintained by the reassuring relays of, once upon a time, the ties of service, and now by the outlay of modest financial "somethings."

The diary records the rites of access to the country, the technical modification of *psyche* and *soma* these routines effect, and the efficacy of Pepys's weekly punctuation by the countryside. But as a rewriting or revision, much like More's alleged transcription of Raphael's account of the island, the diary marks also a further miniaturization or framing of what was already framed by the windows of the coach: an encounter with the countryside "in small." No wonder, then, that Pepys ventures out every Lord's Day he can, religiously, we might say, taking such steps he must to invest in the production of this idyll. "The word idyll derives," as Terry Gifford reminds us, "from the Greek *eidyllion,* meaning a small picture . . . [or] short poem of idealized description."[25] We could say then that Pepys's autobiographical urge, just like my own earlier fictive turn in front of the television, and Holbein's image of More, Giles, and Hythlodaeus in the garden, serves to miniaturize or remediate an experience still further, replaying the sound of the stones flying before him, recording and revising the scene in his diary and so reconstituting the day as a kind automatic archive that installs an authentic Sunday experience among the literal flocks whose shepherds' Bible study enables Pepys to render them figural. Reread the diary and the idyll materializes all over again—a teletopical Sunday outing for those weeks when work or gout or the weather precludes Pepys from getting away.

If, for Williams, television serves as a key actor in what was for him the most recent chapter to the story of pastoral, it is because, in the early twentieth century, "there is," as he writes, "an operative relationship between a new kind of expanded, mobile and complex society and the development of a modern communications technology," replacing Pepys's pocket watch, coach, and diary with their fossil-fueled inheritors, the train and car, and such other miniaturizing technologies that come to replace the paper, pen, and ink with which Pepys and More crafted their accounts.[26] Although at no point does Williams say so explicitly, in *Television,* he revises and expands the analysis of pastoral forms as they relate to the organization of material life that he undertakes in *The Country and the City* (1973), but now transformed by the teletopical figure of the broadcast. Television figures as a material, semiotic, and rhetorical event. It has a technical and political history all its own, but its effects, as in the passage from letter and humanist table to pocket watch and coach, and beyond, amp up the already installed humanist script for "making up people" and constituting collectives.[27] Channel surfing in California, Williams gives himself over to this unevenly regularizing flow. I picture him sinking into his couch, glued to the box, in a "look, no hands" idyll of idling remote control.

WHAT ANIMAL *OTIUM*?

Welcome, you might say, to the writing machine. Welcome, Giorgio Agamben might say, to the alliance that obtains between writing systems and the "anthropological machine," ancient and modern, and so to the routinized procedures by which the "human" is produced as a "space of exception," theoretically distinct, that is, from any other form of life by way of a calculated subtraction that yields up something called "animal." Pastoral, as it were, unfolds as an elaborated ecology or environmental representation keyed to particular technologies of self. Its vistas, its landscapes, its staffage, its flora and fauna, stand in precise relation to the titration of individual bodies (human and otherwise) that result. *Otium* designates the mechanism of exchange or crossing that occurs for both technologies of self to function (to make up people and populations) along with other polities of actors (sheep, grass, and so on) that we take as "world." *Otium* manifests, then, as a curiously doubled, empty, and overfull state of being, recessed deep within our discourses and worlds—the enclosed garden of Holbein's engraving, the emptying but not yet empty garden of vice in Mantegna's painting. It hovers within everyday concerns, appearing as if outside or beyond them, from which it derives its privilege in humanist discourses that seek after an immanent mode of knowledge that may be applied to the optimization of collectives and Commonwealths, to recall the Latin title to More's "truly golden handbook."

Otium corresponds, then, or serves as one instance of the state of being that Agamben regards as key to the functioning of the anthropological machine in its modern iteration: the generalized experience of what Martin Heidegger calls "profound boredom" ("*tiefe Langweile*").[28] This boredom prompts a "being left empty" or "abandonment in emptiness" ("*Leergelassenheit*") that constitutes "nothing less than an anthropogenesis, the becoming *Da-sein* of living man" (68). As Agamben notes, for

Heidegger, the *locus classicus* of this experience was the "tasteless station of some lonely minor railway" (63) that will replace Pepys's coach, where the world of disposable objects (train timetables, magazines, chocolate bars, cigarette stubs, fellow passengers—if there are any) fails to captivate. For Heidegger, then, as Agamben observes, the conception of *Dasein* "is simply an animal that has learned to become bored; it has awakened *from* its own captivation *to* its own captivation" (70). "This awakening of the living being to its own being-captivated," he continues, "this anxious and resolute opening to a not-open is the human" (63).

Ever eager to burst what he takes to be a mode of Heideggerian idyll, Agamben prefaces his account of Heidegger on boredom with the story of a very famous, now iconic, tick. For coincidentally, while Heidegger was metaphorically cooling his metaphysical heels at that lonely country station, just a few hundred miles away, in a laboratory in Rostock, zoologist Baron Jakob von Uexküll had managed to keep a tick "alive for 18 years without nourishment . . . in a condition of absolute isolation from its environment" (47). As Agamben notes wryly, under precise conditions, this tick had "effectively suspended its immediate relationship with its environment, without, however, ceasing to be an animal or becoming human" (47), and he wonders whether, if Heidegger had known of said tick, this knowledge would have made a difference in his modeling of *Dasein*. "Perhaps the tick in the Rostock laboratory," concludes Agamben, "guards a mystery of the 'simply living being.'"

But just as Agamben seems primed to insist that we inquire into the status of this tick, he opts out. He ends his analysis of the anthropological machine instead with a figure of a further suspension that attempts "to render inoperative the machine that governs our conception of man" to assert, as he does at the end of the book, "the central emptiness, the hiatus that—within man—separates man and animal, and to risk ourselves in this emptiness: the suspension of the suspension, Shabbat of both animal and man" (92). This cessation very quickly morphs into "the messianic banquet of the righteous" in which "living beings can sit" "without taking on the historical task and without setting the anthropological machine into action" (92). And as a series of perceptive readers have remarked, what remains amounts to some order of "postsecular negative theology *in extremis*"—

"an empty utopianism"—that leaves us in some version of the proverbial rock and a hard place hewn by Agamben's philological stacking of terms and flattening out of "bare life" to render it a homogeneous category. As Cary Wolfe points out, by this suspension of the suspension, this figure of inoperability, we lose "our ability to think a highly differentiated and nuanced biopolitical field, and to understand as well that the exercise of violence on the terrain of biopower is not always, or even often, one of highly symbolic and sacrificial ritual in some timeless political theater, but is often—indeed maybe usually—an affair of power over and of life that is regularized, routinized, and banalized in the services of a strategic, not symbolic, project."[29] The problem, however, is that utopia is never empty exactly. On the contrary, its neutralizing or hiatus effect, such as Agamben performs, is the by-product of the crossing it performs, the back and forth of an imaginative or affective labor that risks positivizing the work or movement of thought as such, or in Frederic Jameson's words, "over-emphasi[zing] . . . the power of rationality in general and a basic and constitutive overestimation of the functional role of rhetoric and persuasion in the historical process whereby an imperfect world may be transformed into a more satisfactory one."[30] And so Agamben ends up shuttling back and forth between lethal forms of the anthropological machine—so readily and catastrophically entered into evidence—and figures of utopian or messianic cessation. But, however compromised, however illusory or impossible the demand, something (small) subsists in these figures of cessation that remains of use, even as it may find itself palmed by the calculus that scripts our access to an apparently quasi-Archimedean point, the figure of an "outside," or immanence to knowledge, the immersive effects of Hythlodaeus's description or broadcasting.

Seeded throughout *The Open,* there remain riddling, ambivalent, semiequivalent examples of these deactivations that may constitute their own iconographic as opposed to philological stacking of "boredom" as a category: the therimorphous figures from a thirteenth-century Hebrew Bible in Milan (1–2) with which Agamben begins; Walter Benjamin's "At the Planetarium" in *One Way Street,* in which we find the "anthropological machine" at a standstill (81–84); Titian's pastoral painting *Nymph and Shepherd,* in which "in their condition of *otium* . . . the lovers who have lost their mystery contemplate a human nature rendered perfectly inoperative—the inactivity . . . of the human and of the animal as the supreme and unsavable figure of life" (87)—"unsavable" because it inheres to a rhetorical staging of the nonhuman that has now ceased to function, ceased to provide an effective transport back to the concerns of *negotium,* of making persons or worlds; and Georges Bataille's exegesis of the animal-headed effigies from the Gnostic Basilides, which he reproduced in the journal *Documents* (89–90). These examples constitute a minimal archive of deactivations or, more precisely, a gallery of idyllic, pastoral scenes that splice together image and text (images cut from the English translation) all of which, by the splicing of word and text, by their elaborate framing, register the presence of something that Agamben does not know how to name other than by and as a mode of waiting or idling.

Of course, as Jacques Derrida offers in a very pointed critique of *Homo Sacer,* it may be objected that it is not that difficult to imagine forms of nonbare *zoë,* of life "backed" differently than by a human animal, forms that do not then accord with the abject sovereignty of a *homo sacer.* The distinction between *bios* and *zoë* cannot so easily be settled. Derrida offers the admittedly overdetermined figure of the incarnation as a counter, of Jesus as Jew, a "zoological Jew, since he unites in his person, as son of God, both *logos* and *zoë.*"[31] "He is zoological," he elaborates, "not only because of the sacrificial lamb, because of the Paschal lamb of the Jews or the mystical lamb that erases the sins of the world," but because he "unit[es] . . . in one body, or one and the same concept, *logos* and the life of the living, *logos* and *zoë*—a zoo-logy or a logo-*zoë* imposes itself." This logistical incarnation serves as a point of capture, a punctuation, or a routing mechanism that truly makes Jesus "count." Indeed, it's what enables the Agnus Dei to serve, as we saw in my last chapter, as the sovereign, universalizing substrate or point of conversion between human and other animal blood—a point made affectively by Arthur Coga as he sought to make sense of the strange gift of sheep blood he received at the hands of the Royal Society's experiments with xenotransfusion. For Derrida, this incarnation serves to put us on notice that all manner of translations and transformations are continuously at work within the field of biopolitics—a field that it becomes difficult, or at least highly problematic, to qualify—for "'bio-power' itself is not new." And while there may, today, be "incredible novelties in bio-power . . . biopolitics is an arch-ancient thing . . . bound up with the very idea of sovereignty."[32]

It is on these terms, I think, that we should approach Agamben's gallery of idyllic deactivations or *otium* archive, as a storing up or pooling of the contradiction that his programming of *bios* and *zoë* produces, a conservation of this problem. And so, for me, this gallery stands in relation to the way, at the opposite end of the incarnational spectrum, *The Open* plays host to what Derrida might have called a "prodigious archive" (in small), that refuses or reknits this "cut" between beings. I refer to Uexküll's now canonical tick, a tick who, it must be said, fails to note that it inhabits a pastoral enclave, who fails, as it were, to take any notice whatsoever of pastoral even as it plays at a literal, blood-sucking, bio- or zoopolitics as part of its reproductive cycle—a point Uexküll underscores but that Agamben does not register.[33]

"The opening has the tones of an idyll" (45). "Every country dweller who frequently roams the woods and bush with his dog," Agamben quotes, "has surely made the acquaintance of a tiny insect," "but here the idyll is already over," he continues "because the tick perceives nothing of it" (46): the "eyeless" tick, this "blind and deaf bandit," comes to know of the "approach of her prey . . . only through her sense of smell." What then was the fate of that tick in the laboratory at Rostock, who waits and waits, waits for a stimulus that never arrives? It's not quite true that, as Agamben has it, Uexküll "gives no explanation" of how he was able to keep this tick alive for eighteen years and merely supposes that during this period "the tick lies in 'a sleep-like state similar to the one we experience every night'" (47). True, Uexküll offers that the sleep this tick takes "during its waiting period" is a "state similar to sleep, which also interrupts our human time for hours." "But," he adds, "time stands still . . . not just for hours but for years, and it starts again only when the signal 'butyric acid' [mammalian blood] awakens the tick to renewed activity."[34] The magnitude or scale differs radically. The tick's *Umwelt* (environment) differs hugely from the one we inhabit and so limits, taxes, interrupts the analogy. By what order of metaphysics, then, may I account for the condition of *this* historically particular tick? Is Uexküll correct in his phenomenological reduction that states, "without a living subject, there can be no time," and the tick merely sleeps? Was this tick, as Agamben might like to have it, "profoundly bored"? Or was it, as a Renaissance humanist might have framed the question, merely "at leisure" *(per otium),* idling, free of the everyday tasks and distractions that come with the *negotium* of seeking out mammalian blood so that it may take its singular feed, reproduce, and die? To what animal, to what species or kingdom, properly, is *otium* reserved? *Otium,* we know, remains silent on the number and quality of its subject's limbs. It requires no hands even if its human devotees produce a succession of handy how-to-do-it and how-not-to-do-it books in its name.

WHAT WAS PASTORAL (AGAIN)?

Given the stakes, the question sounds faintly ridiculous. We know what pastoral is. Not quite a genre, but sometimes a mode, activity, vehicle, or poetic device—idealizing or ironic—pastoral treats high themes in low costume, enabling its producers to explore issues of patronage, poetic servitude, "life," set against a world that appears static, innocent, or, as Pepys might put it, "pretty." As Robert N. Watson writes in *Back to Nature,* "pastoral is thus another cultural phenomenon explicable as a response—often simultaneously as an inscribed banner of protest and a blank flag of surrender—to the burdensome knowledge of mediation—in other words, the knowledge that there will be no unmediated access to some 'thing' we call 'nature' or the 'natural.'"[35] In his hands, the Renaissance vogue for pastoral, which dominates the mechanisms by which they attempted to dream away the time, becomes a vision of a society backing into a Nature that everywhere was felt to be receding from them.

Famously, for Raymond Williams, pastoral and its kitschy, downgraded "equivalents" from the sixteenth century onward are understood as a perpetual stasis machine that sustains a murderous environmental and social fiction. An "escalator," Williams calls it, "moving without pause" forever backward into an ever-receding image of the Golden Age du jour.[36] Rhetorically, it may seem that I should disagree with Williams and assert that there is something worth saving, something about pastoral qua pastoral that actually matters, but it's not quite that simple. One thing I have learned is that tactically, it is best to agree with everything everyone has ever said about pastoral. This is not because I am some happy shepherd advocating an idyllic, pastoral pluralism but rather because everyone, in her own way, tends to get his particular section of the rhizome correct. The extended or general text of pastoral, pastoral as archive, manifests in different times and places folded in strange or seemingly contradictory ways, as its core tropes are successively performed.[37] Williams flags this point early on in *The Country and the City*: "the initial problem," he writes, "is one of perspective" (9)—where you find yourself distributed, that is, in relation to pastoral. He begins the book with a cannily lyrical reconstruction of the two "networks" he has inhabited, or fallen between, familial and academic, which parse out loosely into "country" and "city"—but which already, by his description, and by the transport technologies (but not yet the television broadcasts) that interpenetrate, fail to sit happily in either category. His rendering of the history of pastoral as the gradual localizing of "a dream and then . . . into a description and thence an idealisation of actual English country life and its social and economic relations" (26) takes aim at the process whereby "the charity of consumption" (a feast) occludes and defers the possibility of a community that may have existed but that remains still to come: "a charity of production—of loving relations between men [and women]" (30–31).

In *Television,* this charity of production finds expression in Williams's valorizing of those unscripted programs that "open themselves towards people not assumed in advance already to be represented"—programs that run live and therefore risk the possibility of "dead" or dull "air."[38] The issue here is not Williams's desire for a utopian figure of unalienated labor—though that is present—but instead the way this occlusion requires that all analysis of social relations occur belatedly, *post festum,* to use Karl Marx's figure from *Capital.* For Williams, then, deeply concerned also by the scandalous ways in which generations of readers misconstrue literary references for historical reality, the story of pastoral yields a theory of ideology—what Paul de Man would call "a confusion of linguistic with natural reality, of reference with phenomenalism."[39]

Read this way, Williams's figure of the escalator seems much less hostile to the likes of a William Empson and his *Some Versions of Pastoral* (1935) than Alpers at least implies.[40] Sometimes a "trick" (11) but, by turns, a "myth," a "process" (22–23), a "machine" (30), a "double plot," an "organism" (144–45), or a "tap root" (261), "pastoral is a queer business . . . permanent . . . and not dependent on a system of class exploitation." It works by what sounds a lot like taxidermy, "putting the complex into the simple" (22–23) so as to "imply a beautiful relation between rich and poor" (11), "giv[ing] the impression of dealing with life completely" (29). "In pastoral," writes Empson, "you take a limited life and pretend it is the full and normal one"—it's a refolding of surfaces that creates the appearance of depth, of "life" or "liveliness," not so much by what it adds as by what it leaves out or subtracts. No wonder pastoral endures—"takes refuge" in a "larval Alice" (269) is how Empson puts it—the activity comes to stand for a "predatory" or "pro-active" mimesis, not exactly the ambient or ambulatory quality of nature writing that Timothy Morton has named "ecomimesis" and that informs Pepys's, Williams's, and my own autobiographical self-staging—but a more fundamental, un-Romantic "stuffing."[41] For Empson, then, pastoral names a metaplasmic operation by which writing touches itself and by that touching, which is in truth a folding, traces the skin of the world, producing a set of effects that may be taken for phenomena, spun off as so many worlds or pasts.

Track back in the company of David Halperin's *Before Pastoral* to the literary world of Theocritus's Idylls and it is possible to isolate the "trick" of pastoral, as Empson called it, still further. For what we find is not a writer of what Virgil will codify as "pastoral" but instead of a "bucolic poetry" defined in terms of its formal features as opposed to its thematic concerns. As Halperin observes, "at the time Theocritus was composing the Idylls, the principle of classifying poetic genres according to meter and the doctrine of fixity and separateness of the poetic genres had long been powerfully established."[42] Moreover, in describing the apparent disunity of subject matter in the Idylls, Halperin argues that this thematic breadth is indexed to a programmatic rereading of Homer and the function of epic in Theocritus's present. "An examination of Theocritus' poetic technique," he writes,

> as well as his treatment of themes reveals a pattern of contrasts or oppositions between bucolic *epos* and heroic *epos*—or perhaps between the heroic and non-heroic registers *within* the tradition of *epos.* Theocritus does not in general introduce totally alien material into the epic genre; rather, he elaborates the non-heroic alternatives to the traditional heroic mode, which are already present, albeit inchoately in early Greek epic. (238)

Halperin finds this recoding or resignifying of epic enacted in the form of an extended examination of the workings of the *ekphrastic* figure of the ivy cup which is given a famously lengthy description in the First Idyll and which stands in relation to the singing competition that the description defers.

Tracing the philological origins of this wooden milking bowl or *kissybion,* and comparing it to the way similar words are used in the *Odyssey,* Halperin concludes that it "seems to be a large bowl, used by rustics—in short, a humble implement belonging to a primitive economy" (169). But whereas in the *Odyssey* such bowls are made from metal and are used for drinking wine, Theocritus has transferred the decoration of such bowls to an object of much more humble origins (172–73). Halperin reads this transfer or translation of decorative motifs to material of more humble origins as a statement that makes readable Theocritus's aim both to outdo his material and also to "provide . . . a picture of the life and feelings of little people, portrayed in situations that are not earthshaking" (177). Reading the successive survivals of the First Idyll in papyrus as evidence of its programmatic nature, he agrees with those critics who read the Idyll as providing "the poet's *sphragis,*" or signature.[43] For Halperin, then, what the question of origins yields is a moment of reading or rereading where Theocritus's "static vignettes [idylls or] *eidyllia*" (literally paintings with captions) (186) constitute themselves as a rerunning or relayering of epic themes and techniques—a way of reckoning or coming to terms with the latent elements of epic by extending epic's techniques (both comically and seriously) to broach subjects of everyday concern. Their root in *ekphrasis* serves to emphasize the way the idylls operate as a doubling or refolding of origins, creating tiny, augmented worlds or pictures in words.

What was pastoral (again)? Ideology? Mimesis? *Ekphrasis?* What about all those shepherds and sheep? Are they then ancillary to its operations and afterlife? "No shepherds, no pastoral"—to which, following Theocritus's lead, we may add sheep. Sheep remain the given, or are produced as the given of pastoral—a point so seemingly redundant that playwright John Fletcher gets a bit nasty when compelled to explain such things to readers in the preface to his not terribly successful *The Faithful Shepherdess* (1608). "You are ever to remember," he instructs, "[that] Shepherds [are]... owners of flockes and not hyerlings"—which was to insist, in a sense, that poets are their own men and emphatically not themselves sheep.[44]

Summarizing numerous readers of pastoral, Montrose observes that the genealogy of the poet-shepherd that Fletcher invokes here stems from the way "literary celebrations of pastoral *otium* conventionalize the relative ease of the shepherd's labors."[45] Sheep largely, so the story goes, take care of themselves, leaving our shepherds with a good bit of time on their hands, time that quickly morphs into pastoral *otium*. He credits George Puttenham's *The Arte of English Poesie* (1589) with a strategic "conflat[ion]... of the attributes of the pastoral world with those of the Golden Age," but also with a careful rearrangement of labor practices such that the "'ease' and 'idleness' typical of the Golden Age" derives "from the organization of pastoral society itself, rather than from the spontaneous fertility of nature."[46] In Puttenham's hands, pastoral becomes a way of placing poetry at the foundation of the world and of crafting a genealogy of different poetic forms keyed to the progress of the social order and its technological advances. In the beginning there was pastoral, sheep, and shepherds. Then follow castles, towns, kings, citizens, tragedy, and comedy. Literary history keys genres to the story of land use, social forms, and technological development, to the production of an entire infrastructure of beings.

Here we may be tempted to deploy an anthropological fix that would posit pastoral, much as did Puttenham, as a discursive exaptation from a primary organization of agricultural resources and animal husbandry that defines the geography, history, and metaphors of the built worlds we live. As Fernand Braudel writes, our metaphors are indexed to an alliance of human, animal, and plant resources or actors, demonstrating "an ancient choice or priority [in the West for ruminants and wheat] from which everything else descended."[47] Certainly pastoral trades on a putative agricultural settlement and mobilizes metaphors that accord with the dynamics of a human, dog, sheep, goat, grass multispecies. Transhumant husbandry is etched into the land. It built the circuits that paper and electronic pastoral will and now travel. And the animation of the "human" derives, in no small part, from the apparent stability, docility, and malleability of sheep as they have been selectively bred and rendered plantlike.

We may even be tempted, as Terry Gifford did or claims to have done, in his excellent introduction to pastoral, to travel to those places where the Golden Age seemingly endures today, accepting "an invitation to observe real pastoralism in practice." Get up early, hop on a plane, and, like some latter-day Pepys, we could pay a visit to real shepherds "of a village in the mountains of Crete," talk with them, listen to their songs, and time travel back to the world that Puttenham envisaged.[48] When he arrives, Gifford finds shepherds, shepherds who even still talk about Arcadia "and still have witty song competitions too, each singer trying to outdo the other." But they also save their money to buy "cattle trucks" and brandish Kalashnikov rifles made in China. Gifford makes the exemplary observation that here we see "from the beginning of its long history [that] the pastoral was written for an urban audience"—or, in the case of these Cretan shepherds, we discover the interpenetration of localities with global networks of exchange. But, as Puttenham's already pastoralizing genealogy of poetic forms demonstrates, the larger point might be to observe that all moves to phenomenalize the metaphors we find expressed or elaborated in pastoral, to go *there,* as Pepys and Gifford, in their different ways, sought to do, will tend to positivize the "past" via a retroprojection of those very metaphors that persist *today,* the past become a relay in the ratification of our various "presents." Tempting as it is, therefore, to derive pastoral from pastoralism proper, and so to invoke sheep as a literal, material basis, inquiries into questions of origin tend simply to transform themselves into this or that invocation of a physicalist base. Shepherds may count sheep; they may sing songs; they may write poems; but they and their flocks remain the rhetorical occasion merely for such a stuffing of the world that is pastoral. They do not provide an origin based on a diverting division of labor.

When, instead, Halperin tracks the discourse of pastoral back to its putative origins in pastoral societies in the Near East (where human persons, sheep, and cows first domesticated one another), what he discovers is a semireadable coding of different types of human speech, in which the voices of herdsman, cowherd, and shepherd were heard by and in their relation to an "inhuman liminality" that derived from their relative remoteness from the city and so proximity to beast and the divine.[49] The story that Halperin discerns of shepherd as "go-between," one in a series of *homo sacer*–like figures, the person who marks the connection and boundary and so who organizes one circuit between the divine, the human, and other animals, and in particular ruminants, anchors a biopolitical field that then proceeds to write the human through and by its un/likeness to sheep and so also to render sheep "sheepish." The sheep, however, remain relative bystanders or casualties to these metaphors, an unmarked catchall to the operation, the captured and stabilized multiplicity that remains still so that all else can appear in movement. Never bored, never capable of boredom, so it seems, these plantlike "sheep" who grow flesh and wool are produced as a residue of what "human" persons are not supposed to be, even as they are. And what matters here, what naturalizes the scenography so that it comes to anchor our routines, is the mixed media, the splicing or folding of forms (word and image) that produces the appearance of fullness or depth. The discourses we name "pastoral," by their media oscillations, condition our access to *otium* as at once a scarce commodity and a diluted biopolitical "fix"—coming soon to a sheep theme park, mattress store, TV or media platform near you.

Given, then, what I think pastoral is (and was), it seems much too early to be moving on, much too premature to think that we can move on or beyond its constitutive metaphors. They remain coterminous with the anthropo-zoo-genetic or multispecies impression whose terms we live out. Pastoral as mixed media or media ecology, as a layering or folding of forms, designates one mechanism by which beings are loaded into discourse so as to authorize and maintain certain significant routines. At the same time, its programming of relations between beings produces a residual reservoir of sameness, a zone of creaturely indistinction—call it sleep; "barely living being"; call it *otium*—through which we all pass, losing our hands or handedness as we do. All that follows, then, will remain pastoral, or, in Empson's modest phrasing, shall be "more versions" of pastoral—different versions, perhaps, decoupled even from sheep and shepherds, so it might seem. But boxed up in the mode or activity, and the worlds they spin off, hiding, as it were, in plain sight, remains this fugitive point of crossing between beings (plant and animal), a point of crossing that surfaces each and every iteration and that might become susceptible to recutting merely by posing it as a question.

For, as Thelma Rowell would be the first to tell me, the metaphors I have traced thus far derive not from the essential "sheepiness" (plantlike docility) of sheep but from their modeling and manipulation as livestock or living capital. As she argues, the traditions of primatology and animal behavior studies have dictated that those animals who lead interesting lives (that is, lives deemed interesting to us) have tended to serve as privileged experimental subjects—especially if they may be grouped as among the relatives of a certain *Homo sapiens.*[50] Accordingly, animals (and that is "most animals") who "spend the majority of their time doing nothing" tend to be neglected or asked only the most boring of questions. "Sheep behavior studies are mostly to do with what they eat, and sheep are not, generally, permitted to organize themselves," she writes.[51] What would happen if, instead, we modeled sheep as what Rowell, along with pastoral policeman Rick Deckard, calls "all gregarious long-lived vertebrates capable of mutual recognition"—a group to which, of course, we also belong? What might happen, in other words, if we allowed sheep to be a bit more interesting or even a bit bored?[52] If the word "interesting" feels a little bland on the ears, then it might be best to hear it according to its Latin origins, "to be among" or "between" *(inter-esse).* If things prove interesting, if we find them interesting, then we wish to dwell or "be" and become with and among them. Heard this way, the word discloses the ontological choreography at work when certain questions are posed and others are not. Our questions function as propositions or projections; they compose worlds.[53]

OTIUM FOR SHEEP

The key to Thelma Rowell's observational protocol—her way of composing or projecting "sheep"—lies in a brilliant, now seemingly obvious leap of fiction. She presents the solution very straightforwardly and somewhat laconically as a "problem of method" that, for a good while, manifested as a problem of accreditation and no little professional frustration. As Rowell puts it, "I decided to watch sheep in the same way I have been watching monkeys, and tried to publish the results." The problem was that "since the sheep were in the title, the manuscripts apparently went to sheep experts to review." And these experts on sheep "were generally appalled by what they saw as my anthropomorphy, and had difficulty understanding why I might be interested in 'our' sorts of questions about social organizations."[54] Watching sheep according to protocols with which primatologists watch baboons or chimpanzees posed basic problems of sense to ethologists of sheep. Rowell's work smelled wrong. It ran the risk of being deemed un- or improperly scientific, of falling outside the confines of the "true" and so out of discourse.[55] Her work vibrated with the disciplinary noise of a category violation that seemed only to yield up a chimera: weird, hybrid sheep-chimps. Of course, it's precisely this charge of anthropomorphy and its attendant problem of fiction making—treating a sheep as if a primate—that renders the likes of Vinciane Despret and Bruno Latour giddily happy. "By importing the notion of intelligent behavior from a 'charismatic animal,'" Latour chimes in, "another of her treasurable expressions! . . . [Rowell] might modify, subvert, or elicit, in the understanding of sheep behavior, features that were until then invisible because of the prejudices with which 'boring sheep' have always been treated."[56] A little, inspired, tinkering with the protocols of observation, treating sheep as if they were chimps, enables a field science to short-circuit the routines by which several thousand years of botched or abusive ethology (selective breeding) came to write the discourses of pastoral and pastoral care. Primatology serves here as what Bruno Latour calls a "trading zone between anthropology, zoology, evolutionary theory, ethics, conservation, and ecology."[57] Its protocols of observation enable a reworking of our constitutive metaphors.

Rowell refutes the routinized, same old logic of "boring sheep are boring sheep" because she "artificially and willingly imposes on sheep another resource coming from elsewhere [so] 'that they could have a chance' to behave intelligently." She tries "to *give* [her] sheep the opportunity to behave *like* chimps, *not* that [she] believe[s] that they would be like chimps, but because [she is] sure that if you *take sheep for boring sheep* by opposition to intelligent chimps they *would not have a chance*."[58] The emphasis here makes explicit the deliberate canniness of the strategy alongside the moral philosophical quotient of the language of gifting and the openness to surprise that characterize Rowell's attitudes. The installed comparativism between sheep (assumed to be boring) and chimps (granted intelligence by their status as avatars of the human) posits and so possibly makes a difference. Rowell operates, then, by positing the following sentence or proposition: "sheep are intelligent chimps."[59] This formulation is premised on a constitutive and considered artificiality that foregrounds the way the experimental protocol serves as a trope, operator, or switch that makes possible the unfolding of other worlds than that with which it began. "It is because of this very artificial collage between unrelated animals—charismatic chimps and boring sheep," Latour writes, "that [Rowell] can reveal what sheep *really* are" or have now become.

Citing Uexküll"s "canonical tick" for comparison, Latour offers that the beauty of a well-articulated proposition is that it "designates a certain way of loading an entity into another by making the second attentive to the first, and by making both of them diverge from their usual path."[60] Like Uexküll, Rowell constructs an experimental protocol that requires her to wait upon or to attend upon her sheep, which may or may not then tell her which questions interest them and which do not. Her sheep are invited to offer an interpretation of the questions they have been posed, but the protocol in no way assumes that their response may be readily processed or understood by the human observer. Perhaps the sheep shall be bored with us, bored by our questions? Rowell therefore has to remain attentive as her proposition plays out. She has to watch and wait. She must run the risk of failure, a failure that it may be hard to own or process as it might take the form of a silence or static, the failure of two only sometimes compatible orders of finitude (human and sheepy) to connect. She keeps open therefore the possibility that an altered choreography of beings, a re-forming of life and forms of life, might teach us something entirely new about what, once upon a time, had seemed so very familiar (sheep, chimps, humans, laboratories, sheep farmers, shepherds, and so on).

The issue here is precisely not some assumed commonality that might manifest under the sign of "speech," "presence," "response," the "organic," "life," but instead the constitutive "as if" that assumes that animals that appear boring to us do so for very good reasons. Recalling the reactions of her red-tail monkeys in Kenya, for example, who, for Rowell, seem closer to a sheep than they do to a chimp, she remarks that they "would make quiet alarm calls and freeze" when their jungle was invaded by a troop of baboons. The baboons "were always squabbling among themselves, with screams and threat barks and grunts, and the calls of temporarily lost infants," not at all unlike the "large groups of school children that are occasionally brought to the forest."[61] "To be distracted is very dangerous," she observes. "I imagine the sheep I studied allowing themselves to be so preoccupied with intragroup squabbles and their resolution, and the thought experiment produces instant mutton. They would be wiped out by predators taking them by surprise."[62] Treating sheep *like* baboons or chimps requires the creation of a formal relationship that removes the threat of predation or that factors that threat into the mode of questioning constructed. The constitutive "as if" fabulates or fictionalizes "sheep," rendering the word a question or a query. Unhappy with the way sheep have been modeled, Rowell switches genres; she invites sheep to respond to questions reserved to primates, chimps, that is, not as they are but as they too have been written or constructed. Accordingly, the assumed "content" to the propositions primatologists have crafted for chimps becomes the ground (still partial, still open to revision) for the inscriptive practices by which Rowell will invite sheep to write with her. Primates and primatology serve as the substrate or occasion for sheep to behave differently as Rowell crafts what Empson might lead us to consider a "double plot" that renders sheep more "fully," more lively, more interesting by way of their likeness and unlikeness to chimps.

The force to Rowell's troping of sheep *as* chimps depends on and derives from the artificial, formal, or media-specific conditions under which differently animated entities interact. The conditions of the scientific protocol make it possible for sheep to manifest as intragroups and, in effect, to historicize and particularize their own behaviors. She invites sheep to manifest as Historical creatures, in the Hegelian sense of the word, though as they do so, History fractures into multiple, competing chronologies and plots.[63] The protocol or proposition Rowell crafts constitutes a type of *différance* machine, an open system, much more so than a system in which sheep are able to say no and so falsify a hypothesis.[64] The protocol constitutes a zone of interaction that is *historial* in the sense that it proceeds "without invoking origins or grounds," even as it may produce something like them (inputs to future acts of making) by and through the anthropo-zoo-genesis, or, in plant terms, "grafting," of sheep, chimp, and human ethologist that the protocol projects.[65] Here it worth recalling that, as philosopher of science Hans-Jörg Rheinberger reminds, the word *graft* shares a "simple etymological coincidence uniting the graft and the graph (both from the Greek *graphion*: writing instrument, stylus) but also the analogy between forms of textual grafting and so-called vegetal grafting, or even, more and more commonly today, animal grafting."[66] Hence "research processes" and "experimental systems" such as Rowell's "concatenate into a constantly changing signifying context." They "oscillate around epistemic things [sheep treated "like chimps"] that escape fixation by transplanting and grafting new methods, instruments, and skills into the set up; or by altering the location of their embodiment." Such grafting needs must prove hospitable. It must "keep . . . alive, as a support, the system upon which one grafts." The resulting form corresponds closely to a mode of parasitism or symbiosis, the cowriting of several now interrelated forms of life. "Sheep" shall emerge, then, as they will have been cowritten with and by and through chimps, and the animals we name human, but this writing remains open to revision, as does "Thelma Rowell" and the chimps or primates that provide her with her ground or given.

Rowell's observations usually start in the morning, with the same ritual. As Vinciane Despret observes, "she takes each of her 22 sheep a bowl of its breakfast. But what puzzles any outside observer is that there are not 22 but 23 bowls, that is, always one too many." "Why this extra bowl?" asks Despret. "Is the researcher practicing a kind of conviviality?"[67]

The twenty-third bowl is, as Despret hints, in what seems like misdirection, about politeness, about offering to sheep the chance to transform the protocols of the observation. The presence of the bowl and so the surplus of food transforms the questions that Rowell poses, removing or suspending an automatic question concerning competition. The bowl "is intended," Despret continues, "to expand the repertoire of hypotheses and questions proposed to the sheep . . . [but] to leave them the choice" of answering other questions than those posed to them. As Rowell offers, "the problem is that you can watch an animal eating very easily. The whole business of food and the competition for food has been much exaggerated because that is what is easiest to see."[68] Alter the protocol; change the scenography so as to render food a background or a given (as it was for More, Raphael, and Hythlodaeus in their garden); allow the question of food, of necessity, to idle, and it becomes possible to imagine sheep differently, for they have been rezoned within our perceptual and cognitive maps, invited, in Williams's terms, to share in a "charity [but not a community exactly] of production."

As Rowell shows, in the place of all those interchangeable plantlike sheep that served as some serial, self-same unit that might be counted, there unfolds a much more complicated social world of shifting intragroups (daughter–ewe relationships; preconciliatory affirmations of "friendship" among competing males of different generations before they enter into aggressive behavior; a sheep collectivity in which individual males work to create a "huge, spectacular noise" that constitutes a "sort of sound and visual display intended to ensure the group's cohesion").[69] It all proves "very exciting to the ewes," Rowell observes. "And they eat together." The twenty-third bowl offers sheep more than choice. It offers them the chance to manifest as rhetorical creatures, as entities that might manipulate codes or perform them to their own ends. It brackets off one routine set of questions concerning food and competition to allow sheep the opportunity to manifest within a broader field of behavior. As Despret remarks, "if a sheep leaves its bowl, shoves away its neighbor to take its place and immediately returns to its [own] bowl . . . did this sheep want to show its fellow creature, and all the others, that it could supplant it?" (368).[70] Does said sheep have a concept of performance, of rhetoric? Can it, does it, lie?

What was pastoral (again)? Will the results of Rowell's observations be pastoral? Must they be pastoral? Or has pastoral now been so irremediably altered that it will be hard to know what a sheep or a human has become? For the reader tuned to the idylls of pastoral poetry, it is tempting to suggest that Rowell's protocol further transforms the mixed media or *ekphrastic* wager of Theocritus's ivy cup or *kissybion,* the layering or folding of media to create the appearance of depth that is pastoral. This time the cup materializes in even more humble garb as a feed bowl for a sheep, indeed, as a feed bowl that is not even used, allowing these twenty-two historical sheep to refigure themselves and the routines that rendered them and us "sheepish." Like Theocritus, Rowell prepares the bowl, but it is her twenty-two sheep whose actions she records that co-*write* the scene it depicts. It is they, provided with a sufficiency of food, who enter that rhetorically complex world named *utopia,* and it is Rowell (and we) who play remote controlled John Clement, running in and out, and waiting on them, trying to catch what we can of their interactions and the world they imagine and live out. Rowell rewrites the relation between *otium* and *negotium* across and through the boundaries of species difference. She and her twenty-two sheep constitute a circuit in which *otium* serves as an input to the writing machine they collectively coauthor. Rowell idles; she waits on and for the sheep. They eat. And then they move to other concerns/ things that interest them, disclosing thereby a "sheepiness" whose substance unfolds less for us than through us. So it is that the human comes to wait upon these historically particular sheep as pastoral *otium* finds itself rezoned, no longer a prospect of utopian cancellation or surplus but instead a common property in a writing machine whose authors prove multiple.

Despret calls the twenty-third bowl "emblematic." "The twenty-third bowl is . . . [Agamben's] open," observes Haraway. It could be described, more technically still, as "idyllic," recalling the scenography of pastoral *otium* that I have described, but refolded now to produce a multispecies writing machine that owns its anthropo-zoo-genetic basis and that offers sheep the chance to migrate from being a plantlike constitution as living stock to become something else entirely—"gregarious long-lived vertebrates capable of mutual recognition," a category to which, obviously, we and sheep both belong.

But as promising as Rowell's choreography of beings may be, it is not time to rest quite yet. For, still, we are not quite finished with pastoral. Nor is it finished with us. Indeed, such rest that might come now that we seem to be moving on might prove oddly wakeful.

SLEEPING WITH ONE EYE OPEN

At the end of *Quand le loup habitera avec l'Agneau,* ethologist Vinciane Despret pauses to explain the full force to the biblical text that gives her book its title. She has taken her readers on a tour of what she regards as the most significant examples of scientific research in the field of animal behavior studies. These examples rise, in her view, to the level of self-conscious anthropo-zoo-genetic practices. Thelma Rowell and her sheep take top billing because their case represents the most pointed example of an attempt to overturn a species hierarchy or hegemony—to offer sheep, if not charisma, then something on the order of a mode of practical sovereignty. This sovereignty extends to other groups of animals also. The more charismatic primates, we learn, respond well to some questions but not to others. And the raven, poor old bird, "stigmatized since the Deluge," almost no one has ever thought to pose him a question at all.[71] Posed more interesting questions, questions that enable them all to take an interest, animal subjects and fellow animal investigators find themselves remade. Sheep cease to be sheepish. Baboons become both less and more than they had seemed. And the raven, so it turns out, has always had volumes to tell us if only we had thought to understand his not-coming-back-to Noah's Ark as a meaningful signal as opposed to a betrayal of natural law.[72] The investigators find themselves transformed also. They run the risk, initially, as Rowell showed us, of their work being deemed un- or improperly scientific, of falling outside the confines of the "true." But such are the risks to be run if we are to alter "the conditions for the production of knowledge" for us that are, as Isabelle Stengers puts it, "equally and inevitably the *conditions of production* of existence for," in these instances, sheep, baboons, and ravens.[73] Like Rowell and company, we shall have to risk failing to produce forms of knowledge that are immediately vendible, forms of knowledge that interrupt the narratives that have conditioned their making in order to make something else possible.

So it is that, when reading Despret, we encounter historically particular bears, baboons, sheep, pigs, ravens, and more, all of whom have been posed questions that enable them to presence as local polities or groups of interest. Or, more precisely, we encounter a set of propositions concerning variously mediated hybrids: data sets or observations produced not by single beings, or even representatives of a particular group or species, but by the questions posed, altered, and so answered by the human investigators and fellow animal subjects as they are folded together by their shared protocol. Divorce the two, the investigator from her responding subjects, the findings from the scenography, and we shall be left empty-handed. Together, the investigators, their subjects, and the media ecologies crafted to pose their questions constitute different instances of multispecies writing machines. In grander terms still, Despret asks us to consider that both the animal subjects and their animal investigators participate in the formation of an altered historical subject, a governing "we" that narrates a story that by its constant posing of questions to its increasing variety of selves comes to accord with the world that we collectively co-make. To read Despret is to watch as she assembles the lineaments of a single story in which, as in the biblical text, the lamb might be said to lie down with the wolf, the leopard cohabit with the kid, by and through the differential questioning of each that allows both to manifest by and in their own particularized forms.[74] The verse from Isaiah 11:6 provides her with a set of instructions that she attempts to decrypt, prying them loose from their eschatological underpinnings to take them more literally or more figuratively still as a desirable voiding of taxonomic laws in the service not of a holy universalism so much as a holey, aproetic, and therefore capacious hospitality premised on what might be said to matter or to "count" for all—no longer the figure of a single messianic banquet or Shabbat but a multiplicity of discontinuous feasts (and deaths).

Despret reads Isaiah not as the description of some achievable future but as a set of instructions or as a recipe that gestures forward, a projective set of rules for assembly that might constitute a different, differentiated "us" and so produce this new historical subject. "If the prophet was right to trust us," she writes, "the 'we' that designates this trust remains still to be constructed." This "we," she elaborates,

> that enrolls more humans and non-humans in one story and a common world remains to be negotiated still, gradually. With confidence and vigilance; with good and bad habits. To inherit the prophecy is also to inherit what has been learned from those who have begun to implement it: what counts [*compte*] to us counts [*compte*] otherwise for those we invite to construct this "us." . . .
>
> Well then? Well then, one day the wolf may dwell with the lamb, but the lamb will probably sleep with one eye open. If Isaiah invites us to be confident, he reminds us at the same time with this wise precaution, vigilance: do not forget to ask the child to lead them by the hand.[75]

I like to think that this project would sound very familiar to the likes of Jack Cade, Dick the Butcher, and company, and now also to Rick Deckard, who learned that his actions were constitutive of but also inadequate to the world he then had to live. I picture nods of recognition and agreement. Like Haraway's multispecies, Despret's "we" might never be said to constitute itself or close on itself as a fully realized object, other than through its performance, an ongoing performance we saw mooted in *Henry VI, Part 2,* when Jack and fellows produced a multiplicity of voices in place of singular, royal speech. Despret ups the ante further as her "we" runs athwart the lines of kingdom and kind. The very basis by which beings are named, the very basis of reference itself, alters, for this "we" no longer refers to a collective of theoretically like-minded or like-bodied beings but comprehends a potential cacophony of different forms of finitude, which it does not know and may not lay claim to but still attempts to accommodate. The trick lies not in some putative universal, then, but rather in the posing of a successively particularized set of questions to each entity as we encounter it such that they are led to take an interest or pronounce themselves utterly indifferent to us, as are "we."

But there are difficulties. Jack, Dick, and company's nodding might slow as they recall that the verse from Isaiah has so frequently served as one anchoring point for a putative cancellation of all those antagonisms that constitute the world as we live it. They might "English" the lines as follows, and speak them back to Despret as a caution, unsettling thereby the way they preface and title her book:

> The wolf shall dwell with the lamb, and the leopard shall lie down with the kid; and the calf and the young lion and the fatling together; and a little child shall lead them. The cow and the bear shall feed; their young ones shall lie down together; and the lion shall eat straw like the ox. And the sucking child shall play on the hole of the asp; and the weaned child shall put his hand on the cockatrice's den. (Isaiah 11:6, KJV)

Come the end of times, come the next world, this is how things shall be. Whichever eschatology you choose, what is imagined represents an end to predation and a universalizing cancellation enabled by the removal of an automatic question concerning the competition for resources, all the forms of information processed as "food." All generations of mammals, nurslings along the weaned, shall share in a veganism so basic that it's like, well, mother's milk, or the Eucharist, or the flesh of a Leviathan and Behemoth that we now share so as not to have to eat one another (depending on whose sacred book you find yourself reading). Moreover, this emerging mammalian "we" shall make its home on the doorways or threshold spaces that lead into the worlds of reptiles, whose bites lose their venom. The verse gestures forward to the quasi-messianic moment in which no one goes hungry and so none must therefore be eaten, when eating constitutes a simple good and sustenance comes less as a gift than a given, a common pasturage—all living beings bioreactors whose tummies digest a world of plants or who eat beings that are understood not to "count." To eat, under such circumstances, might no longer be to eat at all.[76]

Obviously, the zoology of the verses remains complicated, the significance of each animal tied to divergent historical particulars and only partially retrievable vectors that incline toward variously local political commentaries and animal fables. "Lions," kings that they are, shall not eat calves; the reinstated outlaws, formally known as "wolves," shall not prey on the flock that constitutes the collective or the Commons. "Man," "wolf" that he is to "man," shall find that, come the end time, the reasons for his unwholesome zoomorphic parasitism shall be removed. But, however evolved the metaphors, however convoluted the layering of beings, the language of political fable remains intact as it primes a theology that alludes to the cancellation or deactivation of its terms. This is all to say that Jack, Dick, Rick, and their sheepy company will take some convincing. For, as they know only too well, sometimes the teeth that do the biting belong not to another human but to a historically particular flock of sheep bought and bred by a group of fellow Christians who grow grass in the place of the crops that would feed you. The interrelations between rival multispecies as they constitute a common world, if not a world in common, fold together assortments of different animal and plant actors in ways that render the verses from Isaiah both yet more appealing and yet more difficult to parse or pry loose from figures of a desired cancellation and an imagined hospitality. It will take a lot of work, in other words, to deprogram the eschatological underpinning of the verse and pry loose from it a technics or protocol that produces a mode of hospitality that does not manifest under the sign of some infinitely prorogued and so filled-in future or as the cancellation of antagonisms through some mammalian–reptilian occupation of the supposedly indifferent world of plants.

Some of this work, as I have argued in this chapter, is accomplished (in small) by the likes of Thelma Rowell and her sheep as they cowrite something that looks like or plays like pastoral but that might prove to be something else entirely (epic?). Best then, for now, to stick with what the perspicacious Despret regards as key: the verse as an invitation to play with the laws of arrangement that keep beings corralled in this or that category such that the prospect of cancellation comes to reside in the particularizing questions we pose to them—something at which Rowell excels. Perhaps, then, Rowell and her sheep will enable us to trope the languages of biopolitics, in turn, such as Despret does when she offers a quasi-parental reminder to the child in the verse from Isaiah, the child shepherd, but also *the* lamb or Agnus Dei to be (in belated, Christian typology). She reminds this hybrid sheep-child to "lead them by the hand," affirming not the governing figure or presence of the divine so much as a governing attitude, orientation, or gesture of care and concern. What matters, what may be of use (still) in this verse, decoupled from the messianic figuration to come, is the gesture, the pose of this sheepy, shepherding child, a child Despret both tropes and demotes as she takes on the voice of the child's shepherding teacher or parent.

Otium, cancellation, utopian deactivation, are red herrings. What counts are the careful questions that must be crafted to render each creature its own, the inventorying of particularizing differences that enable the production of this differentiated and partial "we." It's a lot of work. There will be no easy equivalences. "What counts to us," as Despret writes, "counts otherwise for those we invite to construct this 'us.'"[77] The work remains ongoing. No wonder, then, that Despret counsels us, like the lamb, to sleep with one eye open (no matter how comfortable the mattress). To shepherd the danger, we shall have to inquire into the particularities of all and each.

"*Omnes et singulatim,*" we might find ourselves mouthing to our surprise and perhaps dismay, all together but also one by one, but now by and through the difference of each. For in the register of biopolitics as it derives from Christian pastoral power, you and I, along with Despret and her investigators, remain shepherds (and sheep). But, by Despret's troubling of the laws of classification, we shall have to become shepherds also to ravens, shepherds to baboons, to pigs, to sheep; shepherds to this putative "we," for which we serve as an occasion or a host but not a referent. Or, more correctly, the very essence of what it means to "shepherd" changes. Shepherd effects, shepherd relays, the shepherd finds itself rezoned as a gesture or technique, unpersoned by the "human," as we find ourselves shepherded also by whatever it is that counts for sheep, ravens, pigs, and ticks. A common world becomes somehow synonymous with a general shepherding of the great variety of finitudes we come into being with and to which we flock.

"Who told you that man was the shepherd of being?" asks Bruno Latour, seemingly out of nowhere in *The Pasteurization of France.* "Many forces would like to be shepherd," he continues, "to the others that flock to their folds to be sheared and dipped."[78] Latour cites Heidegger's famous "Man Is the Shepherd of Being," from "The Letter on Humanism," with the aim of sidelining the adage, of subjecting it to the vertigo of the purely neutral parasitic cascade that he traces through his own analysis of Pasteur's hygienic war on microbes. "In any case," he corrects, "there is no shepherd," venturing that "there are too many of us and we are too indecisive to join together in a single consciousness to silence all the others." He invites us all to shut up and let all these others speak.

What I have proposed in this chapter marks something less immediate and more difficult but true nevertheless to both Heidegger and Latour—a project to which Rowell and Despret, in different ways, contribute key concepts. For in their hands and between the hooves and mouths of Rowell's sheep, shepherding ceases to designate the scenography I have traced in this chapter and instead points toward a generalized bio- or zoopolitical activity that conserves and redistributes shepherd effects across the range of beings invited to contribute to the description. The scenography of care remains—as it did for Heidegger and does for Latour, however restricted or botched that scene—but it finds itself generalized, radicalized, as the terms deterritorialize athwart the supposed lines of species and shepherding finds itself endlessly reexpressed as a necessarily inadequate gesture—inadequate because it aims not to capture or corral but to offer the shepherd as screen or medium for another. *Singulatim et omnes* (one by one and so for all): what counts for sheep counts differently for me, re-makes me, and so convokes this "us" that makes no sense without (all of) you.

But let's not get carried away and miss what now serves as a given or a ground. Bees, according to certain traditions, hate sheep.[79] According to others, sheep hate hyacinths—hate them with a passion that renders them floricidal, eating "so greedily thereof [that they] will swel till they burst." As the mythic ecology or parasitology of Ovid's *Metamorphoses* has it, sheep revenge the violence done to them by Ajax of old. Driven mad in his rage at Ulysses, Ajax kills an entire flock of "knott ewes."[80] So, now, sheep waste their "old grudge" on Ajax's metamorphosed remainder every chance they get, gorging themselves on hyacinths, even though they risk dying as a result of their chronic overeating. How then do we reckon with the plant stock or vegetal being from which Rowell releases sheep to become animals again? What place do we give to this vegetal being that serves as terrestrial grounding, that this putative "we" takes as a given and so eat as food?

II
ORANGES

3

Invisible Inc. (Time for Oranges)

> Here's another exceedingly interesting subject for analysis: the Prison Regulations, and the psychology which develops among the prison staff—a psychology deriving from the Regulations on the one hand and from the contacts between warders and prisoners on the other. I used to think that two masterpieces (I mean this quite seriously) concentrated between their covers the thousand year old experience of man in the field of mass organization: the Corporal's Manual and the Catholic Catechism. I now see that one must add to these the Prison Regulations . . . because they contain real treasures of psychological introspection.
>
> —ANTONIO GRAMSCI TO TANIA, Milan Prison, April 11, 1927

> Practical deconstruction of the transcendental effect is at work in the structure of the flower, as of every part, in as much as it appears or grows as such.
>
> —JACQUES DERRIDA, *Glas*

On the morning of October 5, 1597, Sir John Peyton, lieutenant of the Tower of London, wrote the following letter to the Privy Council. It is the kind of letter that no prison warden ever wishes to have to write:

> This night there are escaped two prisoners out of the Tower: viz., John Arden and John Garret [Gerard]. Their escape was made very little before day, for on going to Arden's chamber in the morning, I found the ink in his pen very fresh. The manner of their escape was thus. The gaoler, one Bonner, conveyed Garret into Arden's chamber when he brought up the keys, and out of Arden's chamber by a long rope tied over the ditch to a post they slid down upon the Tower wharf. This Bonner is also gone this morning at the opening of the gates. Mr. Beling, the attendant in the council chamber, is his brother and assured me of his honesty. One

> Chambers, a gaoler at my coming, finding him negligent in his office and knowing Anies (whom he kept) to be a dangerous prisoner, after the recovery of my sickness I displaced. But not having time to discern the condition of this Bonner, being generally commended, I let him continue. I have sent hue and cry to Gravesend and to the Mayor of London for a search to be made in London and in all the liberties.[1]

What Sir John Peyton doesn't know, or fails to remark, is that, in among the paper and ink in John Arden's chamber, there were also several small piles of orange rind or peel. Closer inspection would have revealed that these piles were not merely the remains of some early-morning snack or dietary supplement but rather carefully cut and pieced crosses strung into a rosary. The flesh of the oranges may have been eaten, and the crosses may be no more than an act of pious recycling. But, had Peyton or one of his subordinates paid stricter attention to the eating habits and handicrafts of his prisoners, he might have saved himself the embarrassment of putting even fresher ink than he had found in Arden's pen to paper too late on the morning after the escape to inform the Privy Council of this lapse.

Ironically, if Peyton mucked up his clothes while he or one of his subordinates was leafing through the papers and ink left in Arden's cell, or if, when he or his secretary (if he had one in this case) were composing this letter, he got ink on his shirt, then whoever did his laundry might well have reached for an orange or a lemon to remove the stain. In his translation of a Dutch book of household remedies printed first in 1588, and then twice more through 1605, Leonard Mascall counsels his readers on the finer points of removing "spots and staines" from clothing:

> If there chaunce by fortune, to fall a droppe of ynke, or any other staine, upon any cloth dyed or coloured, or being white, woolen or linen, Ye shall doe as hereafter followeth, that is take the juyce of rawe Lemons, or the juyce of a great Orange Apple . . . which hath a hard pill or skinne . . . and with the juyce yee shall all to rubbe and chafe the sayde spottye places, and then with luke warme water, and so then scrape the filth thereof, with some spoone.[2]

Spoon in hand, the remnants of an orange or lemon by her side, Peyton's laundress cum chemist was potentially a better reader of events in the

Tower than the forensic paleographer who employed her. What is more, the next time Peyton went to put on the shirt he wore on the day he potentially stained his career, he might perhaps have detected the faintest scent of oranges. Oranges and orange peel, orange juice, as well as or even more so than ink, were the true subject on this particular morning in 1597. It is they that enabled the escape, they whose remains littered the scene. Now why might this be so?

In my last two chapters, I examined a series of anthropo-zoo-genetic figures or tropes that derive from the long story of our co-making with sheep. My aim was to reconstruct this ovine substrate to our discourses and to demonstrate the way our categories of being, our bodies, our texts, are populated with sheepy remainders that constitute one fragmentary multispecies impression. It was relatively easy to trace these figures because within the discourses of pastoral care that give us our daily biopolitics, sheep and not-sheep humans exchange properties: shepherding provides the grounding mode and metaphor for the care of the flock, human and otherwise. But at the same time, the two multiplicities (sheep and not-sheep humans) found themselves differentiated with regard to a grounding plantlike *phusis* or figure of pure growth. The animate existence and labor of each were managed in reference to a more basic still vegetal being to which, in fact, all creaturely life might be reduced or expanded. Access to this terrestrial grounding, this vegetal sameness, went, on one hand, which is to say for the human, by the names of sleep, *otium,* rest, leisure, idleness, whose careful management renders us busily and productively animate but whose abuse renders us supine, handless, divine, and bestial. On the other hand, which may mean, as in the case of sheep, no hands at all, a degraded *otium* or "sheepishness" describes the plantlike state of livestock or animate (plant) stock selectively bred to "grow" wool and flesh. The sheep's liveliness, its existence, becomes accidental to its successive use and commodity values. Enter the capital fellow that is the animate, agentive, human subject leading a parade of abjectly sheepish sheep sloughed off from this calculus, this counting, that projects and then eventuates the vegetal equivalence it posits.

It was against this disanimating calculus that writes sheep (and humans, from time to time) as sheepish that primatologist turned sheep observer Thelma Rowell invited her historically particular sheep to cultivate their own forms of boredom and interest. Modeled as if primates, posed questions usually reserved to chimpanzees, Rowell's sheep were momentarily promoted up a species hierarchy that allowed them to become more "interesting." Rowell's protocol rezones *otium* as a common property of both sheep and humans—both of whom become, as a result, differently interesting, differently animate, differently bored and boring representatives of an altogether new category: "all gregarious long-lived vertebrates capable of mutual recognition."[3] Yes, the phrase is a bit of a mouthful, but that is part of its charm. It plays with the laws of taxonomy in ways that subject categories such as "primate" or "human" or "sheep" to an alternative, more sheepy or sheep-centric set of criteria. It demonstrates the way a multispecies writing machine, such as Rowell and her sheep craft, might produce concepts as they emerge from determinate practices, concepts that prove hospitable to a greater range of beings because of the scenography they propose or entertain.

Rowell's work offers one example of how an altered technique of observation may yield an altered ethical orientation to specific beings. Interfering with the protocols governing how sheep are observed or posed questions writes "sheep" and sheep observers differently. And this altered sense of what sheep (and we) are or may become offers us the opportunity to rethink and rewrite the genres and tropes that shape our forms of sociality, our ways of being. *Otium,* for example, finds itself resignified, gone mobile across the apparent divide of species and kind. So also, as Vinciane Despret demonstrates, figures of messianic cancellation or cessation, as in the verses from Isaiah 11:6 that give her *Quand le loup habitera avec l'Agneau* its title, become invitations to play with our laws of arrangement. The eschatology to such figures remains (*things* end or unravel; I shall die; you shall die), but their affective force or impetus finds itself rerouted as input to successive imaginative experiments and acts of making. Given, in other words, that all our taxonomies have shown themselves to be unlawful or lawful only through a law-making violence, these figures of cancellation become, instead, invitations to begin again, to own the fact that we know so little of the beings we have written and now take as given.

Sleeping within this rearrangement of newly animate and "interesting" animal beings (mammals, birds, and reptiles), indifferent, so it seems, to this ongoing calculus and its recalculation, the world of plants subsists still in the mode of some universal substrate or matter more material. For Rowell the plant world appears in the form of the emblematic or idyllic twenty-third food bowl she sets out for her twenty-two sheep, removing thereby an automatic question concerning competition for food as the governing orientation to their observation. Food becomes a given for these historically particular sheep. And by this given, sheep find themselves also no longer modeled as if food, as if a set of resources to be variously assimilated to other beings. Sheep become historical beings. Within this scenography, plants manifest as the given Despret inherits from Isaiah's prophetic positing of a feast in which all are nourished but no one (nothing that counts) is eaten—in which eating has ceased to be what it is for you and me, and for sheep along with their predators. The world of plants funds Rowell's protocol and Despret's renegotiation of creaturely categories. Plants continue to function as they have, clinging or winding their way through our discourses, metaphorical epiphytes, root stock to our animations. But this vegetal given poses no disabling scandal. Rather, it suggests the stakes to differently distributed givens or grounds. Both Rowell's protocol and Despret's ethology of practice remain thoroughly adequate to the task of inquiring into what might be said to interest plants, into what "counts" for them. All that is required is a little imaginative rewiring of our regimes of description sensitive to the fact that, as Despret offers, "what counts [*compte*] to us, counts [*compte*] otherwise for those we invite to construct this 'us.'"[4] No easy equivalences. No pastoral, then, for plants. Or if there is, it will manifest in some still other, perhaps attenuated or only partially available mode that will require us to imagine other genres, media ecologies, or kinds of writing.

In this chapter and the next, I inquire into the texture of this vegetal substrate and explore how our discourses are marked by forms of vegetal being, calibrated by vegetal temporality. I am not proposing that plants exist as some equivalent category to the abjected sheep that Rowell renders interesting, enabling them to become "animals" again as opposed to plantlike hybrids. That is not the point. Instead, I posit plants as if a semisovereign polity or series of polities, whose being and whose boundaries are, nevertheless, coterminous with our own. Our infrastructures, the worlds we live, our bodies, are shot through with their remainders—sometimes obviously, mundanely, sometimes in ways that are difficult to perceive. "The challenge," as Michael Marder puts it, "is not [to] assert an unconditional right of admission into the vegetal world, which is the world *of* and *for* plants, accessible to them," but, instead, to "let plants be . . . to allow plants to flourish on the edge of or at the limit of phenomenality, of visibility."[5] We remain, as he offers, "strange archives. Surfaces of inscription for the inorganic world, of plant growth, and animality—all of which survive and lead a clandestine afterlife in us, as us."[6] And the differing forms of finitude we come into being with necessarily canalize the cognitive, affective, and so metaphorical capacity of beings to interact. Plant being sponsors certain transversal identifications and disables others. Indeed, by and through their difference from animate creatures, plants offer something on the order of what Jacques Derrida calls an automatic or "practical deconstruction" of our categories. The "structure of the flower," he offers, functions as an always already self-sacrificed or precut entity that scrambles creaturely codes. The flower signifies neither a loss nor a gift. It does not participate in a sacrificial economy or, if it does, it takes no cognizance of that fact. As Marder elaborates, this mode of vegetal *différance* captures the passing of time, or emits time effects, in the form of an ex-scription as opposed to an inscription, a form of marking in and as and by and through the distributed body of the plant as a surface that necessarily allows for no separation between what the animals known as "human" call a "text" and its substrate, backing, or media. No erasure exists for plants exactly as it exists for us. For the flower, the fruit, and the leaf are already cut, precut, as it were, destined to die or, as in the case of an orange, living still as it falls from the tree, to die on, dying into the future, so that the plant may live on *(sur-vivre)*. Vegetal being, orange-being, offers us an "acepholous" or headless discourse, a mode of organization that does not operate by way of a hierarchized notion of the body, of inside and out, surface and depth.[7]

This chapter may be thought of as an exercise in tracing or unfolding the practical mode of deconstruction that an orange sets in motion.

In what follows, I aim to inhabit or "perform" the archive or repertoire of letters, memoirs, bills, reports on security, and so on, generated by John Gerard and John Arden's escape from the Tower of London in 1597. The story, as I tell it, has the flavor of detective fiction, albeit of a strange or self-ruining kind that pulls itself apart, voiding its moments of revelation, as it attends to the various actors that make up this event. My aim is to produce something on the order of a multispecies monograph or singularity writing tuned to the distributed remains of oranges that populate this repertoire of texts, inflecting the relations between ostensibly human players. As I add successive textual remains, the *thing* or gathering that was and is and will become the escape changes, pulls itself apart, its ontology shifting by and through the reference management that my textual collage produces. As the texts turn and I privilege this and not that perspective, oranges flicker in and out of existence. Whole and cut, cut and pieced, whole again, oranges—the individual orange always morphing back into a multiplicity or stream of oranges, differentiated only in the moment of its consumption or use—punctuate the story. The serial repeatability, the iterability of oranges, undergirds this archive that plays host necessarily therefore to fragments, pieces, particles of orange, that appear as if an index or icon to some set of human practices, the orange already an archival entity, whose peel inclines us toward the way our own discourses are shot through with its errant, accidental afterlives.[8]

If it seems that in siding with oranges I have strayed from the world of sheep and shepherds, and so leave behind the biopolitical and governmental forms that derive from Christian pastoral care, then it might be useful to recall that the occasion for this chapter is a prison escape precipitated by the existence of a Catholic mission to bring the wayward heretical flocks of a reformed Protestant England back to the fold of the Universal Catholic Church. In 1570, the Jesuits spearheaded a mission to reconvert England, ministering to its Catholic community, known as recusants, because they refused to attend Anglican services.[9] This mission depended on the steady recruitment of English gentlemen to become priests. English Catholics, like John Gerard, left England for colleges or seminaries on the Continent to take orders, from where, depending upon your point of view, they returned as hybrid missionary shepherd priest sowers of true seed or as clandestine foxy invaders, wolves in shepherd's clothing, who sowed seeds of rebellion and diverted good Protestant sheep along the crooked paths of Catholicism. In 1585 it became treason for a priest ordained since 1559 to set foot in England and treason also to harbor a priest.[10] So when priests arrived in England, awaiting them were two opposed networks. Each proposed mutually incompatible lines of emplotment: one led to the safe house, the other to the prison. One sought to maximize the yield from the seminaries abroad. The other committed its resources to channeling the returning "seedmen," as they were called, into its own receptacles—the prisons that ringed London.[11]

Omnes et singulatim. We remain within the precincts of biopolitics even as certain of the actors to this story seem intent on escape or redirection: Jesuit John Gerard, Catholic gentleman John Arden, and the ambiguous Bonner, their warder. Still we count sheep—all together and one by one. The scene remains the same, a scene of differential and differentiating care, however recalcitrant, however seemingly wayward, however boring or interesting, the sheep. The prison milieu makes a difference—something Antonio Gramsci avers as he expresses his frustration with the experience of writing in prison, the difficulties (technical and personal) in obtaining pen, paper, and ink, eventalizing the act of writing itself in ways he refuses to find disabling. "Writing has become a physical torment to me," he writes home to Tania, "because they give me horrible pens which scratch the paper and make me pay a maddening amount of attention to the mechanical side of writing." He thought he would be able to obtain "the permanent use of a pen," but "I didn't obtain permission, and I don't care to press the matter."[12] Gramsci extrapolates from these difficulties and offers the Prison Regulations as a supplement to the British Army's Corporal's Manual and the Catholic Catechism as documents to "the thousand year old experience of man in the field of mass organization." Beyond anticipating an analysis of the penitentiary or the elaborated forms of discipline, beyond fracturing notions of "bare life" into a whole phylum of forms of life that still must be lived and therefore have meaning, Gramsci puts us on notice to the way a prison ecology routes the relationships between its inmates and warders through a world of objects that take on urgent and unexpected importance: pens, paper, and ink (obviously); rope, rivers, and boats (in the case of a waterborne escape from the Tower of London); but so also to a host of things that appear, at first, unmarked, something as seemingly accidental as, say, a taste for oranges.

Gramsci withholds the circumstances that make it hard for him to "press the matter" of obtaining a pen that works—those circumstances go unrecorded—but in the case of Jesuit John Gerard and his warder, we know more, that is, if we take Gerard's recollections, for now, on trust. For the two men's relationship, so we learn, was built through a mutual liking, an inclination, a fondness or foolishness, a taste for oranges, oranges that also come to enable Gerard's access to pen, paper, ink, to rope, a boat, and escape. So it is in this chapter that I choose to write under the sign or impression made by these oranges, oranges that leave no phenomenal marks on Peyton's letter but whose encrypted agency saturates it, nevertheless. I'll show you what I mean by reading the oranges for which Peyton seems not to know to look back into his letter a bit further.

BURN BEFORE READING

Now, Peyton was no fool. It should be remembered that the rhetorical occasion for his letter was not to inform the Council of Arden and Gerard's escape but to demonstrate that he had not been derelict in his duty. Close inspection of Arden's chamber has been made. An inventory of the contents taken, revealing that the "ink in his pen is very fresh."[13] The method of escape is known, as is the agent, the warder, "one Bonner," who has fled the scene. A search has been ordered and the Council informed. Subtext: no time has been lost; they cannot have got very far; I have done my best; and, anyway, it was an "inside job." The gaoler did it. Beyond all of this, as the Privy Council knew only too well, Peyton had only been in the job for three months (he was appointed in June 1597); he had also inherited the business of conducting a survey of the Tower's precincts and of supplying a description of its defects. He had also been ill. What is more, in another case, he did dismiss the unreliable Chambers, thus preventing, perhaps, another prison escape. If only he had had more time. If only he had not been ill; the warders not so unreliable; had Bonner's brother not been misled, and so misled Peyton, then, no doubt he would have rumbled

Bonner and foiled the escape. Peyton subordinates his forensic powers to the rhetorical necessity of his position. He's on the lookout for whom to blame and so sifts the debris of the cell to find what he can immediately connect to the escaped prisoners. Paper and ink do the job nicely: they refer to the now past presence of the inmates and to an act of writing, to letters that must have changed hands. Bonner's absence also speaks volumes. Oranges and orange peel mean nothing because they cannot be linked to a person in a significant relation. They exist merely as the static debris of the prison milieu.[14]

Peyton's subsequent proposals for reforming security at the Tower confirm that he diagnosed a series of human failings as the root cause of the escape—failings that stem from what he then models as systemic labor problems at the Tower. His report survives in two copies: a draft he was writing on or about the date of Gerard and Arden's escape in October 1597 and the revised and final version he submitted to the Privy Council in September 1598. In the draft, under the heading "Defectes Needfull to be supplied," Peyton notes that the Tower is manned by only thirty warders—much too "weake a watch" given the magnitude of the task and their range of duties.[15] He also complains that the warders are allowed "but eight pence a peece *p[er] diem*, the w[hich] sufficeth not to feede them above one meale in a daye." This means that those warders who rely upon the room and board they receive at the Tower for their subsistence subcontract their free hours to their fellows, who use the time they are supposed to be on duty to pursue other trades or business that proves more lucrative. Peyton worries also that some warders have to "keepe theyr families" in the Tower, which raises concerns regarding their safety given the store of gunpowder and their providing "intelligence to the prisoners."

In the final draft of 1598, Peyton attempts to alter the conditions of employment and so to remedy these defects. He adds several paragraphs concerning the Tower's warders that aim to improve their pay and enhance his ability to recruit capable men for the job. He notes that "ther bee amongst ye warders divers unfit for the place; some of them utterly neglecting their duties in service, others given to druncknes, disorders and quarrels, other[s] for debilitie of bodye unable to p[er]forme their duties, others double officed and cannot attend in two places. By meanes whereof," he concludes, "the guard of this place is muche weakned."[16] Peyton then asks that the terms of employment at the Tower be revised so that in place of "graunting romes in reversion . . . for terms of lief by means whereof suche as are founde negligent, unable or unfaithfull in yo[ur] service cannot be by law discharged," Her Majesty grant rooms to warders "*ad bene placitum*" (for as long as it pleases) so as to "keepe the warders in better regarde of their duties." Peyton begs permission to recode the labor of the warders, to revise the particulars of their tenure and pay, with the aim of recruiting and maintaining a more motivated workforce. He also proposes a series of measures that will restrict traffic through and within the Tower by removing access to areas such as the "brewhouse," "backhouse," and "hackney stable" that had been "comon" thoroughfares. Good shepherd or human resources manager that he is, Peyton aims to rid himself of the likes of Bonner or the man he assumes Bonner to be, cutting him from the flock for the good of the flock.

The problem is that, in Gerard's Latin account written some ten years after the event, Bonner knew nothing of the planned escape. He permitted Gerard to move around the Tower, which was unusual but not exactly unheard of, but, other than that, he was, apparently, not guilty of much else. So, if Bonner's guilt is an artifact of Peyton's construction of the escape, then who was he really?[17] To Peyton, whose letter is the only surviving document from this affair to mention Bonner by name, it is the name that counts, the legal identity and agency he can offer the Privy Council as culprit, and so as evidence that he has fulfilled his role as their representative. Oranges mean nothing because they cannot be linked to a person in a meaningful relation. But, in Gerard's memoir, Bonner is referred to simply as the warder who "was particularly fond of the fruit [*quorum esu videbatur delectari*]," as he, in other words, whose taste, whose desires or needs, whose physiology, predisposes him to the oranges Gerard also desires, albeit for different reasons.[18] Like all of the players in this event, Bonner is a hybrid—a name, a legal identity subject to laws and to punishment, a corporate representative who locks and unlocks doors, and a go-between, he who, in a different way than Peyton's laundress, bears oranges. His liking or "taste" for oranges is really a *thing* lodged within the zone of the human. It is Bonner who brings oranges into the Tower, who makes all that oranges mean (their taste, their color, their chemical properties, their temporality, their history, their principle of connection) available. It is he who gives legs to the fruit.

What was the exact nature of this shared liking or "taste"? How did oranges calibrate the relationship between Bonner and Gerard? If you visit the Tower of London today, you may begin to understand what I mean.

Translated from the Salt or Caesar's Tower, as it was sometimes called, where Gerard was incarcerated, to the Cradle Tower, you can see a somewhat cozy mockup of John Gerard's cell (Figure 10). Whole, juiced, and pieced, oranges dominate the scene, multiplying so as to supplement the ostensibly blank sheet of paper that hovers above the desk as if we were holding it, as if we had wandered into Gerard's cell (which, of course, in a way we have) and were about to read over what we ourselves or he had written. If you pause in front of the display, an electric lightbulb placed behind the sheet of paper turns on and off automatically at roughly twenty-second intervals, illuminating a still indecipherable secret message that appears only to disappear, revealing nothing more than the presence of something that passes as writing hidden within the expanse of white space. As the curators of this exhibition know and Peyton's laundress might have been able to tell him, one of the properties of a weak acid, such as orange or lemon juice, is that it weakens the paper or parchment where it is applied such that when exposed to heat, the script oxidizes at a faster rate than the surrounding white space, bringing out the script in a dull brown. In his memoir, Gerard goes on to explain that oranges were, to his mind, the preferred medium for secret writing because they offered the most security. "Lemon juice," he writes,

> has this property, that it comes out just as well with water or heat. If the paper is taken out and dried, then the writing disappears but it can be read a second time when it is moistened or heated again. But orange juice is different. It cannot be read with water—water in fact washes away the writing and nothing can recover it. Heat brings it out, but it stays out. So a letter written in orange juice cannot be delivered without the recipient knowing whether or not it has been read.[19]

Figure 10. John Gerard's cell, Cradle Tower, Tower of London, July 2014, mocked up to represent the secret writing practices he describes in his memoir. Photograph by the author.

Gerard had been communicating secretly with friends and associates outside the Tower. That is how the escape had been orchestrated and planned. Oranges are about secret writing, about another kind of pen-and-ink story. Case closed. Gerard's liking for the fruit, so it would appear, refers to a tactical knowledge of substances that offer a hermeneutic advantage and so have some utility to the building and maintenance of an underground movement.[20]

Recipes for invisible ink were not uncommon in the period, and their knowledge was a useful if discontinuously remembered commodity as well as curiosity. The most celebrated source of many recipes as well as rumors was Giambattista Della Porta's immensely popular *Magiae Naturalis* (1558), which ran through some ten Latin editions and sixteenth-century translations into Italian, French, and Dutch before appearing in English as *Natural Magick* in 1658. Book 16 is titled simply "Wherein is handled Secret and Undiscovered Notes" and anthologizes the various means of invisible writing then available. For the pre-Baconian Porta, the efficacy of orange or lemon writing stemmed from the "acrimony" or sharpness of the fruit's "undigested juices." When "they are detected by the heat of the fire," writes Porta, "they show forth those colours that they would show if they were ripe."[21] The fire "cooks" the script or matures it much as the sun would the fruit. In the intervening time prior to its "bringing out" by flame, the script hibernates or winters in the paper until it may be brought out or ripened by the heat that restores and accelerates the passage of time.

Porta runs short on details, however, as well as on technical assistance, something the Elizabethan inventor Sir Hugh Platt modestly remedies in his *Jewell House of Art and Nature* (1594), where, among other ingenious recipes and devices, he advises would-be steganographers or practitioners of "covered writing" to "write your minde at large on the one side of the paper with common inke, and on the other side with milck, that which you would have secret, and when you would make the same legible, holde that side which is written with inke to the fire, and the milckie letters wil shew blewish on the other side."[22] Recipes for invisible or sympathetic ink include various fruit, vegetable, mineral, or bodily matters, ranging from onions to urine, as well as more highly evolved alum or gall-based inks that could only be brought out by the somewhat hazardous application of compounds of mercury. Gerard's detailed description remains highly unusual within what otherwise proves to be a highly discrete archive, whose most typical entries tend to be little more than a few lines in a manuscript recipe or receipt book promising that you may "write so no man may read" but which recommend little more than squeezing a lemon or an orange and using the juice as ink.[23] Such discretion might be reversed for comic effect on stage or in print, as in the cornucopia of fruits and vegetables suited to secret communications that Sir Politic Would-be provides in Ben Jonson's *Volpone* (1606), where oranges appear in among the "musk-melons, apricots / Lemons, pome-citrons" and cabbages.[24] More explicitly still, in Thomas Middleton's *A Game at Chesse* (1624), the Jesuit Black Bishop's Pawn and the black Knight Gondomar examine a letter that is declared to be "Blind work . . . / The Jesuit has writ this with juice of lemons sure. / It must be held close to the fire of purgatory / Ere't can be read."[25]

In February 1587, Mary Queen of Scots had fallen as a result, in part, of the discovery of letters written in lemon juice, and by the time of the Gunpowder Plot in 1605, the Privy Council knew only too well to suspect and even to solicit the citrusy desires of their prisoners. Between March and April 1606, when Gerard's Jesuit superior Henry Garnet was imprisoned in the Tower, letters he wrote in orange or lemon juice were intercepted and read and the text recopied and sent on to their addressees. Garnet's warder had pretended friendship, offering to deliver messages to and from his patron Anne Vaux, but he first delivered them to Sir William Waad, then lieutenant of the Tower, who passed them on to Attorney General Edward Coke. Six letters in all were intercepted, the secret writing discovered, owing to the amount of white space left in the letter and the slightly too posy-like aptness of the visible writing thereon. The first of these letters, written from Garnet to his nephew, is a rough piece of paper used to wrap a pair of spectacles, upon which appear the words "I pray you lett these spectacles be set in leather and with a leather case, or let the fould be fytter for your nose." The enclosed spectacles were the vehicle for the transmission of letters and so messages. Subsequent letters include inventories of goods received—a mundane but necessary acknowledgment that things sent into the Tower had not gone astray—and variously illegible attempts to write in invisible ink, prompting, perhaps, in one letter, Garnet to write that "your last letter I could not reade. Your pen did not cast inck."[26]

Now, while this disclosure seems to explain everything—oranges are about secret writing, about a different kind of pen-and-ink story—it marks only the very beginning to another way of reading these events, or of beginning to read them at all. All I have done is move the threshold of intelligibility slightly: oranges now presence in their relation to writing, secret writing at that, and so to the human agent who writes. All that has occurred, in Porta's terms, is that the "undigested" orange juice has "ripened," disclosing its physical presence by and through its translation into a medium for writing, writing that then signifies not because of its visible contents but because it refers to a suddenly decrypted human agency. The content of the messages, visible or not, does not really matter. What does is the act of appearance itself, the bringing out of an invisible medium, of writing that faded as soon as it was written and that hibernates like some stored potentiality or dormant *kairos* awaiting reactivation. It is the interruption of one order of writing as it is backed by paper and ink by another order of orange or lemon writing that signifies, creating thereby the appearance of presence, agency, and so revelation. The bringing out or ripening of the message reveals only the operation of a medium-specific technique indexed to the potential that there was, at some point, and still may be, something else to be known, something both entirely past and of the future, illegible to the pen-and-ink present of the letter that is its medium, ground, and ruin. This is all to say that in a strict sense, the effect of invisible ink, of orange or lemon writing, is to render writing both hyperpresent and virtual—something the iconic, redundant blankness of the white paper in Gerard's mocked-up cell at the Tower and the illegibility of the message that appears there (Figure 11) captures vividly in its solicitation and frustration of our desires for historical truth.

In a strange way, then, Gerard's mocked-up cell reprises the tropic work performed by the only instance of onstage orange or lemon writing in the period, the moment in Christopher Marlowe's *Dr. Faustus* that sends

Figure 11. Detail from John Gerard's cell, Cradle Tower, Tower of London, July 2014, focusing on the blank sheet of paper that serves as the focal point for visitors. Photograph by the author.

Mephistopheles running to "fetch . . . fire to dissolve" Faustus's congealing blood. The action pauses or fits. The play seems to invite a metadramatic stall or even failure, something Faustus himself seems to remark when he wonders what the "staying of my blood portend[s] / Is it unwilling I should write this bill?" But before the scene turns into pantomime, or as it does so, the fire gets things moving again; makes his blood run "clear"; serves as ink. Faustus signs. "*Consummatum est.*" But as he does so, the words "*Homo Fuge!*" flicker in and out of view on his arm, a message that Faustus cannot quite construe. "Whither should I fly?" he asks, missing the point, blanking out the moment of ironic revelation of what might be termed a given, that his soul and body are not his own, that he may not sign them away, even as he does so.[27]

Mephistopheles's fire does double duty. It catalyzes two incompatible orders of writing or stages their collision. Within the diegesis, the fire liquefies Faustus's blood so that it may serve as a form of bodily ink enabling the delusional sovereignty he takes over his flesh. Onstage, it functions either as a winking and perhaps topical reference to the practice of secret writing and/or as a theatrical and so diabolical technique that enables writing to appear and disappear on the arm of the actor playing Faustus, much in the way nineteenth-century table-turning spiritualists would divine messages from the beyond by way of skin writing or dermatography with lemon or orange juice.[28] The fire enables Faustus to sign and so assert a sovereign control over his flesh. But it also enables the revelation of a divine precontract or tattoo that trumps or tropes his signature. Faustus cannot sign even as still he does; the pact proves both impossible and, paradoxically, effective, for his flesh, our flesh, reveals itself here as already archival, prewritten, host to variously time-bound remainders that presence even as we put it to our own use. Faustus signs, but his signature remains haunted by indications of the agency or, in Lowell Gallagher's words, the "incipient meaningfulness" of his blood itself.[29] And such meaningfulness or divinity that resides in Faustus's flesh and blood appears or is hosted by the zoomorphic presence of the fruit (an orange or a lemon) that enables the transformation of his life/blood into the ink with which he signs. The divine, the inhuman, "ripens," as it were, by way of an orange or a lemon cooked by Mephistophelean fire.

Playing a different game, trading on the amatory ingenuity to be had from the homonyms *lemon* and *leman* (lover) in Early Modern English, Royalist poet, secretary, and sometime cryptographer to the exiled Queen Henrietta Maria in Paris Abraham Cowley comes closest of anyone to pondering the efficacy of the fruit pressed to use in secret writing. He begins "Written in Juice of Lemmon" from his poetic miscellany *The Mistress* (1656) by remarking the strangeness that "what I write I do not see" (the poem disappearing into the page on which he writes) along with the forlorn foolishness of a muse who knows that her efforts shall end up "read by, [the letter's] just doom, the *Fire*."[30] The poem takes the act of composing in lemon juice as its occasion. The speaker addresses the paper on which the poem, or something else entirely, is written and into which the script disappears. As the poem oscillates between its moment of composition and the culminating "bringing out" of the message, it renders the imagined story of its transmission as its content. The poem trades on or is generated by the shifting sense of writing that appears and disappears and the cascade of associations that the use of fire to bring out the script permits, recoding the tropes of Petrarchan fire and ice as well as those of Catholic martyrdom in the process.

Early on, the speaker laments that the paper, twinned with the body of the mistress, "think'st thy self secure / Because thy form is *Innocent and Pure* . . . unspotted." Its "blotted" nature will be revealed, however, by the "last Fire[s]" of Judgment Day and the literal application of heat to reveal the poem that is "scrauld o're thee." But because the letter may only be read (as) if burned, barring the mistress "pardon[ing]" it and so leaving it unread, it may "like a Martyr . . . *enjoy* the Flame," even as that act of reading risks a misapplication of the flame that might reduce the whole project to ashes. Such is the plight Cowley's poem voices, the plight of an intermediary whose existence must be transformed, wholly altered, to render its message readable. As the moment of reading or transformation beckons and the heat is applied, the poem stages the appearance of writing as a change of season. Vegetal growth irrupts within the snowy whiteness of the paper. Nature's characters, the divine signatures that "write" the world, the Book of Nature, manifest, in small, ripening or "deinscribing" not merely the lemon or the lemon tree but figuratively a whole orchard:

Strange power of heat, thou yet dost show
Like winter earth, *naked,* or *cloath'd* with snow,
But, as the quickning sun approaching near,
The *Plants* arise up by degrees,
A sudden paint adorns the trees,
And all kind *Nature's Characters* appear.
So, nothing yet in Thee is seene,
But, soon as *Genial heat* warms thee within,
A new-born *Wood* of various Lines there grows;
Here buds an A, and there a B,
Here sprouts a V, and there a T,
And all the flourishing Letters stand in Rows.

The fire appears as the source of a "*Genial heat,*" a principle of life, a giver of forms, such that letters "bud" or "sprout," the facticity of their posy-like, occasional appearance amped up by the fact that the moment of revelation produces no words, no message for us, just time and space-bound letters in the "now" of reading that reactivates the past "now" of writing.[31]

Lemon writing proves hyperoccasional, then, a technique that enables two otherwise unconnected moments in time to come into contact. Reading becomes thereby a telepathic or telegraphic act of joining with those short few moments before what was written disappeared—the mistress (but not the reader) reading the whole as yet unread message seemingly for the first time. Cowley invites us to join in wonder at the process, at the efficacy of the heat, and at the sympathy of the lemon that joins the speaker to his leman, despite the intervening distances or obstacles. Cowley's poem might be said to radicalize the poetic form of the posy, then, the poem as flower or flowering, a time-bound, occasion-specific form of writing that unfolds in and as space in a specific medium. As Juliet Fleming offers, "the posy is a form of poetry that takes in its fully material, visual mode, as it exists in its moment, at a particular site." "Paradoxically," she continues, "such poetry is portable ('something to be carried away') precisely because it has not achieved, and does not hope to achieve, the immaterial, abstracted, status of the infinitely transmissible text."[32] With the lemon, this portability, premised on the virtuality of the script, produces the ability to cross and close distances instantaneously—with no apparent loss.

If the "seely *Paper*" feels a bit left out at this point and objects that "all this might as well be writ with Ink," the poem ends with a further corrective and a caution. What's crucial, we learn, what makes lemon juice the requisite agent, is the fact that "as *She Reads* . . . [the mistress] *Makes* the words in Thee." The bringing out of the message by the mistress makes (or mars) the message. It transforms the act of reading into the quasi-divinity of a mode of sovereignty for which reading has become writing, for which to read is to write, to bring into being, and for whom Mephistophelean fire has become mere technique. All hail then to this sovereign mistress who shall read-write this poem, she who remains absolutely singular in her power. But the poem does not end here. It makes a further set of recommendations for the likes of you, and I, for the paper, and all the other mediators that enable the Mistress to read-write the poem. For, unlike the mistress who makes or mars the message, we must remain cautious with regard to what we learn from such moments of revelation or mere appearance that come with the bringing out of lemon writing. Intermediaries or substrates that we are, the mode of reading reserved to the mistress is not for us. Like the paper, we may get burned, end up on the fire or aflame. "The *Gods,*" the poem ends, "though Beasts they do not Love, / Yet like them when they'r burnt in *Sacrifice.*" We might want to go carefully, then, Cowley advises, as we play with fire and attempt to read what is written in orange or lemon juice. For such sovereign highs of revelation that come do so only on the basis that *we* burn before reading, mistaking an archive and its ash for what may or may not have occurred. By the end of the poem, the only thing we know is that it claims to have been written in lemon juice, a claim belied and rendered ironically superficial by the fact that we read the poem at all. Make the mistake of knowing more and we may come to look a little like Faustus signing his life away as he blanks out the fine print of theological time. We may even end up in the fire, like Cowley's paper (and possibly Faustus also), sacrifices to a desire that would collapse reading and writing, being and time.[33]

All the while, of course, indifferent to our plight, the presacrificed lemon or orange winks in and out of being, cut and pressed to use by an infrastructure or contexture that it both enables and disrupts, synching up or differentiating these and not those polities of human animals; rival shepherds and their sheep: Royalist and Republican; Catholic and Protestant; paying audience and playing company; paying tourist and past-purveying museum professional; whose pleasures and pains, like Cowley's lovers, they calibrate. In the process, in their use as media, the orange or the lemon becomes something else entirely, something, as Michael Marder writes, like "Spirit . . . at once the voice of Reason and that of Revelation . . . and so ceas[e] to be . . . a plant."[34] Or, to put it in terms to which Marlowe, Cowley, Gerard, and their readers might subscribe, for whom words were still *things,* orange and lemon writing designate one process by which plants funded and framed acts of inscription, literally (and figuratively) growing, living out their own forms of being, in and by and through their use as media that bound human expression. When words appear on Faustus's arm and groves of lemon trees sprout as letters in Cowley's poem, something of the plant qua plant remains, albeit in translated or grafted form. And as Cowley seems at pains to point out, the mere fact of writing in lemon juice to begin with, regardless of what is written, already binds thought to matter in a way that shall decide everything.

While we may wish to assimilate orange and lemon writing to technologies of forgetting and reuse in the period, such as the whitewashing of walls in Elizabethan houses, the razing of writing from paper or parchment by a knife, or the use of erasable wax tablets, which retain always at least a trace of what once was written, the ephemeral or superficial quality of the orange or lemon writing works differently. Unlike writing in ink, in which the pen might cut or scratch the surface of the paper, lemon or orange juice is painted or washed over the blank sheet in between lines written in ink. The hand that holds the pen must be light of touch, taking its cue from the painter or limner of portrait miniatures as opposed to the copyist, for lemon or orange writing makes no impression as such, it leaves no marks, other than those that occur through its weakening of the substrate (in our terms) or subsequent cooking by the heat that brings out the message (in theirs). The process remains entirely superficial—a matter of sympathy or faith. Nothing is hidden in the papery depths of lemon writing. The message glosses a surface on which, within which, whatever was written will appear, taking root, fruiting as it appears, in the manner of some waking parasite or symbiont, the orange or the lemon some autograft or epiphyte that makes its home within the media ecology of pen, paper, and ink.[35]

This is all to say that we are still a long way from understanding the events that took place at the Tower. And our understanding now hinges on what counts as an archive, what counts, in the broadest sense, as "writing," and so as evidence that can be motivated in relation to what will then become the story. As long as oranges figure as an expendable or "undigested" material, something I narrativize, ripen, cook, and so exhaust in the process (finally eating them four hundred or so years later), I shall blind myself to their share in producing the escape. Exit the multiplicity that is orange, the stream of oranges, cut and whole, whole and cut; think you have eaten the orange; and you will have decided everything, constituted a world and a "past" by and through your performance of it. Reverse this arrangement and consider a reciprocal process in which persons serve as bearers of different declensions of orange, in which oranges and persons exchange properties and so form an asocial, only partially cognizable, time-bound series or differently scaled multispecies, and much more remains to be said.

TIME BOMBS

Let's assume that we are novices at the College of St. Omer in the Louvain. The year is 1609. Our teacher is Father John Gerard, an Englishman, who was deputy to the Jesuit Mission in England during the 1580s and 1590s. Imprisoned for months in the Tower, he is famous, among other things, for his escape and for remaining in England on the run thereafter, fleeing to the Continent only after the discovery of the Gunpowder Plot in 1605. Or, perhaps, in his absence, we are having his story read to us by one of our peers. Gerard wrote a description of his time in England, titled simply *Narratio Joannis Gerardi,* most probably as a memoir, self-justification, and pedagogical guide.[36] The memoir is also readily recognizable as an early tradecraft manual, as evidenced by its inclusion in the CIA's online library of tradecraft history.[37] In this memoir, Gerard tells us how he was inserted into England in November 1588 and describes the various cover stories and aliases that enabled him to move around; the dos and don'ts of setting up an underground circuit of safe houses; and how to behave in the event of capture, imprisonment, interrogation, and torture.

Gerard was apprehended in April 1594 after a four-day search of Braddocks, a safe house in Essex, and sent to London for questioning. Upon his arrival, he was confined closely at the Counter and then the Poultry before being transferred to the much more loosely organized Clink later that year. There he remained for almost three years, before being transferred to the Tower in April 1597, where, on suspicion of having received letters from abroad containing treasonous designs, he was tortured and interrogated by Sir Richard Topcliffe, chief priest hunter to the Privy Council, then lieutenant of the Tower Sir Richard Berkeley, Attorney General Edward Coke, future chancellor Francis Bacon, Solicitor General Thomas Fleming, and Sir William Waad, then secretary to the Privy Council. Suspended from above by his arms and hands, which were placed in manacles so that his feet were a few inches above the ground, Gerard experienced "gripping pain . . . worst in my chest and belly, my hands and arms. All the blood in my body seemed to rush up into my arms and hands," he writes, "and I thought that blood was oozing out the ends of my fingers and the pores of my skin" (109). He was subjected to three bouts of this torture in all, which his warder witnessed, "wip[ing away] the perspiration that ran in drops continuously down [his] face and whole body" (110). Gerard thinks Bonner stayed out of "kindness" but notes that he "added to my sufferings when he started to talk . . . begging and imploring me to pity myself and tell the gentlemen what they wanted to know."

We join Gerard as he recovers from these bouts of torture and attempts to remake his body:

> Left to myself in my cell I spent most of my time in prayer. Now, as in the first days of my imprisonment, I made the Spiritual Exercises. Each day I spent four or sometimes five hours in meditation; and everyday, too, I rehearsed the actions of the Mass, as students do when they are preparing for ordination, I went through them with great devotion and longing to communicate, which I felt most keenly at those moments when in a real Mass the priest consummates the sacrifice and consumes the *oblata*. This practice brought me much consolation in my sufferings (116).

Prison turns back the clock. It returns Gerard to the seminary, and he becomes a student once again, passing his time practicing St. Ignatius's

Spiritual Exercises. He performs a dry Mass (Missa Sicca), the Mass minus the Massing Stuff, finding consolation nevertheless in the partial repetition of a form that to him is second nature. He goes through the actions with "great devotion and longing to communicate," which he feels most keenly at those points when his gestures mime the transformation of the Host, the palpable presencing of the inhuman. But Gerard's hands are empty. His gestures are exactly that, gestures. He performs a ritual minus the substance, the observance of pure form. Time passes, or is made to pass. And in the place where communication should take place, Gerard experiences a longing, a longing that transports him elsewhere. The gestures of the incomplete Sacrament, which effect no transubstantiation, serve as a curative *technē,* allowing him to mend his body. At the end of three weeks he is able to move his fingers, hold a knife, and feed himself. He asks his warder for a little money and a Bible, which his friends get for him. He then asks him to buy him "some large oranges [*poma aurea magna*]," and, "as he was particularly fond of the fruit [*quorum esu videbatur delectari*]," Gerard tells us, he makes him a present of them, "thinking all the time," he adds, "of another use [he] could put them to [*at ego aliud meditabar cum tempore*]" (116).[38] The two men discover a mutual liking or inclination for oranges. Gerard asks him to get him some and then voids the act of exchange or completes a different one by making a gift of them. The sharing of oranges necessitates a second, and then more frequent still, shopping expedition.

Gerard first puts the oranges he receives to use remaking his body. "My finger exercises," he writes, "consisted in cutting up the orange peel into small crosses; then I stitched the crosses together in pairs and strung them on to a silk thread, making them into rosaries. All the time I stored the orange juice from the oranges in a small jar." "My next move was to ask the warder to take some of the crosses and rosaries to my friends in my old prison" (117). Thinking that no "harm could come from this," the warder agreed and continued to do so for the next five months. "I had no pen and I did not dare ask for one," adds Gerard, so instead he asks for a short quill to pick his teeth, from which he makes a pen, careful to make sure that the toothpick "looked long enough for the warder not to suspect that I had cut a piece off." When Gerard's hands are sufficiently mended, he is permitted to write a few lines in charcoal on the papers that wrapped the rosaries. "All this [the warder] . . . allowed," writes Gerard, "suspecting nothing, but, in fact, on the same sheet of paper, I wrote to my friends in orange juice, telling them to reply in the same way if they received the note" (118).

So on Gerard goes for the next six months writing secret letters to friends and fellow priests, confining his letters to small matters and establishing thereby a network or postal system literate in the ways of invisible ink. "I write to my friends in orange juice," he advises, "telling them to reply in the same way if they received the note, but not to say much at first, and to give the warder a little money, promising him something each time he brought them a rosary or cross and a short written message from me telling them that I was well" (118). The secret messages do not matter (at least at first). What does is the formation of a shared literacy, the sharing of the codes of orange writing, a reconditioning as to what counts as writing and where writing may be found. No wonder, come Gunpowder Plot, the likes of Coke and the Privy Council might be frustrated in how little orange writing revealed. It was the fact of orange writing itself as much as the messages it concealed that mattered. No wonder also, therefore, that they monitored Garnet closely as much to learn his network of associates as the content of the letters they received.[39]

Provided with oranges and the means to purchase them, Bonner delivers the letters, first vetting their visible contents, which Gerard confines to spiritual matters while writing the true message in the white spaces between the lines. He stands behind Gerard while Gerard reads the letter aloud to him. "One day while he was looking over the paper," Gerard tells us, "I read something quite different from what [he] had written down" (118–19). "I did it four or five times," he elaborates, "and when he did not correct me I turned to him with a smile and said frankly that there was no need for him to go on watching me." Gerard explains that his warder "liked listening to what I read out." Bonner could not read. And, having confessed this much, he carries whatever Gerard wants to whomever he wants, even allowing him "some ink" and taking "sealed letters to and fro."

Inside the Tower, meanwhile, Gerard was conducting another correspondence with John Arden written in orange juice. Or, he would have been, had Arden realized that the rosaries he was receiving from Gerard were ancillary to the pieces of paper in which they were wrapped. The two men shared a common line of sight between their cells in the Salt and Cradle Towers across the privy garden. But "I dared not call to him," writes Gerard, for he might easily be heard given the distance. So, instead, Gerard mimes "squeezing the juice out of an orange," "dipping his pen in the juice," and "holding the paper to the fire to bring out the writing" (129). When Gerard's letters still go unanswered, he sends another inquiring why and discovers that Arden misunderstood the signage the first few times and has been dutifully burning Gerard's letters—as instructed. So, had Peyton or one of his subordinates chanced to take a stroll through the Queen's Gallery that overlooked the privy garden on any given day in October and looked up, he might have seen a man charading or telegraphing a lesson in secret writing to another, miming the gestures of a writing technology that to Gerard but not yet to Arden was second nature.

After more than a month of clandestine exchanges of letters written on the wrappings for items such as rosaries, Gerard persuades this warder to allow him to visit Arden in his cell in the Cradle Tower in the hope of the two saying Mass together. And with everything arranged, early on the morning of October 4, from the roof of Arden's cell facing the Thames, the two men slid down a rope to the quayside and onto a boat waiting for them there. This was the plan at least, when Gerard and Arden watched from the roof as their rescuers were forced to paddle back up toward the bridge to avoid a night watchman who had spotted them on the wharf. During the delay, the tide turned, and their boat became stuck on one of the piles driven into the riverbed to break the force of the water near the bridge. Gerard and Arden looked on as help arrived from the bridge in the form of a "powerful seagoing craft . . . with six sailors aboard," pulling the would-be jail breakers to safety moments before their small boat capsized and sank (132–33). (The river Thames is also an actor in this story.) The next night: better weather, better timing, a bigger boat, and the escape goes like clockwork. The two men make it away. Arden disappears we know not where, finally out of the Tower after being imprisoned ten years earlier for his alleged part in an attempt to assassinate Elizabeth I known as the Babington Plot. Meanwhile, Gerard goes into hiding in London, spending the next nine years moving clandestinely between safe houses in and around London on a missionary circuit, before crossing over to the Continent.

Running in tandem with this use of orange writing is a more conventional but no less subtle pen-and-ink story. With their escape planned, Gerard writes three letters that he intends to leave in Arden's cell. "The first was to my warder," he writes, "justifying myself for contriving my escape without letting him know. . . . The second was to the Lieutenant," Peyton, and

> in this letter I made further excuses for the warder, protesting before God that he was not privy to my escape and would never have allowed it if he had known. . . . The third letter was to the Lords of the Council. In the first place I stated my motives for regaining the freedom that was mine by right. . . . Finally, I protested and proved that neither the lieutenant nor the warder could be charged with connivance and consent. They had known nothing about it: my escape was entirely due to my own exertions and those of my friends. (134)

Read alongside the correspondence in orange juice, the production of these virtual pen-and-ink letters—virtual because the only guarantee we have of their existence comes from Gerard's own facsimile of his experiences—represents an even more uncertain set of signals. They attempt to exert some control over the reception and interpretation of the escape, and they do so by addressing and so choreographing different and disconnected audiences.

To the Privy Council, Gerard apologizes, carefully outlining that he, and he alone—with the help of nameless and mysterious friends—engineered the escape. He apologizes also to Peyton, whom he takes pains to exculpate. He is also careful to delete the role of Bonner completely and so produces the fantasy of the Jesuit as miraculously effective agent, as he whose attachments are multiple and invisible and who draws on unseen reserves and affiliations in a time of crisis. But these letters spoke also to Gerard's fellows. Given that martyrdom was the preferred and generically more predictable outcome to imprisonment in the Tower than escape, and that escape might indicate that he had become a turncoat or double agent, Gerard's report of these same letters in the memoir serves as reassurance to the contrary. They serve also to suggest a new plot or significant pattern for understanding the role of Jesuits such as Gerard in England. The key, here, lies in the seeming redundancy of addressing a letter also to Bonner, who, as Gerard takes pains to tell us, cannot read.

When Bonner receives his letter the same morning that Peyton puts his own pen to paper, Gerard reports his reply to the messenger. "But I can't read," says Bonner. "If it's urgent, please read it for me" (137). Richard Fulwood, the messenger, obliges and, narrating the contents of the letter, Gerard says, "I briefly explained why I had escaped, and then, though I had no obligation in the matter, told him that yet I would see him to safety. He had always been faithful in his trusts, and I would stand by him now. If I wanted to save his skin I had a man ready with a horse to take him to a safe place a good distance from London" (137). Gerard goes on to say that he promised him two hundred florins a year, but all with this caveat: "he must settle his affairs in the Tower quickly and go off at once to the place to which the Jesuit would lead him" (138). So off Bonner goes, and with great satisfaction, Gerard adds that within a year, "he became a Catholic," and ends by saying that "my escape from prison was, I hope, in God's kind disposing the occasion of his escaping hell" (138). Gerard ensures that Bonner makes it away, that he will be gone when the empty cells are discovered, that Peyton will put two and two together to make four and so conclude that Bonner is to blame. Gerard compensates Bonner. He provides him with new financial and spiritual attachments—no longer a warder with a "taste for oranges" but a good Catholic saved from the ultimate "bringing out" that is the fires of hell. The incremental process by which oranges become pith, juice, hands able to write, crosses, rosaries, paper, pen, the sound of a man's voice reading aloud, a man listening, a shared taste or liking, an escape, is finally all folded into the larger, divine plan, which was to make a convert. Had Bonner not liked oranges, he might have remained a Protestant and found himself in the fire—along with Faustus and perhaps the rest of us. We have been reading not the story of Gerard's escape or failed martyrdom but the story of Bonner's conversion.

Maybe. For what exactly does Gerard's text teach? His memoir discloses the secret writing technologies and practices of the Jesuit underground and so informs on matters of tradecraft—but it is hard to know whether this disclosure is not itself a further encryption of the actual ways in which the escape was planned. Was Peyton, in fact, correct? Was a "particular fondness for oranges" itself already code for Spanish sympathies—Claudio, in William Shakespeare's *Much Ado about Nothing* (1598), is described as "civil, civil [Seville] as an orange"?[40] Was Bonner, indeed, a sympathizer already? Does the text, in other words, teach its content (techniques and substance) or its form (a way of moving)? Or does it play an even more subtle game of disclosure and obfuscation, mixing fact and fiction in an exquisite blend to produce holes or gaps in the narrative where the inhuman, the divine, tropes the story and where that presence may be remarked or even felt as in the sacral or sacramental space Gerard creates by his performance of the dry Mass, a Mass in which the missing Eucharistic feast is trans*signified* into the orange-bearing warder? Probably the answer is both. For what Gerard installs in his listeners and readers is the fundamental, defining importance of the Ignatian *Exercises* themselves as a foundational set of routines for the remaking of self and for being in the world. Gerard's text teaches us a way of moving, of reading persons and events, with an eye to the oblique, with an eye to the *things* persons carry lodged within them, treating people as vectors, prisons as turnstiles. His memoir teaches us about rival modes of translation, alternate ways of declining *things,* informing us on which things to keep, to hold on to still, to create our own gatherings within situations that seem entirely precarious.[41]

Here it is worth recalling that the Tower remained a powerfully contested site of discursive production and inscription. It was and remains today a stony archive of competing networks of surveillance and dissent, a place where different forms of writing, broadly conceived, faced off in a hermeneutic contest that could, as it did for Gerard, become harrowingly physical in its application.[42] It is to the fact of torture and the possibility of slander that Gerard's memoir responds. Written abroad, the text exists as one anchoring point or relay within a generalized Jesuit writing machine that challenged the writing practices of the Tudor and Stuart state at every point—something Gerard makes viscerally clear when he describes his first night in the Salt Tower, which at one time had held the martyr Henry Walpole as an inmate. Gerard recounts that he made out Walpole's initials among a series of names and dates "cut with a chisel in the wall" (104).

He finds "his little oratory where there had been a narrow window," even though it was now "blocked with stonework." Gerard's recovery of these marks sets in motion a string of associations that transport him elsewhere. He recalls that Walpole "lost the use of his fingers" because he was "tortured more often than they wanted known" (105). He remarks that the marks on the wall are barely legible: they "looked like the writing of a schoolboy, not that of a scholar and gentleman."

The graffito comforts; it leads Gerard to recall some verses that Walpole wrote on the occasion of Edmund Campion's earlier execution and martyrdom in 1581:

> Why doe I use my paper yncke and pen,
> And call my wyttes to counsell what to say?
> Such memoryes weare made for mortall men
> I speak of sayntes whose names shall not decay.
>   An angels Trumpe weare fytter for to sound
>   Theyre glorious deathe yf suche on earthe wear founde.[43]

Gerard's discovery of the graffito and reclamation of the sacral space of the oratory or diminutive chapel that Walpole constructed dislocate the cell from the precincts of the Tower, write the space into a divine narrative that requires mediation only for "mortall men," whose actions and whose suffering, whose martyrdom, are already a remediation of the lives of the saints "whose names shall not decay." No graffiti for saints, then, or, more precisely, the physical marking of the Tower by the likes of Campion, Walpole, and Gerard, their annotating or rubricating countertexts, "bring out" or decrypt the truth they find in the Tower, a truth written in blood and with bodies that no longer remain.

Of Bonner, it is harder to write. He hovers on the margins of these texts even as he seems central to the plots his rival shepherds hatch or posit. As harrowed as Gerard's relationship to pen and ink might have been, Bonner seems to have no relationship to them at all. In Peyton's letter, he is the human element, the missing person who reveals that the escape was an inside job, the failure of the Tower's warders, a "defect" to be mended if Her Majesty follows Peyton's recommendations in the survey of 1598. He becomes both the particularized historical person designated and thereby defaced by the name "Bonner" and the exemplary instance that defines the systemic problem posed by the warders in the Tower. He is the black sheep who must be cut from the flock and then selectively bred out of existence. In Gerard's memoir, Bonner is deprived of name and particularized instead by his "fondness" for oranges that binds him to his new shepherd and, in the end, saves him. Transplanted between Peyton's forensic or socioeconomic analysis and Gerard's rendering of the formal engagements of an underground movement, Bonner flickers in and out of view. He remains inscribed by the semantic field of the oranges that go unremarked by Peyton and that exist only in Gerard's Latin text not as oranges but as "golden apples" *(poma aurea magna),* as apples that might still prove to be oranges but that just might jingle in the pocket. Bonner dwells, sleeps, fitfully within the semantic possibilities to be had from these vegetal—or are they mineral?—oranges. He manifests as a double artifact, a discontinuous or interrupted contact zone, waxing Catholic, impecunious, orange eating, Spanish loving, gold wanting, as he oscillates between the differing accounts.

The story and its losses continue today. If you visit Gerard's cell at the Tower of London, beyond being treated to a preview of his orange writing (Figures 10 and 11), you may read about him and respond to what his story means post-9/11 or post-5/7 (Figure 12). "How far would you go for your beliefs?" reads one of the questions on the interactive console that seeks to capture the public's view, engaging them, enfolding them into the archival relay that is the Tower.

The menu of options reads as follows: "They don't really matter"; "I'd go to prison for my beliefs"; "I'd die for my beliefs." They do not ask you whether you might torture or kill for your beliefs, but they may as well—though to access that part of the story, you must exit the Cradle Tower and descend underground where they keep the rack and the manacles, separating out competing forms of writing that, at the time, could not be kept so neatly or nicely separate.

The exhibit mentions Gerard's "warder," but it does not name him. Its focus remains on the visitor whom it invites to identify with Gerard or Arden, who escaped. One of them is pictured climbing down from the Tower. The scene of writing on which we enter in the Cradle Tower takes place in an emphatically empty cell (Figure 10). And, if you turn

Figure 12. Exhibit. Cradle Tower, Tower of London, July 2014. Photograph by the author.

to look out the window overlooking Tower wharf, you might just catch Gerard now as he climbs down. There he is in silhouette, transducted to a perspectival grid and illuminated by the all-seeing solar eye (Figure 13) —brought out or "ripened" by the heat of the sun.

In the exhibit, Bonner seems to disappear entirely, or perhaps not. For in asking us how far we might go for our beliefs and so quizzing us on whether they do and do not matter, the exhibit positions us as if this warder, this not exactly innocent bystander. What would you do? Who shall you be? But make that decision here, where the past revisits the present instant and not elsewhere. For *here* the stakes are low. The cell is empty. Gerard, we can be sure, has gone, and with him goes any need to decide or delimit the course we might ourselves take as torturer, killer, turncoat, or convert. Press whichever button you like. Press all the exhibit's buttons like kids do because pressing buttons is fun. We are paying guests, after all, and this exhibit only momentarily detains us before we continue our visit, stop for tea, visit the crown jewels, refrain from feeding the ravens, stay within and behind the ropes, and, generally, "keep off the grass."

Figure 13. I spy John Gerard. Cradle Tower, Tower of London, July 2014. Photograph by the author.

How, then, do I even begin to speak of "Bonner," the warder with "a particular fondness" or taste "for oranges"? How do I recover the texture of his liking and not transform him and it into either Peyton's forensic account or Gerard's story of conversion, each of which captures and so puts to use the vegetal temporality that punctuates this story? What "treasures of psychological description" and news of "mass organization," in Gramsci's sense of things, might Bonner and his oranges still have to offer? Thankfully, the mobile polities of actors that form within the interstices of our built worlds endure. They appear in this exhibit in the form of the ornamental, zoographic figures of the disinterested ravens perched on the Tower's walls, who may simply choose to fly away; the indifferent ivy that clings effortlessly to the outside of the Tower that Gerard and Arden had such a hard time escaping; even the grain of the wood that backs the exhibit speaks back to this place of stone. Perhaps one answer, one way forward, would simply be to take Gerard and the exhibit at their word and inquire further into what these posy-like animal and vegetal presences conserve: the vegetal–mineral remainder with which Bonner remains always twinned or tuned, the oranges or golden apples or gold that he so liked and enjoyed. Perhaps he ate the oranges. Perhaps he sold them on because he needed the money. Perhaps they were in fact golden coins to begin with. Or, perhaps, both Peyton and Gerard have everything right and still they do not account for Bonner, for his liking, his connection to oranges. For, in 1597, to like oranges was, quite by accident, to become a fellow traveler with laundresses in search of cleansers and Jesuits on the lookout for invisible ink. To like oranges was to align one's footsteps with everyone who desired oranges in London, for whatever reason. London was marked by these trajectories. So, let's now inquire further into the nature of this liking or taste. What does it mean to like oranges? What, for that matter, is an orange?

TO LIKE ORANGES, AN EXPRESSION-IMPRESSION

Prior to its assimilation to the human, an orange is a type of berry, part of the reproductive technology of a particular species of plant. However fond some of us may be of them, however much you or I may enjoy their scent or find ourselves attracted by their color, or transported by their taste, the orange remains unmoved. The orange does not discriminate on the subject of species. It addresses itself to the great variety of differently animated entities (human and otherwise) that might render it mobile, whom it rents or recruits as it makes use of their bodily capacity for movement, their physiology, offering itself to be eaten in and as the process by which it moves.[44] The primal scene for this encounter, the coarticulation of orange and animal host, is that of consumption—the moment at which we peel and eat the orange. It is this moment also that prose-poet Francis Ponge inhabits in *Le parti pris des choses* (Siding with things), in his poem "L'orange." Ponge investigates the impression orange makes on his body as sound box and so also the impression we make on it. The poem eventalizes the moment of consumption, animating both the mouth that eats and the orange that has been eaten. The two become mutually animating partners even as the human appears in the poem as "torturer" or "oppressor" and the orange as a much too forgiving or obliging entity that nevertheless makes us taste, smell, swallow, and speak.[45]

The poem begins by lauding the peel, which, even after the orange has been eviscerated by its "torturer," retains, so Ponge writes, an "elasticity" that seems to wish to return to its previous shape, to mimic the *thing* it was before:

> Like the sponge, the orange, after undergoing the ordeal of expression, longs to recover its composure. The sponge always succeeds, though, the orange never: for its cells have burst, its tissues are torn apart. Only the peel, thanks to its elasticity, to some extent retains its shape. Meanwhile an amber liquid has been spilled, which, refreshing and fragrant as it may be, often bears the bitter consciousness of a premature expulsion of pits.[46]

Poor, hapless orange, it never manages to reverse its ordeal, to turn back the clock and recover itself to itself. But still, something of its former self remains in the impetus or responsiveness of its skin, of this remainder that somehow remembers. Cohabiting in the poem with Ponge's namesake and emblem, the sponge *(l'éponge),* the zoophyte or plant-animal that defies categories, that loops the inside and outside of being so as to appear self-grounding, perfectly reversible, the orange figures an expenditure, an irreversible expressiveness or expression that, unlike the sponge, "is too passive." The orange "let[s the oppressor] off much too easily," rewarding him with its "perfume" as the skin is broken and with the "glorious color of the resulting liquid," which, "as is not the case with lemon juice," requires "no apprehensive puckering of the taste buds." The orange makes so few demands. It requires only this of us: that "the larynx has to open wide to pronounce the name as well as to ingest it." Both the word and the fruit go down with an easy equivalence, or rather, the opening of the mouth necessary to eat the orange corresponds to the movement of the mouth necessary to saying its name, to sending its name out into the world. The orange gives itself to be eaten and by that giving comes to be named, recruiting or convoking us as its ally in the process. We eat and name the orange. And the orange, in leading us to do so or in not resisting us, names itself. All praise then to the orange, "this tender, fragile, pink balloon," this "epidermis of thick, moist blotting paper," which gives itself to us and which accordingly we torture, subjecting it to the ordeal of its putting to use, its consumption or continual expression, until nothing but the peel is left, peel that, as we all know, you may candy, steep in brandy or liqueur and so preserve, cut and piece into a rosary, as we "express" the last traces of its expressivity into still other media—our bodies, our words, our worlds.

The poem ends by approaching what it calls the innermost core of the orange, its indigestible pip or seed that will pass through us. At bottom, then, with the pip, the "wood," it seems as if we hit something like a referent or an outside to Ponge's poem, the limit to what he can do as a prose-poet of substance or signsponges. But the poem permits no such exit. Instead, it proposes something stranger still. "In concluding," he winks, "this all too summary study, carried out as roundly as possible," Ponge finds that he has still to

> . . . cope with the pip. This seed, shaped like a
> miniature lemon, is the color of the white wood of the lemon tree,
> and inside a pea or tender sprout. After the sensational
> explosion of that Chinese lantern of tastes, colors, and scents—the
> fruity balloon itself—we here recognize the relative hardness and

> acidity (not entirely insipid) of the wood, the branch, the leaf: a very small sum but unquestionably the *raison d'etre* of the fruit.

As Ponge "copes with" or, more literally, "has to come to" *(il faut en venir),* the pip, which appears indigestible, unusable, and so resists the orange's otherwise seemingly generous or gratuitous assimilation to the human, we encounter not the limit to what he can say but, instead, an unfolding of the orange, of orange as it takes hold of his language and this poem. As with the moment of decryption in Cowley's lemon poem, the pip sprouts, turns tree, grows and sheds its leaves, buds, flowers, fruits, and so exfoliates the multiplicity that is orange within "L'orange," within *this* orange that we have tortured, eaten, and read.

At the moment of conclusion, of culmination, at which Ponge might be said to exhaust the orange, he encounters neither a referent nor an outside to language, so much as the point at which the expressivity of the orange comes to take hold of the poem, of him, of his language, and so of us. The innermost pip of the orange—there must be several—reveals the orange to exist as what we might call, already, the expression or preexpression, prewriting or coding, of the plant, an expression housed in pith and peel, a dormant *kairos* that waits, cut and tortured already, a precut and segmented entity, the reproductive technology of a particular genus of plant that goes mobile in and by its recruitment or rental of those differently animated entities we name animals (human and otherwise). So it is that Ponge, the sponge, the sovereign zoophyte, takes in and expels the orange. As he does so, he passes from active subject to passive, sensing orangey substrate or screen, from animal to vegetable and back, as the looping motion of his muscle for writing passes from animal enjoyment, his heady, violent expression or explosion of the orange, to registering the process by which, to "conclude," to reach an end, or make an "end" of *this* orange and of *orange,* requires that we encounter the *telos,* "end," or vectoring of the orange itself, of the way in which all that we take from it, all that he registers under the term "expression," all that we enjoy, is made possible by the orange as itself a form of writing that impresses itself on us.

Yes, the poem "can be read," as Sianne Ngai offers, "as a figure for a number of personification strategies."[47] Yes, the poem serves as an *exemplum* on the reversibility of *prosopopoeia,* the trope that means to give face and so voice to things, and that, as Paul de Man writes, "implies that the original face can be missing or non-existent," that it exists only because of its being figured or by the program of figuration itself.[48] Yes, the orange may be "cute," an object that both attracts and repels, whose passivity engages the reader, viewer, eater, in a reciprocal passivity that dominates both—it's certainly handy, lends itself to the hand, befriends handedness, naturalizes (itself to) the fact of hands. And, yes, this cuteness functions a little like that of the commodity form in that when and as we consume it, we are, in fact, put to use, made use of, even as, in Ponge's poem, we appear to be in control, our agency maxing out as we eviscerate and torture the orange, using it all up—except for the pip. "L'orange" may serve to set in motion a *mise en abîme* by which to eat becomes to be made to swallow and to be made to speak, but beyond or beside offering a template for either the glamour or *schein* of the commodity that we passively consume or the status of a modernist work of art, the poem registers the existence of the orange as itself a form of writing—a literal posy or flower-poem in which the orange tree exfoliates itself.

My aim here is not to deny the interpretive validity of Ngai's renewal of interest in aesthetic categories (the contours of perception as affectively felt and known) as a way of parsing out our captivation by the object world of *Capital.* On the contrary, in recovering the vegetal efficacy put to use in the serial appearance of orange and oranges in our lives and texts, I seek to delineate the way her interpretive schema runs in tandem with or is built upon the rooting of flows of animal, vegetable, and mineral matter that are already themselves rhetorically and semiotically dense forms of coding or writing. What shall we learn from owning this correlation or cooptation of other forms of "expression" as they impress themselves upon us, mediated or accelerated, as they are, by the irrational mechanisms of exchange value? To posit orange as already a form of "writing" enables us to return, for example, to where Ponge begins his poem and to understand the longing or "aspiration" of the orange's skin to "recover its composure," or "*contenance,*" bearing, attitude, or face. The automatic motility or automimesis of the peel functions as an archival remnant of the orange as already an expression of orange writing, of the "plant's proto-writing." Still elastic, still given to mimicry, a several repetition, on its way to becoming something else—food, excrement, cleanser, invisible ink, a rosary, a Catholic convert—every orange marks the singularity of an event, for each is different, each differently the same. In *Plant Thinking,* Marder offers the iterability of the leaf or fruit as a gloss on Ponge's own frequently stated admiration for the ability of plants to "'repeat the same expression, the same leaf, a million times' and 'bursting out of themselves,' to produce 'thousands of copies of the same . . . leaf.'"[49] Like the leaf, each and every orange "will vary with the material expression of the being of plants, that 'is' the being of plants," a fact which, as Marder points out, is as true of the leaf or orange as it is of every "word or concept that will carry slightly different semantic overtones depending on the singular event of its enunciation."[50] If both the leaf and the orange serve as an "ephemeral register for the inscription of vegetal time as the time of repetition, a register not archived [exactly] but periodically lost and renewed," then, with "L'orange," we encounter a record or inscription of that ephemeral archive on Ponge's language and our senses, the two coming to serve as a substrate for the expression that is the orange.[51]

There is no encrypted message for us to be "brought out" from the orange. It remains emphatically itself no matter how ingenious our efforts and the uses to which we put it. Every orange is disseminated to an unknown addressee. It has no plan beyond its dispersion: the orange is quite precisely a dispersion strategy, whose dispersal produces (for us) a series of orangey flowerings or irruptions. However fragrant, however colorful, however sweet, however joyful its putting to use, the orange remains entirely indifferent. It does not discriminate. But, that's not quite right either. The orange proves indifferent, but that is not to say that it remains uninterested. For it functions as something more than a purely grammatical or material-semiotic program. It inclines toward the trope, offering itself as a mode of hyperoccasional multispecies address that seeks, posy-like, to interpellate you as the plant plays its own form of bio- or zoopolitics. The orange is a thoroughly rhetorical entity, then, always, as it were, in the vocative case, the orange itself an act of nomination that designates the multispecies polity of animate beings that shall eat and so name and so disseminate it.

To whom, then, does Ponge address his poem? Who is it that joins him to become the "we" that expresses the orange? This collective subject need not, from the perspective of the orange, be an exclusively human entity. It might refer just as well to any passing animal or animating vector as to a Jesuit, a warder, a laundress, or a poet. The "we" that the poem proposes proceeds just as well by way of tooth and claw or beak, by way of this or that prosthesis, as by the way of hands and fingers—however handy the orange. And this "we," accordingly, does not refer to a stable category so much as it designates merely the range of beings convoked by the orange, the group of animals, plants, fungi, and bacteria that all "eat" and so, in different ways, "name" and express the orange, each taking a share in what Gerard, who captures this group's essence best, might name an *esu delectari* for oranges, a taste, "fondness," or inclination for the fruit.[52]

"Must we take sides," asks Ponge early in the poem, reprising the title of his collection, between the orange and the sponge, between vegetal being and the particular order of *poiesis* that is Ponge? Must we "prefer one of these modes of resisting the oppression of the other?" "The sponge," he writes, "being nothing but a muscle, fills up with air, or with clean or dirty water, as circumstance dictates: a degrading performance." It is all surface, a continuously animate, sensate thing. Pure receptivity inheres to its texture, but so also a total expulsion, as it expels what it has taken in completely. The sponge boasts a certain order of self-grounding sovereignty that leads Derrida to describe Ponge as "self-remarked," as he who, by the accident of his name, loops the inside and outside of being, his name and his words a Mobius strip that mimes mimesis itself, mimes the orange miming or modeling his own eye and mouth, his language a skin that grows over or that responds to the skin of others.[53] Neither the orange nor the sponge is to be preferred, for the two exist as they are, by and through their mimesis of the other—the sponge, the Ponge, the zone of contact between animal and vegetable being never one, always both.

In their profound neutrality, Ponge's prose-poems stand as points of contact with what he calls the "side of things" and what, in this instance, we might name orange-being, the serialized multiplicity that accounts for or that expresses "orange" and oranges, punctuating the object world with particles of sweetness and color. After this encounter, neither we nor the orange shall recover our former bearing, even as the shared elasticity of our skins mimics the *things* that we once were. Ponge's poem suggests also the way in which what we take to be our archives, archives backed with this or that animal, vegetable or mineral matter, rely upon and are no different from the differently situated ephemeral registers of the leaves and fruits of plants that remember by forgetting. With Ponge, we come to recognize the way in which the "proto-writing" or simply writing of plants undergirds our own poetic acts, marking them and inscribing on us and through us the movements of vegetal time. It is for this reason that I have written this chapter under the sign of oranges and their peel, every text, every turn of the event, a scene of expression-impression, of translation in the fullest sense of the word.

But, at the same time, as in the case of secret writing, of all the various iterations of invisible ink I have described, of invisible incorporated (inc.), the indifferent temporality of plant being, of plant writing, finds itself put to use. The orange or the lemon appears in order to disappear. It is pressed to use as one among many vegetable and mineral substances to produce states of suspended or immanent sense, writing that inheres or inhabits a medium, whose encryption and then revelation differentiate different communities of readers. In the story I have told, oranges fund the technical requirements of an underground movement and the theological mysteries of writing as a quasi-magical mode. The winking instance of orange writing mimes the metaphysical density of divine signatures or theological DNA that inheres to all things. And as with the Eucharistic "appropriation of the mute body of the plant," invisible ink (invisible incorporated) harnesses the iterability of plant writing so that it becomes, parasitically, a medium for human writing.[54] Such techniques function, by definition, as a way of differentiating and maintaining competing communities of human animals, each of whom trades on (so it hopes) a superior knowledge of the properties of substances to seek its own advantage or survival. In the Tower, pressed to use as a technique within a highly evolved writing contest, Ponge's orange consumption becomes a scene of multispecies competition that crosscuts or inheres to associations that we parse more readily in terms of confessional difference (Catholic and

Protestant), differential literacy (the ability to read and write or to enjoy being read to), the nostalgic joys of Latinity (listening in on a Mass). The orange both convokes and delimits these rival polities—polities for whom its efficacy, its effects, transmute into a sense of belonging, into gold, divine presence, conversion, treachery.

It is tempting, then, to end by suggesting that even as this story of invisible ink, of orange being "incorporated," remains bound to its time and place, this technical appropriation of the plant gestures forward to the "ontological exhaustion" of plant life that their engineering as genetically modified commodities may threaten. But against this plotting of orange, there remains still the brute facticity of Bonner's alleged liking for them, a liking that it seems impossible to know other than through or by its instrumentalization in the competing accounts of Gerard and Peyton. Perhaps this liking, this inclination, might gesture toward a different territory in which life effects, liveliness, being, is generated always with reference to an ecology that, as Jeffrey Nealon puts it, "cuts across all strata as we've known it."[55] Perhaps, then, it is possible still merely to like something, and so to give yourself to a way of being one among the many who take a share in an *esu delectari* or taste for oranges. What would be the boundaries of such a polity or merely territory of orange tasters—only some of whom may belong within the group Rowell names "all gregarious long-lived vertebrates capable of mutual recognition"? With the notion of a shared taste as the basis for group belonging comes the possibility that we may be able to transform our sense of the common beyond or athwart the misnomer of species, taking our cue, as it were, from the orange so as to render our polities more *thing*-like, self consciously calibrated by the historically particular plants, minerals, and other animals with which we come into being and whose own zoo- or biopolitics enfolds us in their worlds. Who and what else share in a taste for oranges?

Perhaps. But, let's wait and see. For to inhabit the structure of this liking or taste, the liking of a "warder who was particularly fond of the fruit," means having to inquire also into the way the formalizing event that is "orange," this mutual naming and nomination of fruit and host, cohabits with the experience of a rival, dematerializing object relation—the fetish of the commodity.

4

Gold You Can Eat (On Theft)

> Everybody knows that some international trade is beneficial—nobody thinks that Norway should grow its own oranges. Many people are skeptical, however, about the benefits of trading for goods that a country could produce for itself. Shouldn't Americans buy American goods whenever possible, to help create jobs in the United States?
>
> —PAUL R. KRUGMAN AND MAURICE OBSTFELD, *International Economics: Theory and Policy*

> The story of the glorious labor of Hercules, when he stole the golden apples, has suggested to me the framework on which I might develop my story of the golden apples. . . . One of Hercules' labors was to go to the gardens, kill the appalling dragon that guarded the golden apples, and bring out to the woods of Media and to the savage regions of the outskirts of civilization the apples of edible gold. So I, as he, must labor and struggle.
>
> —JOHANNES BAPTISTA FERRARIUS, Frater, *Hesperides, Sive de Malorum Aureorum Cultura et Usu*

It is difficult, so I am told, to grow oranges in Norway. And this difficulty, while not exactly insurmountable, turns out to be one of the favorite examples for illustrating the benefits of global capitalism. As anthropologist of the fetish extraordinaire William Pietz informs, it's this dearth of home-grown Norwegian oranges that those "secular clerics of the latter-day House of Orange," economists Paul Krugman and Maurice Obstfeld, reach for "to bring home the point that in global free trade everybody gains, even those who are exploited in the Marxian sense that the goods they receive 'contain' less labor time than the goods they sell."[1]

Climactic conditions in northern Europe (for now) preclude the cultivation of citrus fruit other than in greenhouses, the cost of which remains commercially unviable—hence the good sense it makes to import them from a warmer climate instead. "Norwegians can get more oranges for the same amount of labor by making some other commodity (refrigerators say)," observes Pietz. "Their natural resources and technology give them a comparative advantage." It just makes better sense for "some underdeveloped nation in the tropical South (Indonesia say)" to grow the oranges. According to this feel-good logic, everybody wins. Norwegians get their oranges, and they get more of them and of a higher quality than they could hope to grow at home. Indonesians, likewise, get their refrigerators, or, actually, wait a minute, they don't. They trade with still other nations than Norway, whose refrigerators remain entirely remote to them—something encountered only by way of the global structure of a market that tells them the price of their oranges and the cost of those refrigerators. Never mind the differential in labor time. Never mind the scale according to which both countries measure their "comparative advantages." Things will even out eventually. Or they won't. In the meantime, the very idea of growing oranges in Norway comes coded by a ridiculousness that almost demands that you set out for the supermarket—on principle.

Picking up the oranges that Krugman and Obstfeld deploy, Pietz diagnoses the sleight of hand by which this "Disneyfied picture of simple commodities circulating across cleanly drawn national borders" attempts to rationalize the inequalities of global exchange. Against this logic, he pits the singularizing overinvestment of the fetish, which interrupts the flow of equivalences in a market constructed as if "free" by insisting on your overweening fondness or taste for, say, *this* orange or *that* refrigerator, irrational though your attachment may seem. It is as if the local and historically particular investments of the fetish serve as a material-semiotic tariff on the abstract spaces that the self-predicating logic of global capitalism posits, creating ripples, unpredictable zones of stasis, where the "free" movement of certain commodities suddenly finds itself arrested.

In a series of now classic essays published in the mid-1980s, Pietz provided a genealogy of the fetish as "basically a middleman's word," a "concept-problem," that, from its emergence as the "pidgin word *Fetisso*" combining Portuguese and native languages in the "cross cultural spaces of the coast of West Africa during the sixteenth and seventeenth centuries," through to its recasting in Enlightenment discourses and eventually in the hands of Freud and Marx in the nineteenth century, refers to a category of *things* with a problematic or contested relationship to exchange.[2] Pietz demonstrates the way the "fetish not only originated in, but remains specific to, the problematic of the social value of material objects as revealed in situations formed by the encounter of radically heterogeneous value systems" (native and Portuguese; Catholic and Protestant; fetishist and psychoanalytic; capitalist and Marxist).[3] The fetish names a confusion of categories. It designates a moment in which a concern for matter and how we understand it comes to dominate the scene. Pried free from its retroactive disciplining as an irrational overinvestment or symptom, Pietz's positive estimation of the fetish enables us to understand the way the pull of the object world upon us produces a parade of DIY sacramental objects that vie with the anchoring power of such supra or sublime objects as the Eucharist or the commodity form. Spirit, as it were, bubbles to the surface, pools, and collects in all manner of places that may prove amenable to factoring our relationship to one and other, to the world, differently—be they the co-opted dispersal strategy of a particular genus of plant twinned with the labor of its collection or going mobile (an orange) or the congealing of labor, mineral, technical resources, and energy we name "refrigerator."

Now, the rhetorical occasion for Pietz's debunking of global capitalism by way of Norwegian oranges was his defense of a collection of essays titled *Border Fetishisms: Material Objects in Unstable Spaces* (1998). At the time, Pietz worried that the book might seem "a sort of postmodern *Wunderkammer,* a cabinet of curiosities taken from exotic but unimportant places . . . for the amusement of intellectuals who have a taste for such things" (245). Rhetorically, he appears in the book as adjudicator cum apologist, writing an "afterward," as he whose extra-academic, activist credentials along with the brilliance of his genealogy of the fetish stand surety for the book's good faith. Pietz is there to see off some imagined gainsayer who coughs and sneers, "Isn't *Border Fetishisms* just another of Krugman and Obstfeld's oranges, and its readers the beneficiaries of a different order of the gains of trade?" Or, worse still, "Isn't the book a veritable hamper of ethnographic exotics transposed from East to West, from South up North, in the 'free' market of intellectual exchange?" "Where's the use-value?" "What's to be gained?" Here is Pietz's answer: he ventures that the collection brings to the reader's notice a series of "heterospaces" where capitalism and noncapitalism coexist. "Rather than adapting their cultural forms to a 'hegemonic' or global economy or 'resisting' it," "indigenous" people appropriate global commodities, transforming them back into objects in ways that defy exchange value (250). Such "border fetishisms" stand as a witness to a productive reversal of fortune "that might be termed the social appropriation of capitalism" to produce collectives that arise on the basis of and in relation to *things,* as opposed to the dematerializing and so empty fetishism of the commodity form. Good-bye to the bad fetish of the commodity; hello to the good, sustaining fetish relation of the retrofitted or otherwise altered objects collectively remade by the labor of these localizing polities whose forms of sociality have become *thing*-like.[4]

But Pietz is not content to end on this note. He remains cautious with regard to the way the collection positions these heterospaces on the edges of capital, on the periphery of its movements. Surely, such ripples, such points of arrest or pooling of rationality-defying overinvestment, must exist everywhere? Surely they inhere to the fabric of our infrastructures themselves? Accordingly, Pietz ends by saying that, in his eyes, "the notion of border fetishisms . . . can provide a useful framework for thought . . . only if one comes to terms with" something else—"the value of shoplifting" (250). For him, the "compulsion to shoplift" raises the question of "whether consumer desire is simply a utilitarian desire for more things or whether there is a deeper truth about the gratifications enjoyed in shopping?" (251). "Is some secular version of the sacred," he asks, "in play within the compulsive pleasures of shoplifting?" Does shoplifting, in other words, rezone the sacred in a way that, if still violent, may "by some perverse logic renew the social"? "If there is an important truth here," admits Pietz, "it is one I cannot yet articulate," and so he ends, instead, by revealing to us the words whispered to him by the "ghost of a surgically altered old man [he] once met in Nicaragua during its brief attempt to hold back the capitalist tide" (251). This ghostly double offers that, if

Pietz really wants to "understand the real 'surplus values' represented in the gains of trade now being reaped in the real Norway (whose national income derives primarily from its legal control over rapidly depleting North Sea oil reserves)" and "in the real Indonesia (where multinational mining firms like Freeport McMoRan and Barrick Gold are chewing up 'resources' to the local benefit of no one outside Indonesia's corrupt national government)," then he "should end" his "afterward with" what he calls "an unconscionable recommendation," a recommendation that coincidentally reprises Abbie Hoffman's iconic, countercultural demand that we "STEAL THIS BOOK!" (251).

It's an odd, wonderful moment. The ghostly figure of a (is it) surgically altered and expatriate Abbie Hoffman recalls Marx's spectral figures and dancing tables in *Capital* but in a way that addresses the reader, calls us out, calls for the question. Suddenly, we all have bodies again, have to take up our relation to the world of ideas and *things* as we register or respond to this bellowed imperative: "STEAL THIS BOOK!" "Constitute your own heterospace," Pietz seems to say. Inhabit the interstices of the built world in ways that try to imagine others; steal from within the flows of matter in which you find yourself. Don't delay. For it's done and been time to turn thief, to "survive," "fight," and "liberate," to reprise Hoffman's 1970s idiom, however those words play now in our respective times and places or translate into the emerging idiom of resilience and precarity. Who knows what is yet to happen?[5]

I ended my last chapter by taking sides, not with oranges exactly but with the structure of a feeling, a relation, a liking or taste for the fruit. Not my liking; not your liking; this liking was attributed to the long-dead warder who took the blame for a prison escape from the Tower of London in 1597. My aim was to discern the orangey remainders that punctuated texts generated by the escape and so to discern the way the vegetal time of orange-being calibrates the discourses that grew up to explain it. In the story, as recounted by the different players, the warder's liking or taste manifests as an unstable artifact to be variously performed. For Sir John Peyton, lieutenant of the Tower, this warder's liking becomes a rationalized set of investments or interests. He constructs him as an impecunious, gold-wanting, and so derelict, Catholic sympathizer, the likes of whom, he hopes, his recommendations for improving security will eliminate from the Tower's labor pool. For Jesuit escapee John Gerard, the warder appears more prosaically as he who likes oranges, as he who has a "taste" ("*esu delectari*") or, in Pietz's terms, perhaps, a "fetish," for all things "*poma aurea magna,*" for oranges or golden apples, for gold you can eat, fruit that jingles in the pocket.[6]

Of Gerard's liking we know much. The oranges he gave, then shared, and then received from his warder were pressed to use as a physical therapy exercise following bouts of torture. He squeezed the oranges; saved the juice; cut and pieced the peel into rosaries. These rosaries provided him with the occasion to ask for paper with which to wrap them to send to friends and so writing paper for secret messages written in the juice. Gerard offers his Jesuit readers a lesson in the quasi-Georgic economy of an underground movement that husbands all the resources to be had in the Tower, human and not. In Gerard's hands, then, the reproductive technology of the orange tree does not grow oranges but instead repairs his tortured body and psyche. Its cut and pieced remains pass as tokens among a circuit of friends and associates bound by their knowledge of the properties of orange juice as a medium for invisible ink. And, like Peyton, who was tasked with managing the Tower, Gerard shepherds his warder, who converts to Catholicism; wins the wayward sheep back to the fold; does so by way of an orangey (or is it golden?) incentivizing strategy, by way of "gold" the warder might literally or metaphorically eat. Whatever the warder wanted, to the extent that he knew, is lost to us. He exists now as a purely archival creature or territory allied to oranges and to gold. He exists as a liking, an orientation to oranges or to gold, and perhaps as an invitation to another way of thinking about the role of objects or other beings in configuring our polities.

In this chapter, I aim to conjure my own little theft. If both Peyton and Gerard, in different ways, steal the warder's taste or liking from him, turning his relation to oranges or golden apples to their own ends, then I attempt to steal it back. In what follows, I seek to inhabit this liking or taste, refusing to cash the orange in or out for this or that grander narrative—be it socioeconomic, rhetorical, theological. Like the warder, I shall have to remain immanent to the rival streams of matter, orangey and golden, with which he is constructed, lose myself to the flows with which Peyton and Gerard ally him. Don't get me wrong, I am not averse to exchange, to relationality per se, but, like the warder, I have to be content, for now, to serve as a screen that registers the way orange and oranges punctuate our infrastructures. I have to content myself with what remains and mime the gestures of a hybrid orange-man who designates one instance in which an animal and a fruit exchange properties, take their respective shares in a multispecies impression. This chapter constitutes its own order of archival heterospace, then, keyed to the ways in which orange, the dispersal strategy of particular genus of plant, by and through its recruitment of human animals, comes to interrupt acts of exchange, inclining them toward an economy of the gift or accusations of theft, and so litters our discourses with errant, erring, fragmented, time-bound polities that unfold by and through orange. Like all thefts, then, this chapter has to keep moving, actively constituting the territory it seeks to name. For, if there is, as Pietz hopes, some value to be had from a surplus, then that value has to inhere in the act itself, in my orientation to Bonner's liking, that liking itself, already, an unruly surplus of feeling or affective attachment, in excess of a discourse of ends, already an end in itself. The issue here is not necessarily some affirmative order of biopolitics so much as it is a more neutral understanding of life and labor itself as already somehow excessive or "in excess," and so resistant to capture. The question would be under what determinate, concrete circumstances such a surplus might enable us to create other forms of personhood and other modes of sociality or the common.[7]

Naturally, it's a Herculean task. For oranges, or golden apples, or, indeed, sheep with unusually golden fleeces—the Greek word *mala* (apples) is an error or partial homonym for *mela* (sheep)—were, so it is said, what Hercules stole from the garden of the Hesperides.[8] Thankfully I am not alone in this endeavor. I merely follow a host of writers who, in different historical moments, have also given themselves over to the multiplicity that is orange and, after their own respective fashion, follow the pattern of Hercules's labor or theft. Seventeenth-century Jesuit Hebraist, polymath, and art lover Johannes Baptista Ferrarius is the most explicit about this pattern. He styles his two-volume treatise on citrus culture after Hercules's labor, though he claims to outdo the demigod (Figure 14). For whereas Hercules made off with only three golden apples, Ferrarius hands "over to your indulgence the entire Hesperidean crop, enclosed in this small volume."[9] In place of a "club," Ferrarius uses a pen, though as the volumes unfold, and he wanders through the orchard of trees that produce "apples of edible gold," he comes to seem more like an overly literal suprahumanist bee, for the bee is "always very careful in [its] choice of flowers" and "especially delighted with the silvery blossoms of the golden apples."[10]

Like all origin stories, the tale Ferrarius tells wanders. As he parses the paths taken by different writers, the story multiplies. The Garden of the Hesperides shifts location; the guarding dragon becomes an impassible sea, an all-seeing shepherd (Draco); Hercules takes the golden apples directly, tricks Atlas into stealing them for him, offloads the weight of the world upon him; the tree and the fruit change names. In the end, the fruit go mobile, are disseminated still further—or they do not; Minerva restores the apples to the garden.[11] As it moves, the story seesaws between a general economy of pure expenditure—the orange become edible gold, the precious metal become self-laboring plant that gives itself to you to eat—and a more pointed act of theft, a parasitic derivation of interest or the opening up of a restricted economy to its other.[12] The status of Hercules's act shifts as he goes: labor, struggle, contest, trick, or thievery. The story will not settle. The promise of pure expenditure seems automatically to call forth a series of faulty, sometimes lethal exchanges. Though, as Ferrarius jokes, given their allure, the truly "Herculean task" was "limit[ing] oneself" to taking only three apples in the first place rather than making off with the whole garden, which, of course, Ferrarius's book does, in translated form, the Hesperidean hoard become a compendium of "flowers" as he gathers all the lore and images he can.[13]

With Ferrarius, then, we encounter all manner of oranges: "golden," "common," "seedless," "curly," "starred and rose-like," "hermaphrodite," "distorted," complete with their mythic explanations, as in the story of

Figure 14. Hercules accepts the gift of golden apples. Johannes Baptista Ferrarius, Frater, *Hesperides, Sive de Malorum Aureorum Cultura et Usu* (1646). By permission of the Folger Shakespeare Library, Washington, D.C.

the tragic, virtuous Leonilla, "one of the best fighting Amazons of Diana," who sought to top her success in a boar hunt by shooting a golden apple off the branch of its tree. Initially, Leonilla's sister nymphs were rather impressed with her prowess in boar hunting, but soon they ended up in an argument over who was the best shot with a bow. A shooting match seems to provide the answer. And the nymphs are delighted with the idea. Leonilla takes the shot. The orange falls and she wins the contest (Figure 15).[14]

But before the fruit even reaches the ground, "blood dripped from the tree and a voice filled with tears said 'Alas, alas, woe is me.'" Leonilla realizes that this voice belongs to her "mother, entombed in the tree," and that she has "been accidentally pierced by the arrow." She rushes to "embrace and help her," but as she does so, her mother's face appears in the bark, begs to know why this hurt was necessary.[15] And the appearance of her mother's contorted face in the tree roots Leonilla to the spot as she herself turns tree, transforming so as to keep her mother company for all time, a living memory device that mourns by entering into a becoming orange, but too youthful, too reckless, only half-evolved; Leonilla, so we learn, comes to name the tree that sports "strange and abortive [or early] fruit" among its beautiful oranges.[16] Still, this vegetal fidelity stands surety for the bond between mother and daughter, a fidelity for which oranges or golden apples shall sometimes stand. Look closely and you can see that she is depicted in mid-transformation, her feet rooting to the ground, her face fixed on that of her mother.

In effect, Ferrarius assembles a mythic archive of the multispecies axiomatic through which orange enters our world and our imaginations. For him, "orange" eats all other modes of relation, which become secondary to this overwhelming, giddy pleasure in the alchemy of gold we can eat, of gold that goes unmined or that the world of plants mines for us. Tuned to this orange ecology that comprehends its economic tasking, that understands economy to derive from ecology, like Ferrarius, I find myself embarked on an inventory of orangey effects as hosted by this or that object or medium.[17] Choose another fruit, another color, and the story you tell and the ecology you trace will shift; connect different times and places; involve different actors. You have to begin afresh with

Figure 15. Dominicus Zamperius, *Leonilla in the Garden.* Johannes Baptista Ferrarius, Frater, *Hesperides, Sive de Malorum Aureorum Cultura et Usu* (1646). By permission of the Folger Shakespeare Library, Washington, D.C.

each and every color, fruit, or *thing*. Though your ability to do so will be canalized by the extent to which you may be said to share the same *Umwelt* or environment, as, in my case, with sheep, oranges, yeast. For, more fundamentally still, this orientation to orange signals the way the phenomenology of a liking (animal sense perception) comes threaded through different ecologies. The sensory effects that we describe as "color," "taste," "smell," and so on, as a "liking," a "fondness," even a "fetish," might best be modeled as products of a multispecies sensory process or network that generates highly differentiated bio-semiotic-material effects for different beings, effects that then take on a metaphorical life of their own as they are translated to different registers. These matter-metaphors prove reversible. The anthropomorphic experience of color, of taste, of the senses, conjoins with a zoomorphic becoming other than oneself, a biomimetic becoming or, for the purposes of this chapter, a turning "orange" or "orangey." Like Leonilla, women who sold oranges in Restoration theaters in England became "orange-women" or "orange-girls," as they were called, rendering the color as hosted by the fruit mobile, stitching its presence into the visual field of Londoners.[18] Such exchanges prove commonplace. They are constitutive of what it means for us to be—all of us, like Leonilla, like these orange-women or Gerard's orange fond warder, animate and animating everyday *prosopopoeias*—hybrid, walking, talking orange-sheep-people.

In what follows, I begin by imagining a world before "orange," before the advent of the name that codifies the color, rendering it a sphere, a fruit, a *thing*. I then go on to tell the story of "orange's" arrival and normalization in England, imagine its absence in one fictive future, its replacement by a different fruit and so color in another, and examine how the object world rearranges itself to accommodate its absence. Like Ferrarius, then, this chapter appears in the guise of a book of flowers, a *florilegium* keyed to the hyperoccasional gatherings that each orange or liking for oranges engenders. Or, perhaps it is a dance, a tracing and tracking of the steps that our animation by oranges, our taste for the fruit, produces. I am not sure that this little book of flowers or surplus of dance steps will offer much by way of consolation. The movements or gatherings I record may constitute dull routines or generate spectacular surprises. They remain partial, ephemeral, subject to dispersal. Dances end; posies wilt, change their hue, find themselves replaced by still other gatherings. Time passes. And that passage registers differently according to the vagaries of our biosemiotic motors. Oranges, for example, change their color depending on their exposure to temperature. In the tropics, they are not necessarily orange at all, but green. And as they decay, they turn white, blue, black, as the polities of bacteria they host "flower."

Still my hopes are manifold. I hope that modeling an archival heterospace tuned to the way oranges derange rationalized modes of exchange might, in some small way, make good on Pietz's kleptomaniacal imperative. Perhaps siding with oranges (and, on occasion, refrigerators), with certain historically particular configurations of labor, matter, and energy, might enable us to "renew the [concept of the] social" athwart the lines of species such that questions of economy might, once again, be understood to derive from a general, even extraterrestrial ecology, an ecology that refuses the hyperreference of terrestrial grounding.[19] What would a polity or sense of the common generated by or through oranges look like? We can begin to measure its presence and charting its contours by traveling our discourses by way of the disturbed exchanges and disturbance of exchange an orange causes.

BEFORE *ORANGE*, CIRCA 1460

"Richard of York," we know from the early-twentieth-century rhyme cum mnemonic device, "gave battle in vain," but at the time of the Battle of Wakefield (1460), which he lost and at which he died, the word *orange* had only just entered the English language and had only just come to designate the color phenomenon that occurs, in the visible spectrum, at a wavelength of about 585–620 nanometers. Without doubt, Richard of York's supporters sang songs to remember a good many things. But when seeking to name the phenomenon we name "orange," they reached for an altogether different constellation of words to designate the range of tones and shades of things that hosted the color in their object world. White roses sported your allegiance to the House of York, distinguishing you from the red, red rose of the House of Lancaster, but if there were such a thing as an orange rose, it had yet, in English, to become "orange." The words *pome d'horange, oronge, orenge, orynge,* and *oringe,* designating the fruit, appear around 1400. But the extension of the word to include any color resembling the skin of an orange is not widely credited before the 1550s.[20] When met with the phenomenon, Richard of York and friends were not open-mouthed or mute. They merely reached for different color words to describe the sun, a marigold, Herod's hair and beard. Even the oranges that Hercules stole from the Garden of the Hesperides or that Venus gave to Hippomenes in order to beat Atalanta posed no problem, for they were not (and may never have been) "orange" or, for that matter, oranges.[21] They remained *pomum aurantium,* golden apples, or, as rendered in Neo-Latin, in homage to the citrus collections of Italian nobility, *medici.*[22]

Codifying the practices of limners or portrait miniaturists in England, such as might once upon a time have painted Richard of York, in small, from the life, Edward Norgate's *Miniatura or the Art of Limning* (circa 1627–28) looks back over the techniques used to craft the jeweled worlds of fifteenth- and sixteenth-century English portrait miniatures. He synthesizes the techniques of his artisan forebears, Isaac Oliver, Nicholas Hilliard, Lavinia Teerlinc, and the color palettes of his time. "The names of the Severall Colours commonly used in Limning are these," he writes: "white," "yellow," "red," "green," "browne," "blew," "blacke."[23] These colors sport a host of medial hues, each of which has its own peculiar temperament according to the nature of the ingredients from which it is mixed. Yellow subsumes "Masticot, Yellow Oker, English Oker"; red, "Vermillion, Indian Lake, Red Lead." Each derives from a working of plant and mineral and insect ingredients, such that the limner had to do more than simply know her colors. She had to know their temperaments, their likes and dislikes; she had to know which were "friends" and "foes." "English Oker," writes Norgate, "is a very good Colour and of very much use. . . . It is a friendly and familiar colour and needs little Art or other ingredient more."[24] But contrast this friendly color to the treacherous "Ceruse" (a white), which "will many times after it is wrought tarnish, starve and dye" and, weeks after it is applied, "become rusty reddish, or towards a dirty Colour."[25] Norgate goes on to describe a lengthy process by which ceruse might be rendered semistable by enlisting the aid of another, more friendly substance—the "hungry and spungie chalk," which will "suck out" all "the greasy and hurtfull quality of the Colour."

Less schematic in its organization, Nicholas Hilliard's short *Treatise Concerning the Arte of Limning* (circa 1600) codes this practical knowledge of the properties of different substances within an explicitly moral philosophical framework. For Hilliard, still the artisan, working hard for the gentrification of his art, the temperament of the limner proves as crucial as the temperament of the ingredients:

> The best waye and meanes to practice and ataine to skill in *Limning,* in a word befor I exhorted such to temperance, I meane sleepe not much, wacth [*sic*] not much, eat not much, sit not long, usse not violent excersize in sports, nor earnest recreation / but dancing or bowling, or little of either, / then the fierst and cheefest precepts which I give is cleanlyness, and therfor fittest for gentelmen, that the practicer of *Limning* be presizly pure and klenly in all his doings, as in grinding his coulers in place wher ther is neither dust nor smoake, the watter wel chossen or distilled most pure.[26]

The limner must curb her behaviors; she must practice temperance in all things so as to be able to temper the colors, manage the unmanageable liveliness or deadness of certain ingredients—learning which colors she can rely upon, such as the friendly yellow oker, and which require careful supervision, as in the fickle ceruse. In a historical moment when color fixity was the very emblem of scarcity value and pigments could turn fugitive and fade, the limner required a specialized knowledge of the etiquette of color, learning to pamper and flatter such substances that might provoke sympathetic color effects. In Bruno Latour's terms, we might suggest that here the limner must seek out friendly agencies, adapting her gestures in the hope of maintaining alliances that would keep color effects still.

Hilliard would have to wait for the ministrations of Henry Peacham's successively revised and enlarged *The Compleat Gentleman* (1622) for the ministrations of the limner to her ingredients to be accorded gentlemanly status.[27] But, eventually, his attentiveness or politeness to colors would lead to the inclusion of the art among pastimes deemed gentlemanly. In the enervated lexicon of the *Treatise,* however, such gentility manifests as an almost obsessive attention to the "fineness" and "purity" of the art.

The word "pure" assumes a unique redundancy in the treatise as he seeks to name the quality of attention and behavior required of the limner in attending to the vagaries of colors whose qualities she inventories. The word "pure" comes to describe both the person of the limner and the materials she handles, purity becoming, in essence, a common property or product in the making, to be proved by the effect of the finished miniature on the viewer. The labor of the limner stands in reciprocal relation to the final stability of color in the miniature. Her husbanding of the materials, her artful handling of foes and reliance on friends, records a joint exercise in making that understands *poiesis* as a cascade of competing agencies.[28] You may neglect the friendly ochre, but on no account skimp on your attention to the ceruse. Keep your chalk dry until you need it.

Hilliard and Norgate's expertise speaks to what Gilles Deleuze and Félix Guattari describe as the artisanal devotion to the "prodigious idea of *Nonorganic Life,*" the agency of all the substances called upon in the acts of making and manufacture they set in motion.[29] Their treatises read like strange Georgic manuals, husbandry books, or participant-observation ethological studies of the animal behavior of minerals, all to obtain very particular qualities of color. Belonging to the long tradition of lapidaries, which describe the particular virtues (force or agency) of gemstones and minerals, both follow Aristotle's theory of substances, for which color was understood to be contingent on physical variables.[30] As Bruce R. Smith summarizes, the "color [red] happens only in the presence of fire or something like fire."[31] Color manifests, then, as a form of affect or sympathy between materials that "contain . . . something which is one and the same with the substance in question." What matters for Aristotle and for the limner most is a color's brightness, its saturation of a medium. For Hilliard and Norgate, this saturation, effected by differing processes in every case, leads to color fixity. Such then was the essence of their skill—the art of keeping colors still by cultivating or rerouting the desires of substances.

The salutary lesson of imagining a world before "orange," a world of saffron and ochre and yellows, lies not in a hardline historical epistemology that would cry foul and insist that there was, at that time, no such color. On the contrary, the absence of "orange" foregrounds instead the way color, like any complicated phenomenon, constitutes a "sheaf of temporalities" or a "knot in motion" that connects different times and places and different entities in a structure of only apparent simultaneity.[32] The emergence of the word *orange* represents a formalizing event keyed to the production of an infrastructure that yokes the color phenomenon to the fruit as it works its way from Asia to the Americas and thus sets in motion a set of routines for making persons and worlds by threading together or knotting otherwise asynchronous chronologies (ecological, social, political, sexual, religious, and so on). The word or ethonym *orange* constitutes a partial archive of this process. Remove the fruit, cancel the various substrates that host "orange," or offer new ones, and the effect that is "orange" will alter. This is as true now as it was then. For even today we have trouble keeping our colors straight and are unable, in the absence of some guaranteeing medium, to call up an exact shade with certainty. Keyed to the rainbow as captured by Newtonian optics, the mnemonic "ROYGBIV" aims to keep something straight that is not, to manage the set of effects named "color" as they occur at the interface that is the human *sensorium.* Learn the rhyme and you will remember the order of colors in the rainbow. But, while the mnemonic assumes a post-Aristotelian orientation to color, it retains, in shifted form, the order of labor and politeness toward matter that Hilliard and Norgate advertise via their husbanding of ingredients to their limning. This focus on the unreliability or the drift to color reorients our attention toward those kinds of practices or routines that keep phenomena still, regularize them, and with what order of politeness or hospitality they do so.[33] It invites us to attend to the ways in which certain operations are conserved, dropped, or redistributed into different discourses or disciplines across historical periods.

Both the rhyme and Hilliard and Norgate's older sense of color phenomena speaks to the translational or transactional properties of color effects, effects that today we manage via that strange, grown-up, ring-bound, folded-out-and-up book *The Munsell Book of Color,* which allows us all to keep our colors straight, even in their absence.[34] The Munsell Code assigns each color a number, and as Bruno Latour writes of the soil color guide version used in fieldwork by pedologists, "the number is a reference that is quickly understandable and reproducible by all the colorists in the world on condition that they . . . use the same code."[35] The book works by way of a "stupefying technical trick—the little holes that have been pierced above the shades of color" that enable you to align this or that colored something you have before you with its like and convert said something into its number referent. The book serves as an object lesson in a model of "circulating reference" that at every point in the translation of an effect or entity must attend painstakingly to what it means to keep that effect stable in a new medium.[36] Once upon a time, this desire for color fixity, to render effects permanent, required the limner to embark on a quasi-mimetic sympathizing with her palette. Now, such embodied knowledge migrates to the tactile certainty granted by a book that translates looking into touching, a book that comes allied to the industrial practices and the adumbration of resources necessary to relaying colors on a global scale. The Munsell Code registers thereby the replacement of the moral philosophically coded labor and techniques of the limner who worked with substances by another order of techniques. But the very existence of the code as the bibliographical form of an elaborated infrastructure attests also to the still necessary if differently deployed labor and resources necessary to keep phenomena "still." Hospitality, become pure technique, reigns, but bypasses the substances whose labor and liking Teerlinc, Hilliard, and Norgate once had to cultivate.

But I am getting ahead of myself. Let's return now to a moment just after the arrival of oranges and of orange in England, to a world reordered by the mobility of the fruit that hosts the color and gives it a name. Let's arrive at the moment of Bonner's liking.

ALLURE, CIRCA 1600

Oranges began to arrive in England from Spain and Portugal at the beginning of the fifteenth century. They generally arrived in March or April, staying good until December. They arrived in quantity (tens of thousands at a time). They were not outrageously expensive, but neither were they cheap.[37] They remained an exotic, coded by the soil—Spanish or Portuguese—that grew them and by their status as a "corruptible," one of those many entities, like us, subject to change. Come the seventeenth century, they would be sold in and around theaters as snacks or, on occasion, projectiles; on the street; as well as in bulk for eating, as holiday gifts, and as table decorations for the Inns of Court and well-to-do houses.[38] Seasonally, waves of "orange" and oranges entered the country, oranges coming to punctuate market stalls, dinner tables, finding their way into the hands and mouths of Londoners. And as with any imported commodity, the key to the fruit's naturalization stemmed from the import of the concomitant skill and knowledge of its use (values).[39]

Oranges were not at first primarily an edible, finding themselves pressed to use as medicinal items, a fashion statement, a cleanser, as well as the more rarified kinds of uses I approached in my last chapter. Nor would they be so until the arrival of China oranges in the seventeenth century, which supplemented the waves of Seville fruits that typified the earlier periods. Moreover, their status as a southern fruit of Spanish or Portuguese origin meant that there was debate over whether northern English souls could properly digest them.[40] In an era when geohumoral theories modeled "Englishness" and "Spanishness" as tuned to different ecologies, eating an orange was to invite a potentially all too animate, foreign actor into your body—an actor that could well produce an indigestible "drift" or "flux" if you were not careful. Eat too many, take too little care in preparing them, and you might even find yourself embarked on an involuntary "becoming Spanish."

Attempts were made to cultivate oranges in England, but they were not entirely successful. Shipments of orange trees tended to arrive in the spring and summer months. Upon arrival, they were transplanted to English gardens, grown usually in tubs, and moved indoors to sheds, greenhouses, or "orangeries" for the winter.[41] It proved difficult to grow trees that would flower and fruit, so the migration of orange trees north was accompanied by a bibliography of advice manuals serially translated into French, Dutch, German, and then English.[42] There even exists a translation and abridgment of Ferrarius's *Hesperides,* which, as it heads north, shrinks from its two-volume, illustrated folio splendor to a slim quarto, first in Dutch and then in English, titled S. Comelyn's *The Belgick, or Netherlandish Hesperides* (1683). The book reduces Ferrarius's elaborate treatment of citrus lore to a minimum, focusing instead on the dos and don'ts of where to buy your trees, how to ship them, and under what circumstances you might be able to overcome their "unfruitfulness." Torn between a Protestant dis-ease with Ferrarius's Catholicism and a sovereign desire to transplant the trees, to render them "Dutch," just as the book and then the tree, so it is hoped, were "Englished," Comelyn justifies the omissions by claiming that "unnecessary Narration is Nothing but useless Labour." Lose the art; lose the easy, idle *otium* of an orange grove; and maximize the labor you devote to "profit." Ferrarius's Herculean fetish likewise finds itself reduced to a general psychology or symptom, "whereunto every one must shew himself as an Hercules, and bend all his strength that he may break through by the waking Dragon into the most inward Garden, to satisfy the sweetness of his Invited Desires to this exercise."[43] With no oranges of their own, the entire cultivation of citrus in Holland and England, if not yet Norway, comes coded as theft. But, ingenuity aside, such satisfaction that these northern Herculeses obtained from oranges still came from imported fruits.[44]

There were notable exceptions. But these required an inordinate commitment of time, resources, labor, and money, leading oranges and lemons to become one of the most technologized of fruits. Writing in the 1590s, Sir Hugh Platt describes a process practiced by a "Surrey knight" (Sir Francis Carew) that "extendeth generalie to all fruite not much unlike spreading a tent ouer a cherrie tree about four-teene dayes or three weeke before the cherries were ripe." Platt recounts further that using this method, last Whitsuntide, he was able to present Sir John Allet, lord mayor of London, with "a score of fresh orenges."[45] In 1601, John Tradescant, Lord Salisbury's gardener, bills his employer for "8 pots of orangg trees of on years growthe grafted at 10s[hillings]. the peece" collected while on a visit to Paris.[46] And in 1604, James I welcomes the Constable of Spain with a feast whose table was graced with "a melon and half a dozen . . . oranges on a very green branch" grown in England. A careful reading of that story indicates that Their Majesties split the melon with the constable but may not have eaten the modest number of English oranges, deploying them instead as a diplomatic superlative that signified the transplanting or grafting of Spain's sovereign goods to England's soil in the form of a sweet-smelling diplomatic centerpiece.[47]

By and large, then, the translation (importation and cultivation) of oranges and orange trees to England unfolded as a series of failed, botched, or contested instances of Georgic. In England, this Virgilian mode exerted a shaping influence on the sixteenth and seventeenth centuries, constituting a loose set of discourses or an "informing spirit" that moralized agricultural labor in the service of a growing sense of the importance of industry.[48] The immense popularity of Thomas Tusser's home-grown Georgic *Five Hundred Points of Good Husbandry,* published first in 1557, going through three editions in 1573 alone, and then "eleven more before the turn of the century," testifies to the significance of the mode.[49] Arising from encomia for agricultural practices, in England, the discourse evolved into an "indistinctly loose" celebration of labor as some fungible mass or energy that might be applied to almost anything. Oranges arrived just at the moment in the sixteenth century when writers sought to downplay the socioeconomic particularity of the plow and the plowman as chief icons of the mode so as to make the discourse available for other kinds of ideological work in the making of persons, careers (political and poetic), nations, religious life, and profit. And these oranges posed a challenge, fueling, on one hand, a desire to develop technologies that might "English" them and, on the other, a sense of wasted, "useless," or unthrifty labor—the oranges a delusive, seductive, foreign "corruptible" traded at the expense of staple commodities. Oranges came tainted with the same conflicting mix of signals that made the *otium* I explored in my last chapter so appealing.

Apprehended systemically, as an object of exchange, the fruit and the trees became subject to economic criticism given the reciprocal flow of capital out of the country on which their re/appearance depended. The price of all that orange and all those oranges required a commensurate absence of gold and silver that went abroad. When, for example, in the 1620s Gerard de Malynes updates sections of Thomas Smith's *A Discourse of the Commonweal of This Realm of England* (1549) for inclusion in his *Lex Mercatoria* (1622), he bemoans the fact that some men "do wonder at the simplicite of the Brazilians, West India, and other nations . . . in giving the good commodities of their countries . . . for Beades, Bels, Knives, Looking-Glasses . . . when we our selves commit the same, in giving our staple wares for tobacco, oranges, and other corruptible smoaking things."[50] Fetishists all, he seems to say; irrationality abounds. But Krugman and Obstfeld would be proud, for Malynes goes on to praise the mayor of Carmarthen, who organized a boycott of rotten oranges passed off as fresh by Spanish merchants in his town, driving down the price of this local "fetish." Unhappy as he may have been that the Spaniards would only trade their wares for ready money instead of Welsh goods, still he knew the value of "gains of trade," for the "Spaniard" in question "afterwards sold his Oranges better cheape, and bought commodities for his returne."[51] The rationalizing rewards of exchange value stabilized by consumer action won out. But, in truth, this story is older still. Malynes's oranges are relative newcomers. In Smith's original, the "evil" or false corruptible in question was an apple and the offending merchant an Englishman.[52]

Criticism came also within the registers of good husbandry. The sheer amount of labor required to maintain orange trees in northern climes led Royal Society member, experimental scientist, arboriculturist, and radical cleric Ralph Austen to take the tree to task in *A Dialogue or Familiar Discourse and Conference betweene the Husbandman and the Fruit-Trees* (1676). Austen deinscribes the divine signatures in the trees of his orchard and provides transcripts of what the different trees say. These scripts put into words what Austen's empirical observation of growing habits told him of their "fitness" to England's soil. Austen's husbandman remarks that there "are a sort of Trees that do not thrive, nor prosper, as other Trees do; what's the reason? Seeing yee are as well planted, and preserved, as other Trees which grow neere unto you." To which the fruit trees reply, "We are forreners; this is not our Native Country we were brought from beyond the Seas, from a warm Clymate; where we had a strong heate, and influence of the Sun; but we are here in a cold country, and it agrees not with us, we shall never grow, nor prosper, nor bring forth Fruits to please thee, or for Profit." "I believe what ye say to be true," replies the husbandman, going on to remark that "many Gentlemen of great estates, to please their minds, have sent for, and brought many rare Plants, and

Trees, from forraine parts, out of the South-Countries, of France, Spaine, Italie, and other Southerne Clymates and planted them in England, severall degrees Northward; which though it will beare many kinds of good fruits to perfection, and ripenesse . . . yet it will not beare good Orringes, Lemons, Pomegranates, and such like Fruits of the South Countries."[53] Better to grow English apples, whose "profitability" has been established by the experiments of the Royal Society.

But oranges still captivated. And because they could not be effectively "Englished," they remained sovereign unto themselves, Spanish, perhaps, but twinned with a serial, handy, sweetness, and beauty all their own. Every orange constituted an event; interrupted the flow of stories and sense; warped them with its affecting presence—with a surplus of liveliness.

In the *Paston Letters* (1422–1509), for example, we learn that "Elizabeth Calthorp longs for oranges, though she is not with child."[54] The letter passes over this longing, remarks it as an excess worthy of comment but offers no explanation. Cardinal Wolsey was famous for his predilection for taxidermied oranges or pomanders, "whereof the meat or substance within was taken out and filled up again with the part of a sponge wherein was vinegar and other confections against the pestilent airs; to the which he most commonly smelt unto, passing among the press [crowd or throng] or else when he was pestered with many suitors."[55] In *Much Ado about Nothing* (1598), the falsely jealous Claudio names the guiltless Hero "a rotten orange," claiming that

> She's but the sign and semblance of her honor
> Behold how like a maid she blushes here!
> O, what authority and show of truth
> Can cunning sin cover itself withal!
> Comes not that blood as modest evidence
> To witness simple virtue? Would you not swear,
> All you that see her, that she were a maid,
> By these exterior shows? But she is none.
> She knows the heat of a luxurious bed,
> Her blush is guiltiness, not modesty.[56]

Here he throws this "rotten orange" back, metaphorically smacking the innocent Hero in the face, causing her skin to mimic the color of the fruit, to blush. And in his poem, "Employment 2," George Herbert wishes that he "were an Orenge-tree, / That busie plant! / Then should I ever laden be, / And never want / Some fruit for him that dressed me," inducting the peculiarity of the orange tree's ability to flower and fruit at the same time into his attempts to craft prosthetic prayer machines.[57]

In each of these instances, as with Leonilla, a person and an orange exchange properties: both are remade in the process, taking on the properties of the other. Herbert wishes to become the orange tree, its fruit and flowers. He wishes to take within himself the tree's ability to flower and fruit simultaneously, continuously, as he becomes a living *florilegium* of spiritual posies; or better, his poem shall comprehend the orange entirely, siphon off its vegetal being, pray on after he dies, the poem become his spiritual survivor. His poem would eat Francis Ponge's "L'orange" but retain the "pip," grafting its poetic and religious resonance to the root stock of orange. For Dame Calthrop, desire itself becomes "pregnant" with oranges; it may not be explained by the cravings that come from being with child. Instead, desire itself becomes that child. Dame Calthorp is pregnant with oranges. Wolsey's pomander works a bit differently; indexes his orange fondness to concerns regarding his shepherding; raises doubts about his fitness as a shepherd who cannot stand the smell of his flock. But in each case, these exchanges set in motion a more or less readable material–semiotic transfer by which particles of orange-being impress themselves on our discourses and the forms of life they encounter turn orangey.

Much Ado's Hero offers a particularly complex example. As an uncharacteristically clipped note in one edition of the play offers, Hero is pronounced rotten "perhaps because an orange may look sound but be bad inside."[58] Here the orange functions as a little clock. The temporality of its ripening, its living or dying on, as it falls from the tree, occurs independently of what fate holds for its pith, which may remain intact and beautiful. In human hands, then, the orange becomes that most anti-Platonic of fruits, for its insides bear no relation whatsoever to its appearance. And in *Much Ado,* this "corruptible," physiological peculiarity finds itself indexed to the instability of the female blush as a signifier (all at once) of innocence, shame, and guilt. Though, in this instance, the material–semiosis proves denser still, for in the theater, this "rotten orange" comes twinned visually with the "orange-women" who sell oranges to the play's audience, and who, on occasion, also sold themselves. The rottenness of *Much Ado*'s orange goes mobile across this hybrid, gendered, animal–plant flesh as it is parceled out in differently animated forms. The passing of oranges from hand to hand, hand to mouth, and their contiguity with a similar set of economic and sexual exchanges produces a nexus by which oranges, human bodies, and gold exchange properties. Coincidentally, in Early Modern English, *lemon* and *leman* (lover) were homonyms, automatically inducting citrus into the amatory registers of the period by an accident of linguistic materiality.[59] So it is that the "civil" (Seville), Spanish, orangey, Claudio picks up a metaphorically rotten orange to throw at the "rotten" Hero and we watch a scene choreographed, perhaps, on the model of a pillory, in which the culprit is pelted with rotten fruit.[60]

It is important to recognize that these discursive forms of orange, this orangey archive, registers those moments when the temporality and chronology that are "orange" intersect with the differently mediated worlds of letter writing, life writing, theater, lyric, and still other political, sexual, economic, and spiritual chronologies. The lesson they offer lies in the nature of the associative process by which they come to signify. In each case, that of the hybrid orange-longing Dame Calthrop; the puffed up, hollowed out, sweet-smelling wolf-in-shepherd's-clothing Wolsey; the rotten-gendered-mishandled-orange-Hero; and the plantlike Herbert, we witness a different "performance" of the orange, a different declension or declination of the multiplicity that is orange as it gives itself to be grasped, as it fascinates historically particular pairs of hands and eyes and so produces different worlds. If there is a common script or pattern to these exchanges, it inheres to the way in which each of these orangey figures remarks a misfiring or misuse, an excess or surplus, that manifests in moral philosophical, economic, sexual, or political registers. The orange, as it were, interferes in these discourses, traumatizes or captivates them by and through its promise of expenditure and threat of theft or fault.

In *A Lover's Discourse,* Roland Barthes names this *topos* "*fâcheux/* irksome" and calls it "orange," keys it to orange, or the historically particular, fictive orange Werther gives Charlotte in Johann Wolfgang von

Goethe's *The Sorrows of Young Werther* (1774). He quotes the young man and adds his own gloss, extending his lines in paraphrase: "The oranges I had set aside, the only ones as yet to be found, produced an excellent effect, though at each slice which she offered, for politeness's sake to an indiscrete neighbor, I felt my heart pierced through." "The world is full of indiscrete neighbors," Barthes offers, ventriloquizing Werther, "with whom I must share the other. The world (the worldly) is just that: *an obligation to share.* The world (the worldly) is my rival."[61] Morphologically complex, the segmented orange gives itself to a plurality of consumers. It is made, so it seems, to be shared, even as each orange presses itself into your hand, offers itself as an "event" for you. The fruit gives itself to us but by that giving embeds us in a story of theft, from the world, from each other, as we derive our liking from out of this obligation to give, an obligation that the orange seems both to give and to offer to take away. Orange-being produces this paradox for us. It convokes multiple, incompatible, competing polities of animate or animal actors.

Of what consists an orange's allure? What is the source of our taste or captivation? In one section of *De Sapientia Veterum* (*On the Wisdom of the Ancients*; 1609), titled "Atalanta for Profit," Sir Francis Bacon offers one theory that explicitly writes the orange or golden apple into his early attempt to name and describe the form of the commodity. He does so by retelling Ovid's retelling of the story of Atalanta's race with Hippomenes in book X of *Metamorphoses.* As you remember, Atalanta's amazing speed poses a bit of a problem for her would-be suitors. So Hippomenes relies on a gift of golden apples from Venus, which he throws in Atalanta's path, distracting her, leading her to lose the race. Bacon offers this commentary:

> The Fable seems allegorically to demonstrate a notable conflict between Art and Nature; for Art signified by Atalanta in its work (if it be not a little hindered) is far more swift than Nature, more speedy in pace, and sooner attains the end it aims at, which is manifest in almost every effect: you see that fruit grows slowly from the kernel, swiftly from the graft; you see clay harden slowly into stones, fast into baked bricks: so also in morals, oblivion and comfort of grief comes by nature in length of time; but philosophy (which may be regarded as the art of living) does it without waiting so long, but forestalls and anticipates the day. And yet

> this Prerogative and singular agility of Art is hindered by certain Golden Apples to the infinite prejudice of human Proceedings: For there is not any one Art or Science which constantly perseveres in a true and lawful course till it comes to the proposed End or Mark; but ever anon makes stops after good beginnings, leaves the Race and turns aside to Profit and Commodity, like Atalanta. . . . And therefore is it no wonder that Art hath not the power to conquer Nature, and, by Pact or Law of the Contest, to kill and destroy her; but on the contrary it falls out, that Art becomes subject to Nature, and yields the obedience of a Wife to her Husband.[62]

As Helen Deutsch comments, "Bacon's golden apples make Atalanta a victim of the marketplace, seduced by the allure of 'profit and commodity.'"[63] Art (technics) proves infinitely superior and speedier than Nature. But driven by a desire to derive an interest from the time it invests, it stops short, gets distracted. Nature moves at a different rate entirely. Its plants and creatures simply grow, just keep on growing; it figures a constant. Art (technics) eventuates itself, proves prone to "stopping" or halting, a bringing up short from the "proposed End or Mark." It abandons the race and "turns aside" to ends other than those its designs posit. Nature just keeps on going. It does not recognize that there is a race and therefore wins.

Bacon genders this structure. Atalanta, we are told early on, responds to the apple "with the eagerness of a woman." And so, on his reading of the myth, "Nature" plays the husband, "Art" the wife. But they make an odd sort of couple. For such sovereignty that "Nature" has over technics (and so us) plays as a little theft—our or Atalanta's "obedience" is yielded through an involuntary if eager captivation that steals our ends from us, installing in us, instead, a desire for other ends still, for "golden apples," sweet gold we can eat now. And the existence of "certain Golden Apples," as Bacon's generalizing syntax makes plain, transforms Atalanta's historically particular apples into exemplary forms, cell forms, it's tempting to say, of the "fetishism of the commodity," by which Art is brought up short. For Bacon, then, the commodity form itself mimics or derives from the goldenness of these apples, from the fruits of a plant's labors. The golden apple or orange, the self-rolling, gilded reproductive technology of the plant, already mimes and embeds the distractions of "profit and commodity," which play here not as "the gains of trade" but as a delusive, enlivening, surplus or stunt and stunted *telos* that prevents "Art," the "human" as technical animal, from attaining sovereignty.

Of course, Bacon's little domestic allegory or discursive couples therapy holds only for as long we agree to suspend the presence of Venus, who, in Ovid's version, and in Arthur Golding's translation from 1567, narrates the scene. In these versions, the allure of the golden apples proves more complicated still, less or differently generalizable. In Golding's version, Atalanta's remarking of the first apple is described as "covetous"—a psychologized and religiously loaded term for desire—but the fruit is "rolling," in movement, and it is that movement which seems to distract:

> Then *Neptunes* imp her swiftnesse to disbarre,
> Trolld downe at one side of the way an Apple of the three.
> Amazde therat, and covetous of the goodly Apple, shee
> Did step asyde and snatched up the rolling frute of gold.
> With that Hippomenes coted her. The folke that did behold,
> Made noyse with clapping of theyr hands.[64]

Atalanta steps aside, picks the apple up, examines it. We do not know what she does with the fruit. These moments serve as minimal instances of *occupatio* (dilation and delay) during which Atalanta becomes a spectacle for us and ceases to be a spectacle for the crowd, who cheer instead for Hippomenes. Her stillness, captivated by the movement of the apple, which she arrests, subtends or runs athwart the movement of the race that ultimately drives the story forward. The narrative moves on apace, but we, and Atalanta, do not. The golden apples introduce holes into the narrative, moments of repose or lapse. The action fits; the crowd's attention shifts. And Atalanta's fascination, much like Bonner's liking, remains unreadable.

Roused from the pleasures of the fruit by the "noyse" of the spectators, Atalanta "recompenses" her "slothe" and catches back up to where she ought to be given her superior speed. She makes up for lost time. But the second apple "stays" her again. And, despite her "footmanship," the legs of sense are held once again or suspended. No more details are forthcoming, however. Golding offers the second apple merely as a repetition of the first. As with Bacon, their serial deployment, along with Golding's successive shortening of their name, renders them a generic placeholder for distraction, physiological, psychological, or otherwise. But the narrative also preserves this distraction, allows Atalanta a narrative space aside from the plotting of the race, which she exits and to which she returns after each rolling golden apple comes to rest, its mobility, its peripheral movement, manifesting as if an event. Golding captures, as it were, the rhythm of the commodity and of consumption. Its serial, punctuating calibration of the race registers the qualitative transformation that its finite quantity or titration allows. Atalanta stands fascinated. She is curious. But each curiously mobile apple sustains her attention for only so long. And so there comes another . . . and another.

The third apple tells a different story. Crestfallen that whatever he does Atalanta catches up, Hippomenes prays to Venus for help, and so she intervenes in what he seems to consider a mid-course correction to her faulty "gift." Hippomenes gives the third apple his best throw, "askew," but when Atalanta seems to "dowt in taking of it up," Venus assumes the first person, and tells Adonis that she put a "heavy weyght" to that apple, giving it such "massinesse" that Atalanta could not lift it. For Golding and Ovid, it seems as if Atalanta could refuse this last "golden apple" and interrupt the repetition compulsion. The golden apple's allure might fail even as it continues to move and distract. But before that can take place, Venus alters the physics of the world so as to arrive at the desired outcome. And this intervention renders Atalanta's refusal entirely virtual, a possible outcome, a thinkable but entirely preterit potentiality. For Ovid, then, the third apple appears less golden. Repetition pales. Atalanta learns. She seems less distracted, or has learned by distraction, seems on the point of recognizing the allure of the apple for what it is—"something that," as Jacques Derrida writes, "comes without coming," that mimes a presence or a density that exists only by and through its mimesis, its movement, the hybrid mineral–plant form gold would take if you could eat it, if it were a plant.[65] And this insight, this curiosity as opposed to captivation, becomes the occasion for Venus's course-correcting intrusion.

Venus black boxes the mechanism of distraction or allure such that we may read only the inputs and outputs. Oranges are sweet. Golden apples are shiny. They distract. Yes, perhaps we could refuse them. But no, they address themselves to us, and so, all of a sudden, we feel the weight or mass of their allure. Cognition comes too late to this particular game. Perhaps this is why Bacon feels he can delete the goddess entirely. She manifests as a personification of the governing *physis* of the world. For Bacon, then, the "massinesse" Venus attaches discloses merely the deeper inevitability of what, on his reading, meshes perception with a psychological drive for the false ends of localized and individualized profit. Bacon no longer has need of the goddess, for he derives a psychology stitched to an understanding of the form of the commodity from this scene of our captivation by golden apples, gold you can eat. We never find out exactly what passes between Atalanta and her three golden apples. Ovid and Golding grant us no access to her moments of fascination even as Bacon seems to take Venus's word for it—so much so that he deletes her. Allure, then, cohabits with this absence in the narrative; its efficacy absents the recipient from what can be narrated. When Atalanta picks up a golden apple, arrests its movement, and considers it, the two constitute a time-bound stay, a hyperoccasional, almost event. Whatever unfolds between an orange and the animal with whom it is taken remains localized, historically particular, present only by and through its loss.

Is this structure what it means to eat? Is this the temporality of consumption, liking become a surplus of liveliness? Is this mode of distraction, following Pietz, what we must somehow find a way to steal?

"'ORANGES AND LEMONS' . . . ," CIRCA 1665

"Say the bells of St. Clements." Maybe you learned the rhyme as a child or read it in a novel. Maybe you did not. Here is how it goes:

> You owe me five farthings,
> Say the bells of St. Martin's.
> When will you pay me?
> Say the bells of Old Bailey.
> When I grow rich,
> Say the bells of Shoreditch.
> When will that be?
> Say the bells of Stepney.

I'm sure I don't know.
Says the great bell at Bow.

Then it all gets a bit frightening: "Here comes a candle to light you to bed. / Here comes a chopper to chop off your head. / Chop, chop, chip, *chop*!"[66] The rhyme choreographs a street game in which two children "determine in secret which of them shall be an 'orange' and which a 'lemon'; they then form an arch by joining hands, and sing the song while the others in a line troop underneath."[67] When the rhyme approaches its gruesome finale, marked by the rising tempo, the two children repeat "chip chop, chip chop," and "on the last chop they bring their arms down around whichever child is at that moment passing under the arch. The captured player is asked privately whether he [or she] will be an 'orange' or a 'lemon'" and then joins the team of the said "orange" or "lemon." Typically, the game ends in a tug-of-war between the two arbitrarily composed teams, leading to a victory for either oranges or lemons.

The text of the rhyme can be traced no further back than the eighteenth century. It appears, in a slightly different form, in *Tom Thumb's Pretty Song Book* (1744), though there was a square for eight dance called "oringes and lemons" recorded in the third edition of Playford's *Dancing Master* (1665)—but the text of this dance, if there was one, is not preserved.[68] According to Peg, a character in Edward Ravenscroft's play *The London Cuckolds* (1682), which takes a postfire 1666 London as its scene, "oranges and lemons" is but "one of a great many things [they cry] in London."[69] This is as far as the written record permits us to travel back into the origins of the rhyme, but it attests to the fact that by 1665, "orange" was a word you might expect to hear as you walked down the street in London. Generically, the rhyme belongs to that group of verses that recall what bells *say,* and such rhymes were marketed to parents as "artificial memory" devices for children in the eighteenth and nineteenth centuries.[70] The text of the rhyme itself is subject to vigorous debate among local historians and antiquarians today, who lay claim to phrases from it for different churches that they wish to make more of a landmark. (The rival locations are St. Clement Danes in the Strand and St. Clements in Eastcheap, both of which boast Sir Christopher Wren–designed postfire interiors.)

"Oranges and Lemons" serves as a fitting emblem or jingle for the procedures by which, in the examples I have offered thus far, person and orange exchange properties, by which mobile, time-bound multispecies polities such as Atalanta and her golden apples come into being. The rhyme choreographs a scene of use and naming, the mutual nomination of person and fruit that serves as the privileged scene of encounter. But the rhyme exists now as a mute record of such encounters, a multiply authored, multitemporal archaeology, accreting references to differently timed peals of bells, landmarks, and half-remembered, half-forgotten "events." It is pure anecdote—the archive become anecdote, catching particles of lost speech and practice, lost scenes of use. And yet, it connects, remembers the connection, even if, as with Atalanta's moments of distraction, it enables no further inquiry. Did you know the rhyme already? Have you sung it? Shall you? It inducts its singers and listeners as "vicars of lost causations," avatars of relations now gone which nevertheless it recalls.[71]

The rhyme invites its own order of reciprocal fetish work—all the archival labor needed not to recover but instead to posit the supposed labors, lives, and deaths lived by previous generations of Londoners whom it somehow touches. The endeavor may prove beguiling, inviting us to hallucinate a past, to catch traces of otherwise undocumented lives as we imagine this or that boy or girl, on the very edge of vision, unloading oranges from a barge on the Thames wharf or singing the song and playing the game. Such are the diversions enabled by the *catachresis* of bells saying things, bells that momentarily possess those who sing, those with ears to hear and tongues with which to vocalize the peal. The rhyme renders us mute as we are sung by phrases past whose meaning we can never know. The cries of London come (and go) again. "Chip. Chop." But even as an immersion in ephemera, "a sadness without an object," beckons, the rhyme puts us on notice also to the way these instances of exchange refer to encounters whose historical particularity cannot quite be rendered. For, as with a certain warder's liking or taste for oranges, to render them is to rewrite their *thing*-like specificity, to steal them so that they may *matter* for us.[72]

Heard this way, "Oranges and Lemons" embeds a sense that our infrastructures come riddled with the remnants of these mobile polities, partial, time-bound practices or thefts that momentarily occupy a street for a game of "oranges" and "lemons" or their equivalent. And these polities unfold on and through and by way of an orange or a lemon, by and through their recruitment by the reproductive technology of a plant that interrupts human narratives, technologies of self, ways of being, nesting its own zoopolitical modality within the biopolitical articulation of our lives. The rhyme reveals also the tropic or rhetorical work of language games in managing these "events," in establishing and maintaining routines, in keeping the built world built, even as its buildings disappear. And while the rhyme may invite a certain order of archive fever, it has within it everything we need to recognize that, in ecological terms, language arts, metaphor, rhetoric, remain key relays within a political struggle for citizenship, and for imagining alternate futures, by and through the way we contest what an orange, a lemon, or a refrigerator is, was, and those who join with them shall be. "Oranges and lemons" is a machine, in small, for writing different worlds. If you manage to lay claim to it or part of it today, mundane as it may seem, you may be able to direct tourist traffic to your landmark, sell a few more postcards and tea things, mend your church's roof.

Of course, some might seek to police these spaces of encounter—to model particular moves and denounce others—precisely because they may prove productive, might err, might provoke possibility. In her short story "'The Orange Man" (1801), Maria Edgeworth tells a story that she hopes might help young minds navigate the perilous London streets and avoid giving themselves over to a life of petty theft. In the story, we meet two boys: Charles, who "never touched what was not his own: *this* is being an honest boy"; and Ned, who "often took what was not his own: this is being a thief."[73] Charles manages to resist the allure of the oranges in the orange-man's basket, fights off Ned, gets a black eye; but for his pains, he has his hat "filled . . . with fine China oranges," gives them away to his peers, wins their admiration (46–49). These same oranges fascinate poor Ned. "The *sight* of the oranges tempted" him "to touch them; the *touch* tempted him to smell them; and the *smell* tempted him to *taste* them" (38). That's how desire works. Captivated by these oranges, Ned turns thief, ends badly. He receives no praise. "*People must be honest, before they can be generous,*" reads the moral (49). And perhaps it is so. But the story ends by asking "little boys . . . [to] consider which would you rather have been, *the honest boy, or the thief*" (50). It makes thievery a choice; it encrypts the comparative disadvantages that, for some of us, render it a necessity of survival.

Obviously, I wish that the story had been interrupted by the "oranges and lemons" of St. Clements's bells and that Ned might have had the chance to become an orange or a lemon, stolen a London street for the time the game took. The rhyme resists narrative, refuses its emplotment, orchestrates its own little time-bound theft by seeding the world with half-realized images that we have to fill in. It reminds us that world is littered with little thefts, nongeneralizable occupations, even if it too ends badly: "Chop, chop, chip, *chop*!" Moreover, under certain circumstances, the mere fact of singing "Oranges and Lemons," badly, of giving your voice over to bells you have never and may not now hear, might matter, might actually do something.

CLOCKS NOT BELLS, 1949

"It was a bright, cold day in April and the clocks were striking thirteen."[74] Against these clocks whose striking announces the altered routines and disavowed past of George Orwell's *1984,* Winston Smith is taken momentarily by a rhyme he hears and which he begins to hum, trying his best to lend his voice to bells he cannot remember hearing, perhaps has never heard. He's alive to the way the Party implants false histories, and so he trolls around, acting on a note to self in his diary that "if there is hope . . . it lies with the proles" (85). He slums, eavesdrops on the cries of London, tries to talk to people and fails. He finds himself seized with "helplessness" when he questions an old man about the past—"the old man's memory was nothing but a rubbish-heap of details" (95). Still, Winston comes to sift the rubbish for this or that stray object that he finds "beautiful." The words "beautiful" and "rubbish" circulate through the novel until they eventually connect and Winston owns the fact that he is on a quest for what he names "beautiful rubbish" (104). He is embarked on becoming an antiquarian, becoming a collector—trawling the remains of London for stray objects whose forms, whose fragmentary archives, somehow activate something in him that he does not quite know how to name. We join him as he considers renting a room from the old man we later learn is named Mr. Charrington, a room, Winston gasps, with "no telescreen" (100).

Mr. Charrington asks if Winston is interested in old prints and beckons him to join him in front of an old picture. Winston

> came across to examine the picture. It was a steel engraving of an oval building with rectangular windows and a small tower in front. There was a railing running around the building, and at the rear end there was what appeared to be a statue. Winston gazed at it for some moments. It seemed strangely familiar, though he did not remember the statue. (101)

"I know the building," says Winston. "It's in the middle of the street outside the Palace of Justice." "That's right. Outside the Law courts," adds the old man. "It was a Church at one time. St. Clement Danes. It's name was." "He smiled apologetically, as though conscious of saying something slightly ridiculous, he added, "'Oranges and Lemons' say the Bells of St. Clements." "What's that?" asks Winston—who doesn't know the rhyme. "Oh, '"Oranges and Lemons" say the Bells of St. Clements.' That was a rhyme we had when I was a little boy . . . it was kind of a dance" (102). Mr. Charrington sings it over but he cannot remember the last line.

Winston looks around at the scene in the print, but he is unable to date the buildings, unable to make sense of the overlapping of medieval, Renaissance, and modern forms and painfully aware that "anything that might throw light upon the past had been systematically altered" by the reigning regime. The old man proceeds to go through the rhyme, or most of it, and to describe the churches to which it refers in detail. It all washes over Winston, and he can't quite process what he has heard. At the end of their meeting, Winston does not buy the picture, but he is haunted by the rhyme. "It was curious," he remarks to himself,

> but when you said it to yourself you had the illusion of actually hearing the bells of a lost London that still existed somewhere or other, disguised

> or forgotten. From one ghostly steeple after another he seemed to hear them pealing forth. Yet so far as he could remember he had never in real life heard church bells singing. (103)

In a moment sometimes held to ridicule for its sentimentality, the linguistic materiality of the rhyme captures Winston and transports him not into the past but into what he calls a "lost London." This affective or druglike hit of transport, this blocked immersion in a lost series of objects, leads him to continue in what he names "his pursuit of beautiful rubbish," his idiotic desire to rent the room, buy the engraving of St. Clement Danes, and carry it home concealed under his jacket. On the way, he starts to hum "Oranges and Lemons" to an improvised tune—not quite his own, but some partial mimesis of Mr. Charrington's rendition (104). The moment is shattered by the appearance of the glaring woman from the Department of Fiction walking down the street. Winston "was too paralyzed to move" (104).

As the novel progresses, the rhyme "Oranges and Lemons" reappears at key points allied to an encrypted past that might set all free. Julia, Winston's lover, his betrayer, and the woman he betrays, knows part of the rhyme (her grandfather taught it to her) and even remembers seeing oranges even as she can't quite keep them or their color straight—"they're a kind of round yellow fruit with a thick skin" (153).[75] Winston recalls a lemon—"so sour that it set your teeth on edge"—and he takes their sharing of this partial knowledge or citrus memory as "two halves of a countersign." All comes to founder, however, on the fact that O'Brien, Winston's interrogator, knows the rhyme in toto, including the last line, "'When I grow rich,' say the bells of Shoreditch" (186), which empties the rhyme of any escape or reactualizing possibility, signaling also that the fragments of the past, all its "beautiful rubbish," might equally be deployed by the state to keep tabs on and eventually to ensnare those who stray.

In the end, "Oranges and Lemons" stands in contrast to or is countered by the compromised lyric of another nursery rhyme—"underneath the spreading chestnut tree," revised by the telescreen and intoned by a lobotomized Julia, "where I sold you and you sold me" (307). There is nothing stable or safe about the aesthetic, the novel seems to say; quite the reverse: whoever controls the sites of enunciation, such as the telescreen, the clocks, or the bells, will write the end of the story. Winston's antiquarian rebellion and turning collector of lost forms augur no escape even as it might offer consolation. "Beautiful rubbish," it turns out, was merely one further mode of the state's biopolitical articulation and profiling of its citizens—part of its gathering less of biometric than bioaesthetic data. The bleak beauty of the novel is that it offers no alternative. Colors die. The world is once more without oranges. We do not know now what they were like exactly. Still, as Mr. Charrington tells Winston in passing, "Oranges and Lemons" was "a kind of a dance." Winston's collecting of "beautiful rubbish," of forms and objects that sit mutely still, might yet have within it the kinematics of another dance that redistributes the vegetal being that inheres to the forms of life it curates by taking or apprehending orange differently.

What might happen, for example, if the infrastructure of orange I have described thus far simply ceased to be? What would happen if the fruit were suddenly recoded, cordoned off as a biohazard, designating less gold you can eat than gold that kills, less an orange ecology than a citrusy crime scene? What if oranges were no longer cute but deadly, even venomous? What if there was only one edible orange in the world—an orange on which the whole world turned as if in parody of the great solar eye?

AN ORANGE OUT OF SEASON, 1997

In *The Tropic of Orange* (1997), Karen Tei Yamashita imagines just such a world. The novel begins in Mazatlán, where there exists an orange grown on a tree transplanted from California. It figures a hybrid—the problematic of the deracinated immigrant become emigrant. "The tree was a sorry one," we learn, "and so was the orange. . . . It was an orange that should not have been. It was much too early."[76] Attached to this "abberant orange—not to be picked, not expected, and probably not very sweet" (11), is a "line—finer than the thread of a spiderweb—pull[ing]" on it "with delicate tautness" (12). This line connects to the Tropic of Cancer, synchs space-time to this unseasonal orange. One day the orange falls from the tree and rolls "to a neutral place between ownership and the highway" (13), where it is picked up by Arcangel, a magic realist angel-trickster-performance artist let go from some Gabriel García Márquez story (71). Arcangel's on his way north across the border with a post-NAFTA California friendlier to pairs of Nikes than theoretically free persons, and he takes the orange with him, causing Mexico and the United States to fold upon one another. People's TVs go all wonky. They get more and fewer channels. Spanish replaces English, English, Spanish. The cries of the city as broadcast by news radio in place of bells get stuck, looped in endless repetitions. Time collapses to an instant.

Meanwhile a consignment of lethal cocaine-concealing "death oranges," "spiked oranges" (139–40), produced by a team of Brazilian drug dealers, wrecks the world market for oranges—which are now too lethal to consider eating. An altogether different polity of actors—those human animals who trade in coca derivatives—nests its own structures within the streams of orange that move more freely than human persons across national borders. Arcangel's illegal immigrant orange becomes one hot commodity, finds itself singularized as a piece of "beautiful rubbish," worth a fortune, perhaps, if it is one of the cocaine-bombs, or maybe, even, simply as an orange, given how rare they have become. One of these bombs goes off, causing a SigAlert of epic proportions on the 10 Freeway, turning it into a city where the homeless set up camp in a utopian cessation of happening. The novel merely describes and establishes the situation, builds several worlds (in small) via its parallel and sometimes intersecting character-bound narratives—seven in all, unfolding over the course of seven days. It ends with a knockout, drag-out, Sunday glam-wrestling contest between Arcangel and the cyborg SUPERNAFTA while the LAPD storm the 10 Freeway to restore their brand of law and order.

The climax of the novel comes when Bobby Ngu, who's "Chinese from Singapore with a Vietnam name speaking like a Mexican living in Koreatown," who peppers his sentences with "that's it" (15), stumbles late on to the wrestling match in search of Sol, his son, and Rafaela, his wife, who's in the ring. Bobby doesn't know about Arcangel's orange. And Rafaela's yelling at him to cut it—"Cut what?" (267)—"Cut it now!" Rafaela feeds the orange to a dying Arcangel and the line that was attached to it flies free. Arcangel "chews and smiles," for "it's all over. Crowds rushing in." They pick him up, take "him away with orange peels scattered on his chest, stink of oranges on his lips, like he's floating on a human wave."

They take him home—home "where mi casa es su casa. Bury him under an orange tree." Repatriation, the novel seems to say, happens all the time, everywhere. "Tag it good." Meanwhile the lines attached to the orange fly free and "little by little the slack on the line's gone. Thing's stretching tight. Just Bobby grabbing both sides. Making the connection." And it's hard work—"pretty soon he's sweating it." And it hurts—all the labor of holding on, of holding this infrastructure together, hurts; it kills; "lines ripping through the palms. How long can he hold on? Dude's skinny, but he's an Atlas" (267).

But that's not going to be the end to this story, not this time, not this iteration. There's no Hercules hiding in the bushes ready to pull a bait and switch. There's no sleight of hand that allows a theft to become a gift by and through the obliterated or comprehended labor of another being (orange, animal, or Titan). Bobby does not become Atlas even though he is one. He will not be crushed by the weight, the "massiness" that his forgotten labor enables to keep on rolling. He's not going to get "split in two." He's not going to "hold on 'till he dies, famous-like." Instead, Bobby sees Rafaela and Sol, "his little family," "stuck on the other side" and despairing of all these "lines"—"What are these goddamn lines anyway? What do they connect? What do they divide? What's he holding on to?" (268); Bobby "lets the lines slither around his wrists, past his palms, through his fingers. Bobby lets go. Go figure. Embrace. That's it." Atlas no longer shoulders the world. He puts down the globe. "That's it." Bobby knows no more than he knows, lacks access to the whole even as its lineaments flicker in and out of view. But by this letting go, his hands are freed to other labors than those of co-opted mediation. Embrace. You might say, then, that with its multiple narrators and vanishing point of an ending, Yamashita's novel attempts to wean us off the need for an Atlas, the muscular mediation that holds the world in place or, more properly, that enables a synoptic God-like sense of the whole, a world picture. Bobby lets go and the lines of one infrastructure unknot themselves. "That's it." Who knows what will happen now?[77]

The future that Yamashita's novel proffers remains hard to read, unavailable to view. Instead, time collapses to an apocalyptic orange event that seems to inaugurate or to admit that the common has to be constituted, made, and lived; otherwise it resides in nothing more than a series of instants or disconnected moments that it is the "life" and "death" of finite beings to connect. In the novel, each character's narrative line never exceeds the perspective it imagines. This multiple, character-bound mode of narration rejects the synoptic power of the worldview. It refuses to yoke its diegesis to the back of one character, choosing instead to travel with many and so to limn several worlds, worlds that it understands to be partial, plural, and imperfect. In doing so, Yamashita interrupts and reorients our infrastructures; renders them sensible; demonstrates their relative fragility, their apparent resilience, and recognizes the everywhere existence of possible, if precarious, heterospaces: here, now, then, and to come. And such instantaneous, time-bound occupations of place or thefts sponsor the possibility of a more hospitable commons in which car horns become musical instruments and freeways communities (which, in a way, they already are—if poorly so).

Ursula K. Heise calls Yamashita's aesthetic, the novel's leveling or weighing of the value of *things* and persons, a "junkyard ecology" that weans itself off the conjoined lure of unalienated labor and unmediated access to Nature.[78] But, even in the novel, our infrastructures remain—even as they momentarily buckle. Consumer capitalism is not defeated or sent packing by the terrorist cocaine oranges even if they do momentarily interrupt its mechanisms. All that happens is that the elaborated base of industrial capitalism works double time to substitute passion fruit as the breakfast beverage of choice. "Talk Show [host] Tiffany" doesn't "miss a beat: 'That's right. Passion fruit is all the rage. Minute Made is selling it under the name, *Passion*™. Make the change now. *Passion*™'" (141).[79] Overnight, purple becomes the new orange—rewriting the visual field that we saw built in England in the sixteenth and seventeenth centuries. Everything and nothing changes; juice glasses flush purple instead of orange.

As Pietz might chime in, it takes something more to challenge the mechanisms of global capitalism—something on the order of a radical reappraisal of worth. And it is for this reason that the same networks prove less resilient to metaphysical or aesthetic attack by the one remaining orange in California, the singularized, "aberrant," unusable, or purely aesthetic "un-seasonal" orange, which says "no" to any use-value and proves inedible even as Arcangel still eats it. The collisions, the careering dance moves and tableaux it produces as he carries it northward, choreograph less a series of utopian cessations so much as a disorienting rewiring of affect that produces the possibility that something might happen differently, today, tomorrow, or is happening now, or already has. Perhaps you just missed it as you mourned your missing OJ and took your daily slug of *Passion*™. "STEAL THIS BOOK!" Or, maybe, steal *this* orange, steal orange itself, let go of all that the ubiquity of "orange" become commodity anchors, and reimagine the text of labor, of theft, and of gift it embeds. What would it mean to sacrifice the "for us" that attaches to the mobility of orange and agree instead to be the vectors to which it addresses itself? How shall oranges roll now? Where will all the weight or mass, that Venus attaches, go? What then would it mean to look upon and to like oranges, gold you can eat, gold that grows?

More than anything in this book, thus far, I have attempted to foreground the way forms of language and forms of life prove coextensive. Allow sheep to alter our understanding of them and you allow for the possibility of understanding the animals we name "human" differently, for the possibility of other categories of being and sociality to emerge from the cowriting of both. So also with oranges; reimagine the fruit as an inhuman zoopolitical or biopolitical address, an asocial vector that nevertheless cohabits with and configures different polities of animate beings (human and otherwise), and you alter the stakes to something as mundane and apparently unmarked as a "liking" or a taste, to what it means to eat. One important strand of cosmopolitical thinking begs us to attend to and so to alter the figural, material–semiotic routines by which worlds remain the same or may come to move differently. And as my tracking and tracing of sheep and oranges have shown, these routines come rooted always in historically particular circumstances that play out at radically different scales of being. Each produces a very different scenography keyed to differing levels of perception and affective response that localize particular ethical and political demands or fields of possibility. Attending to these relays allows us to toggle back and forth between the affective modes in which we consume and the objective conditions that render the things we love mobile. Pastoral scenes of *otium* offer us an occasion for rethinking our sheepy co-relations. Troubling scenes of orangey theft or exchange offer the possibility of rezoning our relation to vegetal being.

Winking in and out of view, hidden, as they are in Yamashita's novel, in plain sight, these scenes allow us access to what might be described as a set of connecting forces that animate, that convoke, and that do so by way of keeping some *things* still and by setting others in motion. These connecting forces are the subject of my last chapter, keyed as they are to bread and stones, to yeast, to *things* that bubble.

III
YEAST

5

Bread and Stones (On Bubbles)

> All history is the history of animation relationships. Its nucleus, as certain anticipatory formulations have hinted, is the biune bond of radical inspiration communities.
>
> —PETER SLOTERDIJK, *Bubbles,* vol. 1 of *Spheres*

> I immersed myself in contemplation of the sidewalk [*Pflaster*] before me, which, through a kind of unguent with which I glided over it, could have been—precisely like these very stones—also the sidewalk of Paris. One often speaks of stones [*Steine*] instead of bread [*Brot*]. These stones were the bread of my imagination, which was suddenly seized by a ravenous hunger to taste what is the same in all places and countries.
>
> —WALTER BENJAMIN, "Hashish in Marseilles"

At the end of my last chapter, courtesy of the unseasonal, aesthetic, orange and Brazilian, cocaine-spiked oranges of Karen Tei Yamashita's *The Tropic of Orange,* our infrastructures buckled. Time and space collapsed. The world, or one version of it, seemed momentarily to "fit" or stutter such that its undergirding patterns and routines became knowable, available to view. Structures—the very notion of structure—rebounded. Overnight, the risk-managed routines of consumer capitalism kicked into high gear and substituted passion fruit for oranges. The next day, glasses all over the world flushed purple instead of orange. But, nestled within this reassertion of routines, still other worlds, other yet to be imagined scripts for being, beckoned. The novel did not grant us access to them—far from it. It testified to their existence but offered no more. "That," to reprise Bobby Ngu's laconic tag line, was emphatically "it!" The End. Burst bubbles. The future in Yamashita's novel remains, as it does for Karl Marx in *Capital,* resolutely blank, something not yet, and very precisely not to be imagined, so that it may not be filled in, before the fact, by scripts that turn out to be uncannily familiar. Instead, the novel stages a more literal, terminal decision, or "cut."[1] Such futures that we may imagine for its characters are the matter of the breath they share but that we do not, the worlds or bubbles they inflate.

Just as the novel lets go of its oranges, cuts and eats the singularized unseasonal, transplanted orange that heads north of the border, obliterates "orange" itself, the fruit become a tropic weapon or work of art that exhausts itself in the process, so must I. Good-bye orange. "Orange" names the price the novel pays for the foundations it opens and the futures it does not close. The scale shifts from *things* that move to the relays or routines that keep *things* in motion, the lines of force that Bobby Ngu allows to slip through his fingers as he lets go. Thus far in this book, I have tracked a series of anthropo-zoo-genetic figures or tropes that derive from our co-making with sheep, other animals, and the reproductive technologies of plants (oranges). The story I have told is one in which animals (human and otherwise) find themselves differentiated with regard to a grounding plantlike *phusis,* or figure of pure growth, the animate existence and labor of each managed in proximity to a more basic still vegetal being. Weird or strange archives that we are, the multispecies basis to our lives comes filtered through discourses and concepts (pastoral, *otium,* utopia, cancellation, gift, theft) that titrate this vegetal sameness and animal being so that so-called higher forms of life appear somehow less vegetal, less like the living plant "stock" of the herd or flock. These distinctions prove of shallow foundation, however, for they remain haunted by their residues, hostage to the very beings they seem to disown. What emerges instead, then, is a more differentiated biopolitical field sensitive to the vast array of multispecies alliances, mobile polities, asocial and parasitic (non)relations, that set different groups of plant and animal actors (far beyond and without reference to the human animal) against one another or make them compeers in constructing a world in common.

In this chapter, I examine the anthropo-zoo-genetic figure or trope that, for us, designates the connecting forces that Yamashita's novel seeks to access, forces that underwrite our notions of infrastructure, of shelter, security, sustainability, and that do so with such power that the organic, the time bound, the ephemeral, appears solid, become a ground, or foundation, a not quite but almost inert mass, as if mineral, as if earth itself. My subject is bread, the yeasty confection that we derive from the action of a fungal plant-animal that ferments, bubbles, inspires, or breathes. In our hands, this bubbling causes a paste of flour and water and sugar and salt to become animate, to rise, and, this baking, by its serial repetition, by its cascade of minute variations, its inscribing of time in and as a dough that rises, and that rising's arrest by the application of a heat that freezes, creates the appearance of solidity, mass, and so foundation. Bread serves as a congelation of labor and matter, a translational node that anchors our notions of sameness, predictability, and permanence. Or, rather, bread's serial difference in sameness localizes and punctuates. It transforms the precarious finitude of being into something sustaining. Obviously, such stability that bread affords requires an inordinate amount of labor, an extraordinary deployment of resources. My aim is to inquire into these metrological processes and the material-semiotic figures they anchor, figures that trade on the appearance and maintenance of routines that keep worlds in place and so manifest as if they had the referential heft of stone.

Here it might be worth observing that this chapter inhabits a tension between bread and stone even as it treats their apparent difference as superficial, the matter, quite literally, of how each anchors notions of a ground or grounding principle. If, as Tim Ingold offers, we tend to think of the ground as an inert surface upon which we construct our infrastructures, then the punctuated animation of bread enables us to recover the way what we call the ground describes not a blank slate but an "insterstitial" matting or weaving of strata, an interface, "the most active of surfaces."[2] Indeed, this interactive, process-oriented notion of the ground as a surface that must be continuously generated might best be rendered not as a noun but as a gerundive "surfacing."[3] The story of the West, as Ingold tells it, amounts to the process by which the ground we tread is paved over with a literal and conceptual "hard surfacing" that renders it as if an instrument. He offers the experience of walking, of attuning yourself to the lay of the land, to its motions, and variability, as one antidote to the inertia of pavement.[4] I offer the differently kinaesthetic experience of bread and baking, of attending to the peculiar alliance of plant, animal, and mineral actors that give us our daily bread as another trope or *topos* for inquiring into the status of the material, of what quite literally comes to *matter.* Though, not coincidentally, many of the instances I inhabit derive from an ambulatory touring of city streets, both paved and dirt. The city, the matter of the city, of composing the common, is a *thing* of stone and bread and, as you shall see, bubbles.

Matter itself might be said to consist of differently scaled and differently timed bubbles, swellings, occlusions, semimobile, time-bound, fragile, yet strangely substantial and enduring masses, negentropic eddies, whose movement, at certain scales, appears to have been arrested, but whose animation, if you are able to perceive such things, might, once more, become visible, audible, knowable.[5] Matter consists, as Michel Serres offers, of a "succession of automorphisms," folds, inclusions, pockets—the same is true of stone and the tissues that make up your body. Not coincidentally, the example he likes best to demonstrate this process is bread or bread making. "What does the baker do," he asks, "when he kneads the dough?"[6] "At the beginning," he elaborates, "there is an amorphous mass, let's say a square. The baker stretches it out, then folds it over, then stretches it out and folds it over again." The process continues; the square remains the same but gains in complexity. "Time enters into the dough" almost as if the kneading were a type of writing, the dough a substrate that takes time "prisoner."[7] This kneading designates what Serres calls "an exemplary gesture," a gesture personified by the baker but that is hardly "human," for it describes also the stratification of layers in stone, the folded forms of matter. Even in the kitchen, the process continues in the baker's absence. The dough has both to rest and proof. The yeast activates; eats its carbohydrates; emits or dumps its carbon dioxide; exhales into the dough whose kneaded complexity captures the gas, forming bubbles, expanding and so rising in the process. Wait too long, neglect the process, and the dough will deflate; the bubble will burst, the gas rejoining a general atmosphere. The yeast eats its sugar. We capture its breath. And via the technical application of fire, we turn ceramist and cook the dough. We cremate the yeast, its exhalation become the smell of bread. The oven mimics the sun; both activates and kills; arrests the process, leaving us with the fossilized or sculpted bubble we name bread, a strange archival remnant that captures the breath of a fungus, the action of its exhalation on a medium composed of flour and water. Always different, always the same, that differing sameness describes a set of limits, limits that shape an existence. The yeasty bubble that is bread inflates, and we are buoyed by it. Or, from time to time, some of us mark the precarity of existence by leaving out the yeast, by failing to capture its breath. We bake unleavened bread; the burst bubble biscuit rations of shipboard life or the unrisen, sun-baked matzah of passing over.[8]

If "all history," as Peter Sloterdijk writes, "is the history of animation relationships" and "its nucleus . . . the biune bond of radical inspiration communities," then what might there be to be had from an orientation that gives itself over to the autonomic facticity of breath and breathing (human and otherwise) as a unit of analysis?[9] Such an inquiry operates at a radically altered and altering scale, attentive to the way the multispecies associations I have tracked thus far find themselves managed in reference to the worlds of only sometimes visible, microbiopolitical actors that inspire or breathe life into this "us" that they help also to differentiate.[10] Here bread and bread making stand as one privileged, historically deep structure or axiomatic, in which a yeasty exhalation breathes life into the bubbles or techniques for capturing breath that we name bodies, individual and social, coming therefore to metaphorize what it means for us to "incorporate." Bread anchors our notions of collectivity. As Donna Haraway reminds, in a gesture that discloses the stakes to her multispecies redescriptions of the world, the word "*companion* comes from the Latin *cum panis,* 'with bread.'" Whatever their scale, whomever or whatever they eat, "mess-mates at table are companions."[11] A Eucharistic or sacramental poetics undergirds the figure of the multispecies. Multispecies modeling or description aims to open restricted scenes of the table to consider the array of possible companions. It rezones the convoking function of a Eucharistic sign across the blurred lines of kingdom and kind, decoupling that convoking function from any stable order of shared community as it goes. Convocations multiply. For "bread" (and wine) read information, however coded or mediated—what information, what "breath," do we share with others or do we take from them? On what basis, on what altered sense of foundations, may we configure a shared atmosphere? What exclusions does this convocation require? What are their costs? How might we rezone the metaphors that attend on our capture and captivation by yeast's exhalations, metaphors that underwrite the Eucharistic suturing of matter and signs across and among bodies, human and ovine, that I have tracked in earlier chapters? What would an altered or extruded sense of the universal as an aporetic, holey, ongoing proposition look like? How, then, to retain, perhaps, the convoking function of such a significant object as "bread," but in a way that renders our collectives porous?

The stakes to this wager or exercise could not be higher. As Sloterdijk puts it, the sphere or bubble designates the definitive problematic of the human tie to "morpho-immunological constructs."[12] "Being in the world," he offers, "means being in spheres. . . . The symbolic air conditioning of shared space is the primal production of every society." A precarious ethics and politics seek to unfold from this insight—for if your breath was not your own to begin with, how could you expect to (with) hold it? Breathe you must; breathe you shall. "Living in spheres," he elaborates, "means inhabiting a shared subtlety," a subtlety he seeks to recover, re-reading the second creation story of Genesis (Genesis 2:4–25) as a neutral "statement about a production process . . . a procedural insight" that records a technical schematic on how "man is an artificial entity that could only be created in two installments."[13] First he is molded: "Adam—the clay creature taken from the soil, *adama*— . . . [is] model[ed]" as a "statue." "The creator" becomes "no more than a potter," a ceramist, or, in Serres's terms, a baker, who makes a hollow vessel into which is breathed the breath of life. The story describes a "two-phase process in procedural terms," a recipe or "anthropopoiesis" in which "breath [becomes] the epitome of a divine technology capable of closing the gap between the clay idol and the animated human with a pneumatic sleight of hand."[14] What appears, then, as an "unbridgeable hierarchical divide . . . an ontological difference between creator and creature," becomes, instead, a "synchronously interchanging relationship between two breath poles," in and out, out and in, creature and creator a mutually animating sharing or exchanging of breath—a "double echo" or, once upon a time, a double articulation. And this insight augurs for what Sloterdijk hopes might become a material–semiotic building block for a "radical resonance" between beings that would demonstrate the way "real subjectivity consists [always] of two or more parties."[15] Never one being, animation requires always two or several. How, then, to construct mutually sustaining bubbles that own the air they put to use and attempt to sustain others by and through their exhalations? How to deploy the bubbles we are in ways that enable our companions also to bubble?

Sloterdijk works toward a theo-technical insight that would proceed on the basis that "what we call historicity is nothing but the time required to repeat God's trick [of inspiration] through human ability."[16] The question concerning technology, as it were, might be revisited productively, he offers, by way of an ongoing inquiry into networks of inspiration, into bubbles, or what elsewhere he names an anthropotechnics.[17] What modes of translation do we deploy to tether these bubbles? What congealing of labor and materials is necessary to keeping *things* still or creating the appearance of stillness and foundation? What bubbles might it yet be possible to blow that would not prove so lethal in their enforcement of a border or boundary? I proceed slightly more cautiously, perhaps, sensitive to the way the word *bubble* may come primed by a desire to find a mode of parasitic relation become stable symbiosis that eludes the eventuating of a biopolitical field, the "decision" or "cut" of Dick the Butcher's knife with which I began this book. That said, I do so with the conviction that the way forward lies in inquiring into those bubbles that form between and among differently animated entities under the sign of "air" or a general atmosphere, even as those relations, as in the case of yeast and the highly differentiated field of animals we name "human," may prove purely coincidental, parallel worlds, if not even parallel play, and so might complicate and enlarge our notions of community. It seems important to entertain the possibility of multiple, sometimes compatible, sometimes vying "morpho-immunological constructs" across the wandering lines of species. Hence the yeastiness of my inquiry into the impetus or efficacy of differently mediated *things* that bubble to their own end, and that do so to the benefit of a multispecies that knows no relation but whose points of contact may still prove significant.

It may prove useful, then, to understand this chapter as an attempt to remember something that Luce Irigaray claims Western metaphysics takes pains to forget. "Metaphysics," she writes, "always supposes, in some manner, a solid crust from which to raise a construction."[18] *Logos* unfolds always as a "topo-logic" or topology, a "fabricated shell" that takes itself as if solid (9). Being, on her estimation, presupposes a "groundless ground," something more and less than a substance, something ubiquitous but unlocalizable, something so fundamental as to remain unthought. Don't hold your breath—she means air, that which surrounds and which makes life possible. It's not that metaphysics does away with air completely. On the contrary, air returns in various guises. It has to be continually evacuated from thought to keep our foundations secure. Indeed, from Irigaray's point of view, a metaphysician becomes something of a songster, "a trafficker in airs," or numbers, whose basis "remain[s] unthought by him" (6). What would it mean to think air? "Is air [in fact] thinkable?" (12).

You might say that in this chapter I attempt to recover these lost snatches of songs or lyrics, scripts for being or, better, breathing. I take as my focus the "biune bonds" that form between yeast and the animals we name "human," bonds that proceed by way of the bubbling that yeast activates, an aeration that enables certain doughy configurations of matter to convoke and sustain the sheepy, orangey worlds I have described. In what follows, I enter into something like a pneumatic pact with those writers who have attempted to think with bubbles, who orient themselves to the animating breath of yeast, a breath so powerful that, figuratively, its products take on the referential heft of stone. I make no particular claim with regard to the conditions that led these writers to choose bread (or why bread chose them). Their decisions to do so were for reasons of trauma, joy, boredom. For some bread manifests merely as a seeming afterthought or passing remark not quite worthy of attention and yet still requiring it. The archive I assemble is discontinuous. My method, therefore, may appear a little precarious, a bit like blowing bubbles whose drift renders me giddy, inflated by the scenes of "being" excorporated they sustain; beside myself as they confuse the relationship between the giver and taker of breath; bereft and melancholy as they burst. Walter Benjamin's "Hashish in Marseilles" suggests the *topos* and topology that give this chapter its title. It's he who insists on the convergence of bread and stone or stones as coeval things. I begin, then, in Marseilles, by drawing breath from Benjamin's hashish-fueled, prose exhalations.

"BREAD AND STONES"

Marseilles. July 29, 1932. Walter Benjamin has been out walking. Or, at least, he says he has. "At seven o'clock in the evening after long hesitation I took hashish," he writes, secure, here, "in this city of hundreds of thousands" where "no one knows me . . . of not being disturbed."[19] But Benjamin is disturbed. He lies on his bed. He reads and smokes. He cannot settle. "Three quarters of an hour" pass but turn out only to have been "twenty minutes." "Opposite" him, all the time, the "view of the belly of Marseilles" beckons. Breaking the protocol of his drug trial, he leaves the hotel, the effects of the hashish "nonexistent or so weak" that "the precaution of staying at home" seems unnecessary. Out of doors, he floats through the city; he quickly loses his "feeling of loneliness," which is replaced by a "certain benevolence," an "expectation of being kindly received." "The nausea disappears"; he takes "special pleasure" in his walking stick. Everyday objects exert an absolute fascination upon him.

One episode of several in Benjamin's experiments with drugs and drug effects in prose, "Hashish in Marseilles" unfolds as a series of self-observations that attain an objective cast as Benjamin works through and over the night before. "One reads the notices on urinals," he generalizes, "becomes . . . so tender" that "one fears that a shadow falling on paper might hurt it." Buoyant with the drug, buoyed by its effects on his sensorium, Benjamin exhales sentences as his imagination gorges itself on what he shall name the bread of stones. Each insight passes before his eyes as if a bubble, in prose, and as it bursts, he blows or writes another, and another, immersed, absorbed, in what the drug makes possible, thinkable, just by breathing or writing. High on the ecstasy or joy of inflation, of his being "excorporated," he finds himself outside of himself, in the city, and then in prose. And this animating buoyancy captures the scene of invention, of possibility, of divine-like efficacy, of creation. As Sloterdijk explains, "while exhaled air usually vanishes without a trace, the breath encased in" bubbles "is granted a momentary afterlife. While the bubbles move through space, their creator is truly outside himself."[20]

Benjamin drifts with *things,* his hashish-enabled prose a matter–metaphor slip and slide in which subject and object merge. It does not matter if he is telling the truth or if he even remembers what it was really like being high; his words translate the night before so that his sentences register what it means for language to bubble.[21] The "hashish eater's demands on time and space" are "absolutely regal," he purrs. "A wonderful beatific humor" rises within him that "dwells all the more fondly on the contingencies of the world of space and time." He feels this "humor infinitely" when he's told that the kitchen at "Restaurant Basso" has "just been closed." Still, sitting at the bar, he shall "feast into eternity" on the "view of the old port." The meal comes later. For now, he sits and watches the passersby; considers their faces; plays a game "of recognizing someone I knew in every face" (675). Sometimes he thinks he knows the name. Often, he does not. Benjamin comes over all hungry; orders oysters but is told that none are left; works his way up the menu; finds himself "on the point of ordering each item, one after another," but seizes on a "paté de Lyon" instead. "Lion paste," he muses, "with a witty smile, when it lay clean on a plate before" him (676). As hungry as a lion, he feasts on one, plans to "go elsewhere and dine a second time." Though, when it comes to it, he does not. Instead, he orders a "half bottle of Cassis . . . piece of ice . . . floating in [his] glass." It "went excellently with my drug," he observes; it subdues his hunger, provides a soothing *digestif* to his attack of the "meta-munchies."[22]

Benjamin hums to himself; contemplates the solitude of his drug trance; "meditate[s] on Ariadne's thread"; concludes that "under the influence of hashish," we become "enraptured prose-beings raised to the highest power" (677).[23] And it's then that the desire takes him, welling up first as a "deeply submerged feeling of happiness" that won't quite translate, that won't quite generate within the mimetic stratum or medium of his writing. Luckily, he has a newspaper to hand, in which he finds the following sentence that localizes the feeling, gives it shape or allows it to take root. "One should scoop sameness from reality with a spoon," he writes, lapping up this newsprint elixir that distils the essence of the night before. He remembers another "sentence that had pleased [him] very much which appeared to say something similar" several weeks ago—most probably from an "exotic novel" by Danish writer Johannes Vilhelm Jensen: "Richard was a young man with understanding for everything in the world that was of the same kind."[24] Fusing with the lines from the newspaper, with the memory of hashish, with what remains of the drug itself become rhetorical vector or shifter, this sentence, also summoned from memory, leads Benjamin to the following conclusion:

> Whereas Jensen's sentence amounted (as I had understood it) to saying that things are, as we know them to be, thoroughly mechanized and rationalized, since the particular is confined solely to nuances, my new insight was entirely different. For I saw only nuances, yet they were the same. (677)[25]

That night in Marseille, or the morning after, high on hashish, or on the rhetoric of drugs, Benjamin found his whole orientation to the world of things, to the regime of description by which we render the infrastructures of our worlds, altered. Everything now has nuance; everything stands forth and out, wears its depths outwardly; and yet, everything is the same—this sameness revealed to be a product of all these minute particulars, these nuances. Benjamin "lives the nuance."[26]

Pricked by the details, by the rarified particulars or elements of *things* in their emphatic, familiar, objectile strangeness, a strangeness he renders by way of a certain mobility of language or sliding of words, Benjamin taps into the pulsating, animate stillness of what passes as Marseilles, a general or extended text from which local particulars phenomenalize.[27] He immerses himself in the sidewalk [*Pflaster*] that hashish renders slick with associations, and his vision escalates. He is in Marseilles. But he might as well be in Paris. The stones of the sidewalk vehiculate, their nuanced, metaphorical sameness become an instantaneous transport. "One often speaks of stones instead of bread" (677), he observes, "these stones were the bread of my imagination, which was suddenly seized by a ravenous hunger to taste what is the same in all places and countries." Bread and stones, stones and bread—the two become conjoined figures of "what is the same in all places and countries," their anchoring identity, their medial incompatibility, building blocks of an infrastructure along which Benjamin's drug *otium* glides.[28]

"Two figures (citizens, vagrants, what do I know?)" pass him on the street as "Dante and Petrarch.'" "All men are brothers," he concludes. "'Barnabe,' reads the sign on a streetcar" that stops "briefly in the square . . . and the sad confused story of Barnabas [stoned to death] seemed to [him] . . . no bad destination for a streetcar going into the outskirts of Marseilles."[29] Leave the city; exit the hubbub; and the stones that become bread to your imagination shall turn stony once again, their mass increasing as you leave town. No more feeling of "benevolence," no more quasi-beatific vision or Eucharistic plenitude deracinated from its host to the streets; get on the bus with St. Barnabas, and you shall find yourself caught in the gravity of a reference that brings you down. The bus leaves. Benjamin lingers out of doors, watches the street life; "a Chinese man in blue silk trousers and a glowing pink silk jacket," the "girls display[ing] . . . themselves in the doorway" (678). He thrills at the "thought of sitting here in a center of dissipation," though by "here" he "mean[s] . . . not the town but the little, not-very-eventful spot where I found myself." There he remains, listening to the music "which . . . kept rising and falling . . . the 'rush switches of jazz.'"[30]

The trance "abates," and he ends by offering that "when I recall this state, I would like to believe that hashish persuades nature to permit us . . . that squandering of our own existence that we know as love." "For if, when we love," he continues, "our existence runs through Nature's fingers like gold coins that she cannot hold and lets fall so that they can purchase new birth, she now throws us, without hoping or expecting any-thing, in ample handfuls, toward existence" (678).[31] Part gift, part hazard or risk, if not quite a pure expenditure, the efficacy or technicity of hashish as a substance that plays the senses produces something like love. But, whereas love gives our existence to Nature in the form of a golden currency she cannot quite cash out, that overflows her fingers as well as Benjamin's attempts to personify her, so that it might "purchase new birth," with hashish it is we who are thrown, gratuitously, "toward [an] existence" that comes to seem like an order of pure production, a facticity that both reduces and raises our awareness, so that all has nuance, everything is the same, and *things* turn upon their names to become other *things.* Played by hashish, played by the chemical effect of a plant on the biomedia of the animal we name "human," the "hashish eater" exists as a figure of transplanted awareness—the vegetal efficacy of the hashish as plant derivative growing within the hashish eater as biological host—who births less a creature or a life than a sentient form of vegetal being that ambulates across a field of nuanced sameness, an asphalt from which animate, botanical *things* grow.[32]

Yes. The experience sounds a lot like that of the flaneur who "botanizes on the asphalt," makes lists like a child, collects specimens of city living, or, for that matter, of the collector who gives herself over to her collection.[33] A year or so later, in "On the Mimetic Faculty" (1933), Benjamin will name this orientation the "powerful compulsion to become similar and to behave mimetically," in sympathy with an object.[34] And this insight stands surety for what becomes, in the *Passagenwerk,* an auto-archiving of consumer capital's broken or discarded goods become uncertainly temporal fossils, *things,* that, having lost the sum of their cross-references, offer a perceptual jolt or "shock" to the viewer. But if, elsewhere, Benjamin judges the "task of divesting things of their commodity" status "Sisyphean," the drug high renders that task a rolling stone, animating everything that comes to his attention. The quasi-automaticity of the experience, the efficacy or agency of the drug and the mobility of reference it permits, subjects codes of ingestion and interiority, the idea of an inside and outside, to a continual troping. It sets them in motion. And this motion reveals or renders sensible the general flesh of being whose ovine and orangey involutions or "nuances" I have tracked thus far in this book. No sheep or oranges here though. The scenography Benjamin provides is different. He reanimates the suspended vitality of an infrastructure built from masses whose animate potentiality, whose breath or being, have been stabilized, made to appear as if still.

Bread and stones, stones and bread; Benjamin's winking refusal to know the difference between the organic and inorganic, nuance and sameness, matter and metaphor, designates a point of equivalence that gestures toward the archival density of our worlds, the way the routines that anchor our structures (and shelters) depend upon the manufactured efficacy of mineral, plant, and other animal actors. Our terrestrial grounding relies on this constitutive "as if" that names what is solid and what is not, what endures and what ends. Against these structures or routines, Benjamin posits a mobile, morphing equivalence between stones and bread: the seemingly inert matter or mass of stone that has been worked into a figure of uniform size and so sameness; the sidewalk; and the airy, kneaded, folded, measured, yeasty fermentation and application of fire that cooks the hits of doughy reference we name "bread." And as the range of Benjamin's sometimes oblique reference indicates, this stony or bready translation, if not quite transubstantiation, set in motion by the drug trance, reveals a set of conversions or passages between matter and metaphor that connect or stack terms so that they come to anchor a particular version or vision of reality: stones and bread; Eucharistic feast and feeling of benevolence, fullness, being; but also a stony loss of breath or extinction, an exhalation and expulsion from the city—the death of St. Barnabas become the terminus to a bus route.

Benjamin's drug *otium* or practice of attention tunes him in to the nuance. His "not-very-eventful spot" offers one precarious *topos* from which to begin to rethink what we mean by an infrastructure or ground. But before his bubble bursts, before we get winded, Benjamin's cities have to be built, their streets paved, their bread baked into its reassuring sameness. Marseilles is not the place to begin even if it is the point at which I arrive. Instead, let's rewind to what one city dweller, an ocean away, names a beginning—however "unlikely" that beginning may seem.

ENOUGH IS ENOUGH?

Philadelphia. October 6, 1723, about eight or nine o'clock in the morning, as replayed from some time in 1771. Benjamin Franklin has been out walking. Or, at least, he says he has. He's the seventeen-year-old as recalled by the sixty-something polymath, statesman, republican man of letters, and printer. He's quit his apprenticeship to his brother James and just arrived from Boston by way of New York. It's been a long journey: a succession of small boats; a storm; a lot of walking; rain. "I was in my working Dress," he writes, "my best Clothes being to come round by Sea. I was dirty . . . my Pockets . . . stuff'd out with Shirts and Stockings."[35]

Franklin is "fatigu'd with Travelling, Rowing, and Want of Rest," hungry, and his "whole Stock of Cash" consists of a "Dutch Dollar and about a Shilling in Copper"; but that does not stop him from paying for his passage, even over the refusals of the crew, "on Account of my rowing," he informs. But he "insist[s] . . . on their taking it," offering that "a Man [is] sometimes more generous when he has but a little money than when he has plenty, perhaps thro' Fear of being thought to have but little." Having barely enough, uncertain of this new departure, Franklin assures himself (and others) that he has by giving some of it away.

On his way to the "Market House," the hungry Franklin meets "a boy with Bread. I had made many a Meal on Bread," he writes, winking perhaps at his earlier descriptions of the various diets or regimes he has tried out to save money for books or for reasons of principle. He asks for directions to the baker's; heads to the bakery on Second Street; asks for "Biscuit, intending such as we had in Boston." Though maybe he means to have a thin, sweetened, spiced, yeastless, saltless, twice-baked wafer like bread—not necessarily the iron rations of shipboard life.[36] But Philadelphia is not Boston, and there is no biscuit. He asks for a "three-penny Loaf," but "they had none such." So, throwing caution to the wind, "not considering or knowing the Difference of Money and the greater Cheapness nor the Names of his Bread," Franklin asks the baker to give him "three pennyworth of any sort." In Philadelphia, three pennies, it turns out, buys a lot of bread. The baker gives him "three great Puffy Rolls." Franklin is surprised; gets more bread than he bargained for; does not have enough "room in [his] Pockets"; and so walks off "with a Roll under each Arm," "up Market Street as far as Fourth Street" (29), pockets stuffed out, the third in hand as he munches away.

The lesson in moral economy continues, keyed to this sufficiency or surplus of bread. For what should you do if you suddenly find that you have enough of something, more than enough, too much, in fact, to keep? What sort of ethical burden derives from too much, from a surplus? What should you do if when, as a stranger in a new city, you find yourself the unexpected beneficiary of your ignorance of the local cash nexus, and so with more bread on your hands than you can manage? Well, first things first, eating all that bread has probably made you thirsty. Franklin finds himself back at the "Market Street Wharf, near the boat" in which he arrived. He stops for a drink from the river and "being fill'd with one of my Rolls," gives "the other two to a Woman and her Child that came down the River in the Boat with us" who were "waiting to go further." Franklin is all "fill'd" up, has had enough, and so he gives his surplus of bread to those who remain in transit, still in movement, who are not yet arrived. Or, more exactly, he allows the woman and the child to relieve him of his burden. The gesture, the gift, unfolds as a reflex—what should he do with those two remaining rolls so that he can bend down to drink? Give them away—obviously—give them into other hands whose emptiness enables them to lend him a hand. Reciprocity derives from this figure of the more-than-enough or more-than-would-have-been-enough that threatens to become an inconvenient bit too much: the "three great Puffy Rolls" Franklin receives for the same money that, in Boston, buys a "three-penny loaf."

This anonymous young Franklin looks quite a sight as he walks down the street all puffed up and out like the rolls, munching away—though, of course, he is not yet famous, and so no one really knows enough to look. The only head he turns is that of one Miss Reed, his wife-to-be, who sixty-something Franklin places in a doorway so that he can put thoughts in her head. He tells us that he happened to "pass . . . the door of Mr. Reed, my future Wife's Father," and that "when she standing at the Door saw me," she "thought I made as I certainly did a most awkward ridiculous Appearance." Still, despite appearances, she marries this anonymous, every-Franklin or not-quite-yet-but-still-to-be-Franklin anyway, the figure he cuts good enough.

Courtesy, then, of the fortuitously placed Miss Reed, who serves as witness, Franklin becomes the occasion for the performance of a carefully rationed and rationalizing sensibility that thinks through and with the conversions between labor (the crew that rowed him—even as he also labored in the rowing), the social value of money (the "Dutch Dollar and about a Shilling in Copper" he has on him), and the obligations that come with a sufficiency of bread. That sense of a sufficiency, of having enough, and so of a bubbling surplus, a surplus that courses through everyday actions such as walking down the street and stopping for a drink, become the occasions for a technology of self that understands and proceeds on its relation to a communal whole or, in the ovine rich language of my earlier chapters, Commonwealth. The figure Franklin cuts, tired, far from home, wandering through Philadelphia's streets, thinking as he goes, offering charity as and when the opportunity offers itself—those opportunities become simple matters of fact, rational imperatives that occur with the spontaneity of common sense—offers something like a Golden Mean that balances the incompatible drives of a profit economy or cash nexus with a Christian moral economy whose yeasty icon led social historians such as E. P. Thompson to name it a "bread nexus."[37]

Writing of the food riots caused by the Corn Laws in eighteenth-century England, Thompson coins the phrase "bread nexus" to refer to the seemingly ancient settlement that rendered "millers and—to a greater degree—bakers . . . servants of the community, working not for a profit but for a fair allowance."[38] Such a paternalist system sought to ensure that cities could provide the daily bread that Christians prayed for, risk managing the supply of wheat and the infrastructure of bread production so as to ensure bread's quotidian appearance—the material-semiotic efficacy that, for Walter Benjamin, if not yet Benjamin Franklin, renders it interchangeable with paving stones. Much more than an attempt at economic regulation, the system whose collapse Thompson investigates "derived in part from earlier Edwardian and Tudor policies of provision and market regulation" (such as I investigated in chapter 1), which aimed to stitch economic policy to a theory of the state or Commonwealth that proceeded on the basis of mutuality and shared obligations, on a Christian humanist understanding of "profit" *(utilitas)* as a communal good.[39] In colonial America, this settlement and its index to baking and bread production was the subject of ongoing and tense revision as different cities adapted the customs they inherited from their English and European citizens, each of whom brought with them differing permutations of the social, spiritual, and physical ingredients that made for good bread and so good civics.[40] It is against this landscape or food/scape that Franklin's arrival in Philadelphia unfolds, an arrival that thematizes the shifting scales of value in Boston and Philadelphia, scales whose translation or untranslatability require an ongoing application of rationality, a circumstantial ethics that responds to what an intention to buy biscuit in Boston becomes when that bread is baked, sold, and bought in Philadelphia.

Beyond inculcating a habitus of quasi-automatic care and concern as a principle of rationality, this reflexive charity signifies also that Franklin has arrived—journey's end. No matter what travels there are still to come, Philadelphia, City of Brotherly Love, shall be home. The city becomes synonymous with these "three great Puffy Rolls," whose yeasty puff signifies a captured and so preserved inspiration, a local atmosphere that makes everything seem possible. The future, as it were, suffuses this present past. Franklin's very being, his reputation, the health and wealth of the city, of America, comes indexed to this joyful, unexpected, gratuitous inflation that continues seemingly indefinitely. Philadelphia is emphatically not Boston, the bread its bakers bake emphatically not the "biscuit" on which Franklin expects to subsist. And this yoking of the autobiographical subject with the recording of the life of the city and its objects signals the way the techniques of self that Franklin experiments with over the course of his life—all those diets, the tabulation of his faults or "errata" as though his life were a book that was being continuously printed and corrected—stand in strict relation to the larger issues of moral economy and statehood that were also his life's work.[41] Moral philosophical precepts, the decisions of a life, prove coterminous with the making of routines that build a world. Virtue, rationality, being, turn out to be media specific. Sixty-something Franklin knows this, and his simultaneously anonymous and hyperindividualized arrival in Philadelphia reads something like a signature to that effect.[42]

The episode presents as a special case—an illustrative object lesson from times past. Franklin takes pains to draw attention to his arrival by noting the shift in the scales of narration, the way in which he is now being "more particular" in his "Description of my Journey and first Entry" so as to put readers "in Mind" (28) of him now, in the present future. As commentators on the episode remind, when he left the city in 1764, three hundred people came with him to the port of Chester, where he was treated to a cannon salute and sung a song to the tune of "God Save the King."[43] All that, so it seems, might be reduced to this beginning, however "unlikely" (28), an arrival the text invites us to reenact, alluding, as it does, to a street map or itinerary that we might follow, which, of course, we can, and tourists do, in the spring and summer months, encountering one or more Benjamin Franklin impersonators or reenactors in and around "Old City."[44] Look, here he comes now (Figure 16).

Figure 16. *Benjamin Franklin with a Loaf of Bread.* Engine panel painting attributed to David Rent Etter, circa 1830. 25 × 13 3/4 in. Gift of CIGNA Museum and Art Collection. Courtesy of the National Museum of American History, Kenneth E. Behring Center, Smithsonian Institution, Washington, D.C.

While the reanimation of Benjamin Franklin by the hosting "wetware" we name actors may have to wait until the twentieth century, in the 1830s, if you chanced to be part of a crowd celebrating the labor, Christian charity, and rational economy of a volunteer fire company, say, the Franklin Engine Company of Philadelphia, you might have seen their prize engine decked out with this panel painting, attributed to David Rent Etter, that immortalizes Franklin as the icon Miss Reed's chance presence at her father's door makes of him.

Roll of bread under his arm, young Franklin's arrival in Philadelphia becomes a badge of civic honor, badge of a project and prospect of civic life premised on a sustainable concept of "enough," however "unlikely [that] beginning." That the episode and its attending iconography are the work of fiction, an impossible thought Franklin places in the head of the woman he marries but who does not yet know him, does not matter. For hovering within this vision of "enough," of a bubble, atmosphere, or world that continues to inflate, that endures, is a sense that the primary obviousness or commonsensicality of Franklin's thirst-driven charity derives from a willed decision to act as if "the fabricability," or the "'fixion,' *both* the fixity and the fictionality, and the constant fabricability of the community and of the plural positions that we can occupy or be made to occupy within it," were true.[45] Live this way; act this way; and what shall the world become? As Jacques Lezra writes, the word *enough* localizes the political precisely because "it is not given . . . that what we are talking about when we say, 'That is enough' or 'That is not enough' is a thing of one or the other sort." The word demands an open inquiry into whether my sense of enough matches your own self-perception: "Was that, in fact, enough?" Is enough in fact enough? "No rules are given, or universally-*enough* agreed," Lezra continues, "allowing us to decide, or even to reason out from first principles, that one thing is the object of practical measure and not the sort of thing that is not."[46]

Take care, then, in the decisions and the cuts you make, in how you delimit the "inspiration communities" with which you understand yourself to draw breath. Inquire into what counts as "enough," and if you find that it is more than enough, act accordingly, stitch that action into your routines so that it becomes routine. Whatever political concepts we inherit from Franklin and this era (insufficient as they are), enshrined as they may (and may not) be in the attending documents of that age, the U.S. Constitution, its amendments, the Bill of Rights, Franklin's derivation of a plural "enough" from the redistribution of a singular too much makes an open inquiry into what counts as "enough" both the ratio and rationality of his city living. In this enlightened Philadelphia of 1723 (or is it 1771?), Franklin, the city, and the Commons, as he models them, stand still to be made by their actions, by the decisions they shall make and the limits of those with whom they agree to draw breath. The bubble inflates. And hidden, in plain sight, funding and founding this vision of an ethics and politics of rationalized and self-regulating sufficiency, are "three great Puffy Rolls" of bread that necessarily pluralize a scene of individual consumption—the satisfaction of an individual hunger the anticipation and answering of the hunger in others. Franklin's "three great Puffy Rolls" are, in every sense, quotidian bread, a figure of miraculous sufficiency transubstantiated into a rational and repeatable miracle. The paving stones, of course, remained still to come—a project of civic improvement for which Franklin, despairing of Philadelphia's dirt on his shoes, would advocate noisily (117–18). If, in the future, this paving would transmute into the conceptual "hard surfacing" that Ingold indicts, then it may prove useful to consider the way Franklin's ambulatory, interactive anecdote communicates the constitutive power of routines in making a world. There remains within this notion of ground and grounding as an activity or process a powerful set of resources for renewing our sense of the common.

But Franklin only takes us so far, helps us to recall the giddy cosmopolitanism that the ever-inflating bubble of one version of America sets in motion, a sense of possibility riven by its willful occlusions. We need to go back further and prepare for a series of burst bubbles. Let's see what happens, for example, if we get on Benjamin's bus with St. Barnabas and try to leave town. How might bread, unlooked for, unwanted bread, anchor something like resistance?

"GODDIS-GOODE"

The Tower of London. April through October 1597, as replayed from the College of St. Omer, in the Louvain, in 1609. Jesuit John Gerard, to whom I introduced you in chapter 3, is in prison in this place of stone. He's found solace in his cell in the Salt Tower, where he spent the first few nights, in the graffiti the martyred Henry Walpole scratched there, leaving his name along with a makeshift oratory concealed in an embrasure. Gerard makes his own mark, lost now among all the initials, inscriptions, and impressions made by the cell's various occupants (Figure 17).

Soon he shall be working on his escape, writing letters in orange juice, sending secret messages to friends and associates in and out of the Tower. Soon he will recruit an orange-fond warder to his side, share his oranges with him, and so make him his factor. Or, that's what Gerard says he did. After his escape to the Continent following the Gunpowder Plot in 1605, Gerard writes a description of his time in England, most probably as a memoir, self-justification, and pedagogical guide. He describes how he was inserted into England, the various cover stories and aliases that enabled him to move around, the dos and don'ts of setting up an underground circuit of safe houses; and he models how to behave in the event of capture, imprisonment, interrogation, and torture.

We join him as he recovers from successive bouts of torture at the hands of Richard Topcliffe, chief priest hunter to the Crown, this interrogation the express purpose of his transfer from his earlier prison, the Clink:

> Left to myself in my cell I spent most of my time in prayer. Now, as in the first days of my imprisonment, I made the Spiritual Exercises. Each day I spent four or sometimes five hours in meditation; and everyday, too, I rehearsed the actions of the Mass, as students do when they are preparing for ordination, I went through them with great devotion and longing to communicate, which I felt most keenly at those moments when in a real Mass the priest consummates the sacrifice and consumes the *oblata*. This practice brought me much consolation in my sufferings.[47]

Gerard performs a dry Mass (Missa Sicca), remakes his body by practicing the *Spiritual Exercises.* He goes through the choreography of the Mass with "great devotion and longing to communicate," which he feels most keenly at those points when his gestures mime the transformation of the Host—the palpable presencing of the inhuman. The gestures of the incomplete sacrament, which effect no transubstantiation, nevertheless serve as a curative *technē,* allowing him to mend his body.[48] At the end of

Figure 17. Henry Walpole's graffito in the Salt Tower of the Tower of London, July 2014. Photograph by the author.

three weeks he is able to move his fingers, hold a knife, and feed himself. He asks his warder for a little money and a Bible, which his friends get for him. He then asks him to buy him "some large oranges," and "as he was particularly fond of the fruit," Gerard tells us, he makes him a present of them, "thinking all the time," he adds, "of another use [he] could put them to." Eventually, these oranges enable Gerard to orchestrate a spectacular waterborne escape from the Tower.

It's easy to miss the bread in this story, or to think that the only "bread" is that of the missing but still efficacious Host *(oblata)*. But there was, of course, bread aplenty in the Tower. Bread, bought or baked at the Queen's expense, was given to inmates daily. "The food [*cibus*]," writes Gerard, "was plentiful—every day they gave me six small rolls of very good bread [*sex panes parvuli sed valde boni*]" (117). These six small rolls signify the passing of a threshold, a sufficiency, something on which Gerard can count, on which his listeners or readers, if they were ever to find themselves in the Tower, might count also. Gerard has enough to eat, more than enough. He tells us that, in fact, he never needed to take supper because of all this bread. The Tower was the only prison in England where the inmates were fed at the Crown's expense, and we know from the bills submitted to the Exchequer for the period of Gerard's imprisonment that he paid no additional supplement for extra services.[49] Accordingly, this sufficiency of food (and its quality) meant that Gerard was free to use what resources he had by way of money to other ends, such as cultivating his orange-fond warder, inclining him to his side, an effort, he says, that eventually would lead to the man's conversion. Gerard derives what he can from the streams of matter that routinely entered his cell, puts them to other uses than they were intended, and this serves as an object lesson in how to manage the terror and trauma of confinement and torture.

The stability of these "six small rolls of very [or strongly] good bread" that appear each and every day, relieving him of the burden of paying for his own food, exists in an exact relation to the more animate orange that almost from the moment it enters his narrative is pressed to use as part of Gerard's curative *technē* and clandestine network building, pressed for its juice, the peel cut and pieced into rosaries that serve as pretext for obtaining paper for wrapping and writing, and so the passing of secret messages. The orange morphs into so many *things,* becomes synonymous, as we saw in chapter 3, with the network effects Gerard produces. Meanwhile the "six small rolls of good bread" continue to appear, each and every day, always the same, their sameness, their quotidian reliability, something worth remarking, but only once. These rolls are in that most literal, prosaic sense good bread, whose "disposing" is a matter of the adumbrated circuits of baking and food regulation in London that ensured that Gerard's flock (reformed and otherwise) received the "daily bread" for which all Christians pray. Gerard remarks them only once because they appear always the same, always "six small rolls of very good bread"—though there were, if you are inclined to count them, more than twelve hundred of them in the seven months he spent in the Tower.

The remarking of these rolls is not quite unmotivated, however. A minimal, prison-based semiotics attaches. "The grades of diet in this prison," Gerard continues, "vary according to the rank of the prisoner."

"The scale," he adds, "is purely a social one, and taking no account of the religious state, it puts first what ought to be esteemed last" (117).[50] Gerard scoffs at his reformed countrymen's valorizing of social rank over religious vocation. They value self-aggrandizing sheep (animals with titles) over good shepherds. But still he likes their bread, eats it, despite their misplaced scales of value; though he implies that he eats nothing more even as he leads us to infer that, if the bread were good, so might be some or all of the food he is offered. If there is a note of reserve to his voice with regard to these bread rolls, it may indicate that he is ill at ease with the fact that in this place of stone, surrounded with the relics of his fellow martyred priests, he enjoys them. Or, perhaps, he is concerned that readers might feel that this bread was, in fact, not merely enough, but a bit too much for someone who should, perhaps, himself have preferred to die than to escape.

What, then, was it that made these bread rolls taste so good? Were they a specific kind of bread: "whigs," sweet breakfast or milk rolls that were coming into vogue in the 1590s; or were they simply bread that has been very well made? The bread flour of sixteenth-century England had only 8 percent gluten, Diane Purkiss informs, as opposed to the bread flour of today, which contains as much as 12 to 14 percent. Less gluten means more kneading. It requires much more manipulation of the dough.[51] Good bread, therefore, requires much more congealed sweat and labor, not to mention the labor and resources of fueling and tending the fires in the ovens that baked it. This "work," as Purkiss takes pains to point out, "was heavy and tiring . . . the baking cellars were dungeons, often tiny, sometimes too small for men to stand up in." She reminds that in *Capital,* Karl Marx wrote that "man is commanded to eat his bread in the sweat of his brow, but Londoners do not know that he had to eat daily in his bread a certain quantity of human perspiration mixed with the discharge of abscesses, cobwebs, dead blackbeetles, and putrid German yeast, without counting alum, sand, and other ingredients."[52] Perhaps the bread in the Tower was remarkable, then, because it tasted so much better than the bread baked outside its precincts in the City of London, bread which, unlike these some twelve hundred rolls, even as it appeared each and every day, was not so reliable, nor always so "very good."

If, on a given day, these six small rolls changed in weight, as they probably did from time to time during the seven months of Gerard's captivity, it's unlikely that he would have noticed or thought it worth a comment. Outside the Tower, however, the situation was very different. Visit your usual baker's stall at one of London's open markets and, whatever the day, you might count, as the newly arrived Philadelphian Franklin could not, that "the halfe penny Cocket and the penie Cocket" would, in fact, always cost a "halfe penie" and a "penie," respectively, and that all regulated forms of bread would remain the same price regardless of the rise and fall in the price of wheat.[53] This price stability was maintained by a regulation that allowed the weight of the loaf to vary according to the cost of grain. Entering into statute law in the thirteenth century, *The Assize of Bread* codified a set of regulations including tables showing by how much the weight of a loaf was to be reduced for every six pence rise in the price of a quarter of wheat. The shape and appearance of the loaf could likewise be stipulated by the authorities, as was the provision that ingredients could or at least should not be adulterated by the baker (that prerogative being essentially reserved to the authorities whose mandate of a constant price in the name of good government and civil security could only be maintained in the face of rising prices for wheat and increased labor costs by licensing bakers to make the same-sized bread with less wheat).

Thus, on a given day, in London, the same-sized loaf, its shape and appearance regulated by statute, its price fixed, varied by weight, the regulations ensuring thereby an open and constant supply. The difference in weight might be made up for by the amount of yeast in the form of the ale "barm" bakers derived from commercial brewing added to the dough, making it just that bit fluffier, or, as contemporaries might complain, sour tasting. *The Assize of Bread* mandated periodic trial bakings by all members of the Worshipful Company of Bakers in which bakers would weigh their products and so confirm their compliance with the statutes. It also required the commissioning of officers whose job it was to "perfectly discerne and knowe what euerie sort of bread shoulde waigh in true proportion and rate" in order that they may be able to detect and punish negligent or criminal bakers with fines, public shaming, and, on occasion, imprisonment.[54] The assize was by no means foolproof and was reformed in the sixteenth century in response to the difficulties bakers faced in making ends meet given the strict regulation of both prices and the forms bread should take. Adjustments were made in different localities, adding to the particularity of bread in each municipality. Bread baked beyond the limits of the City of London was not regulated, and the importation of bread from outside its bounds by regatresses might carry a surcharge to protect citizen bakers or itself constitute a criminal offense.[55]

Unbeknown to the civic authorities, to Gerard and his contemporaries, the regulations that calibrated their supply of bread in the name of security, mutuality, and Commonwealth, ensuring that Londoners received the daily bread for which they prayed, led them to encounter the effects of the fungal actor *Saccharomyces cerevisiae* (baker's yeast) that enabled this multispecies relation or food web. They did not encounter yeast as such, as an entity or phase in the life cycle of a fungus, but mediated by and through its effects, what it made possible and what too much or too little of it might make or mar. Gerard and his contemporaries did not know that the "ever-recurring leavening miracle," as one baker and food historian puts it, was the effect of a "plant with a single active cell," the microscopic, animal–plant–fungus we name "yeast" but which they named "leaven" or, on occasion, more archaically and still more miraculously, "goddis-goode," the phrase a contracting sentence that condenses the process of fermentation that a baker captures and puts to work, the resulting bread become self-grounding, both gift and given.[56] The fact that yeast is a living entity and that, when fed reduced forms of carbon, it produces carbon dioxide and ethanol would not become known until the experiments of Louis Pasteur in the nineteenth century with grapes. But, still, the process of brewing and bread making produced a powerful, observable figure of inhuman process that therefore proved divine.[57]

Early modern dietetics worked on the principle that bread "is the staff of life" and that it was accordingly no coincidence that when "Christ would describe himself unto us while he lived, and leave a memorial unto us of himself after death . . . his wisdom found no hieroglyphical character wherein better to express himself (the only nourisher and feeder of all mankind) then by the sight, taking, and eating of Bread."[58] And as this attention to the sensory appeal of bread makes clear, the infrastructure of baking and food regulation yielded a powerfully semiotized and phenomenologically nuanced object, a nuanced sameness, in Benjamin's terms, each and every iteration of bread always the same and always different, different in its recurring, localized phenomenality. "Leaven" designated the process and presence of this agency or divine gift that enabled bakers to keep the material–semiotic fact of "bread" still, to render bread synonymous with their infrastructures, however precarious on occasion they may be. Fermentation or leavening was understood to be a process that required human monitoring, just as the instability of supply had also to be regulated and so secured. "Leaven" was both "the Mother and Daughter of corruption," of change and transformation. Too much (especially of ale barm) might make the bread sour and unwholesome or cause it to over-rise and fall; too little and the texture of the bread would be wrong and not nourishing. Indeed, in one account of the process, the texture of the bread, bread itself, derives from the meal's resistance to and so tempering by the leaven.[59]

We should not smile, then, at Gerard's seemingly idle remarking and then second-guessing of his liking for the Tower's bread rolls. As Joan Thirsk observes, writing in an age of industrialized bread production, "the way contemporaries looked at their bread" may "be a novel experience for us who are accustomed to conformist standards," but Gerard would have "expected every piece of bread . . . to taste differently, and so scrutinized it critically," something she illustrates vividly, citing the debate between Simonides and Polemon in Juan Vives's 1538 *Dialogues* over what counts as good bread. Simonides sits "down to a meal" that includes "fine white bread" that "weighed less than a sponge . . . the flour well sifted," and waxes lyrical on the subject. Polemon joins in and the two debate the value of each other's preferences.[60] Still, Gerard's reserve or concern with regard to his own liking or, to use his own loaded phrase when it comes to oranges, "*esu delectari*" (an inclination, fondness, or taste), marks a tension with these rolls, evidence that in the Tower, given the circumstances, the yeasty, bready hieroglyph stutters or falters.

Not for him the Eucharist, that saturated phenomenon or contested gathering of significance, of signs, of presence (real or commemorative), these "six small rolls of very good bread" constitute a more mundane but no less miraculous and unexpected gift, enabling him to rely on a steady source of nourishment as he recuperates. It is difficult, however, not to feel that these leavened "six small rolls of very good bread," which appear each and every day whether or not he prays for them, stand in relation to the missing, yeastless Host, operative still, whose absent presence Gerard mimes, his Mass mimesis mending his body and bringing him consolation. This bready sufficiency haunts the experience and the narrative even as the rhetorical pattern of Gerard's text suggests that the deferred communication of the Mass materializes in the form of the orange-fond and then orange- and letter-bearing warder. Gerard's empty-handed offering translates into the figure of this warder whose oranges Gerard animates. For by Gerard's celebration of the dry Mass, he crafts his own version of Walpole's oratory, a sacramental, quasi-Eucharistic space that alters the phenomenological density of his cell in the Tower. And this Hostless, sacramental bubble, animated by a divine power or breath, stands in tense relation, pushing back against but also coopting the yeasty rolls that Gerard eats and enjoys. Here, perhaps, with the efficacy of this sacramental economy, lies the lesson that Gerard offers his readers and listeners: the reformed bread of the Tower funds the epistemic, rhetorical, sacramental force he exerts within the confines of his cell, allowing him to expand his presence beyond its confines. His memoir models a regime that teaches us about which *things* to keep, to hold on to, still, and which stillnesses or stabilized forms, such as "six small rolls of very good bread," might enable you to animate still other *things* differently—piecing them together to create your own gathering. Stuff yourself with this gratuitous yeasty bread, he seems to say; it shall sustain you and enable you to derive from this sufficiency still another world, still another bubble of divine breath, breathed into you by the sustenance you steal from this yeasty exhalation we name "bread."[61] Plus, this bread tastes good.

How terrifying, then, it must have been to contemplate a world whose routines were interrupted, whose "foundation," to use the word of faith that Gerard's Protestant inheritors might choose, unmoors from the yeasty puff of the multispecies inspiration community that funds this divine gift, the divine, the inhuman itself an effect, the translated product of the animating breath of yeast. A world without bread would constitute something on the order of a sacramental void, a world not, perhaps, without god but without the contracted sentence become noun that declares, in a single breath, that "goddis-goode."

BREATHLESS

London. 1665–66, as replayed from 1722. H.F., the narrator of Daniel Defoe's *A Journal of the Plague Year* (1722), has been out walking. Or, that's what Defoe claims in this fiction cum spiritual biography, semi-documentary archive and anthology of lost stories. H.F. wanders the streets, caught up in archiving the course of the latest distemper. This modest saddler shuttles between his two apparent options: "whether I should resolve to stay in *London,* or shut up my House and flee, as many of my Neighbors did."[62] Not quite a Crusoe, H.F. lives shipwreck as a constant, lacks the store of provisions that would enable him to hole up and wait for Friday. He weighs matters accordingly and, finding no true course, concludes that, "as nothing attended us without the direction or Permission of Divine Power," to "flee . . . from my Habitation" would be "a kind of flying from God" (11). H.F. practices divination, randomly accesses the Bible for a verse that might guide him, opens the book to Psalm 91 and learns that "the Lord . . . is my refuge, and my fortress, my god in him I will trust" (13). H.F. reads on and the psalm instructs that no "evil [shall] befall thee, neither shall any plague come nigh thy dwelling, &c." Thus ordained as a witness to the "reward of the wicked," mewed into his own little morphoimmunological mobile construct of one, wherever he goes, we might say that H.F. stays put. Trouble is that, as time passes, his witness reveals that this same reward (a horrible death) befalls the not so wicked along with the wicked. The plague does not discriminate on the basis of conduct.

H.F. seesaws, goes out walking, explores this or that part of the City where the plague figures rise; but, "terrified by those frightful Objects" he encounters in the streets, he "retire[s] Home sometimes, and resolve[s] to go out no more." Unable to "keep those Resolutions for [more than] three or four Days" (66–67), however, during which time he prays, confesses his sins, fasts, he ends up back on the streets, where "people fall . . . dead" around him. He encounters scenes of waking oblivion, minor tragedies in which the usual bonds of care and concern between family members, fathers and mothers for their children, lead to infection. "It was very *sad* to reflect," he observes, how such persons might become "a walking Destroyer, perhaps for a Week or a Fortnight" (173), infecting all around them. He despairs at the multiplication of false remedies, medical and spiritual, and remains critical of the wisdom of the shutting up of houses containing plague victims. Such measures prove counterproductive, he thinks, and he documents all the elaborate ruses detainees come up with to escape to the countryside or merely to the streets. No exact pattern or shape emerges from these movements even as the *Journal* records the condition of H.F.'s soul. Instead, H.F. is left to wander, sifting the world, the people he encounters, for some measure of inner or spiritual security or, in his words, "Foundation" (94) or "Pattern" (105). The cast of characters he meets or assembles, named and nameless, stand as practical moral philosophical examples or enactments of how (and how not) to live now, in a world recoded by the plague event, by the parasitic rerouting of a built world by the microbial actor *Yersinia pestis* and its mutations.

Among the papers and documents on his desk that the *Journal* collates, H.F. mentions that he has by him "a Story of two Brothers and their Kinsman" (51), a soldier, a seaman, and a joiner or carpenter, who formed a traveling band, collecting up provisions and people as they wandered. "Their Story has a Moral in every Part of it," he writes, "and their whole Conduct, and that of some they join'd with, is a Pattern for all poor Men to follow, or Women either" (105). The three form a sort of joint-stock company in small. They incorporate, pooling their resources and expertise. The three men talk through their responses to the crisis. H.F. has the transcription of some of their dialogues; he offers them to his readers as templates or even role-playing exercises that might help model the scripts that lead, if not to survival, then to some precarious conservation of personhood as the world disintegrates. But all the time, even as he is impressed by such stories, he remains haunted by the possibility that the security or "foundation" of these souls derives not from who they are but from the way they have been distributed within the built circuits of the city. The men have names—John and Thomas play the two brothers. But the names mean little. It is their trades and skills that matter.

The same holds true of the most inspiring and affecting person H.F. meets, Robert, the waterman, whom he sees "walking on the Bank [of the river], or Sea-wall . . . by himself" while all the surrounding houses were "shut up" (92). Robert tends to Rachel, his wife, and their two children from a distance. He remains entire to himself, to his moving island of a boat, acting as postman and courier between larger ships and the land or other houses, with whom he has no direct contact so as not to communicate the disease. H.F. gives him some money unasked, and Robert gives him a ride in his boat—agreeing to the risk because he feels certain that since "your Charity has been mo'vd to pity me and my poor Family; sure you cannot have so little pity left, as to put your self in my Boat if you were not sound in Health, which would be nothing less than killing me" (96). Robert stands as H.F.'s model: he whose "Foundation" was such that he could continue in his life and work secure in his faith but also entirely rational, "us[ing] . . . Caution for his Safety" (94). The two form their own momentary community on that small boat, breathe the same air, run the same risk. But H.F. finds no lasting community with which to share his air—other than, perhaps, the one with which he began, the City of London itself, to whose citizens he addresses and dedicates his book. For on what does Robert's own foundation rest—his faith or the accident that he is a waterman?

H.F. begins and ends solo. The *Journal* ends with an intake of breath, with the brute facticity of breathing as proof of life, proof of survival. No security, no foundation beyond the intake of air, H.F. ends with this stark envoi:

> A dreadful Plague in London was,
> In the Year Sixty Five,
> Which swept an Hundred Thousand Souls
> Away; yet I alive! (212)

Generations of readers have wondered at H.F.'s panicked but never panicking rationality. Some have even asked whether he is himself, not, perhaps, a ghost, and the journal an "apparitional narrative," its narrator dead or breathless. Much like Walpole's graffito, these parting lines testify to a presence that once was and now is (again). For, when read aloud, the envoi causes the reader to draw breath (in and out), inducting us into the undead air of the inspiration community H.F.'s journal anchors. He lives on—his words bubbling in our anonymous mouths and ears. And we his wetware—breathed by the rhyme he sets in motion. As to whether there remains some otherwise than textual afterlife or Judgment Day, the *Journal* remains mute. H.F. never finds a sufficient "foundation" beyond the everyday heroics of London's magistrates and other civic authorities that he extols—especially its bakers. The security provided by providential narrative, political theology itself, finds itself downgraded or translated into those relays of city living that keep going, quasi-automatically, even in a state of such dire emergency.[63]

For, remarkably, as H.F. points out (if he is to be believed), throughout the crisis, Londoners were able to count on a steady supply of bread at a fair price. "The Price of Bread," he writes, "was not much raised" despite the course of plague (155). "In the first Week in *March,* the Penny Wheaten Loaf was ten Ounces and a half," he continues, "and in the height of the Contagion, it was to be had at nine Ounces and a half, and never dearer, no not all that Season" (155–56). This stability was maintained by the "Sheriffs and Aldermen," who ensured that, for the most part, the streets were kept clear of "frightful Objects" (159) and also the "Master of the Bakers Company" who was "directed by the Order of my Lord Mayor" to ensure that the regulations put in place by the *Assize of Bread,* which obliged the bakers to keep their ovens going "on pain of losing the Privileges of a Freeman of the City of *London,*" were kept. "By this means," he concludes, "Bread was always to be had in Plenty" (159). And were it not for this plenty, this regulated supply of quotidian but still "hieroglyphical" bread, then, perhaps, H.F. would have had to face a different truth, a truth that inheres to the unthinkable thoughts on the margins of the narrative's sense or in his seesawing, rationalized panic: that god is not always "good," and may not be at all.

The plague may parasitically reroute London's infrastructure. It may interrupt its routines. But the city is what endures. The weather changes. The plague diminishes. And all the while the alliance of plant, animal, and technical resources that produce the *thing* we call bread remains intact. H.F.'s foundation resides, finally, in the legislated sacrament of a daily bread, in the baked or dried-out paste that captures the exhalations of yeast, a microbiopolitical actor just as mysterious as the plague, but more benevolently so.

YEASTLESS?

No wonder, then, that in *The Life and Strange Surprizing Adventures of Robinson Crusoe* (1719), when H.F.'s forebear finds himself shipwrecked on that famous island, whose natural advantages Crusoe augments with what he salvages from his ship, he sets about investigating what it will take for him to be able to bake bread. Bread doesn't grow on trees, you know. Or it will not for Crusoe until the publication of a children's adaptation of Defoe's novel in the nineteenth century, in which his initial survey of the island locates a "bread-fruit tree" amid the lemon, lime, and orange trees whose juices refresh him. Crusoe knocks several of these breadfruits off the tree with stones. But, "as it was not very palatable when raw," he cooks it in a fire, testing bits of it as it cooks until eventually it tastes "so like Bread that I could not doubt that it was the true Bread-fruit."[64] Along with the llama milk he obtains, the meal sustains him but offers no long-lasting solution to his ills. For to live by breadfruit alone might run other risks, Crusoe descending to the level of a beast or, worse, as far as certain of his English cousins might feel, an Irishman.

No amount of cooking, you see, will dissipate the "rawness" of this breadfruit or other "bread-roots," as early moderns sometimes named the potato. Both bypass "the whole satisfyingly social and symbolic cycle of planting, germination, sprouting, growing, ripening, harvesting, thrashing, milling, mixing, kneading, and baking which makes wheat into bread."[65] To take them as or instead of "bread" would be to organize the world differently, to enter into a deevolving, becoming earthy that might manifest as a loss of Christian civility or a return to what Crusoe names "a meer State of Nature."[66] The very same endeavor, of course, might also lead us to imagine other modes of association rooted in or animated by an alternative, differently yeasty, or even yeastless, multispecies alliance with other plant and animal actors. *Robinson Crusoe,* the Robinsonade, occupies this double zone of economy and indistinction. Its limit cases designate scenes of policed possibility, moments at which certain convocations or inspiration communities might reinvest their energies in the same old bubbles and their routinized sense of "enough," or become otherwise than they have been.

But Crusoe is not there yet. He's only just ashore; only just breathing normally; given up "hold[ing] fast by a piece of Rock" and "hold[ing his] . . . Breath" (35). Legs back on dry land, he embarks on his salvage operation, "fill[s] . . . [his] Pockets with Bisket" (37) to keep him going—the yeastless, saltless sailor's kind (Franklin had, of course, read Defoe).[67] Crusoe lucks out, finds "a great hogshead of Bread" (42), rations his "Provisions (my Bread especially)" (46) as much as he can so that he has enough for as long as possible. Though at some point, unannounced in the text, the "bisket," just like his ink, which comes in short supply, and which he has to mix ever more thinly, runs out, even though he "husbanded [it] to the last degree" (97). Crusoe ends up without bread for more than a year before he becomes self-sufficient. Exactly when the ink runs out we're not sure, obviously, the narrative impossibly fading into nothingness, and Crusoe, like H.F., something of an apparition. Still, in the meantime, he finds the whole enterprise of bread making "a little wonderful," especially when you pause to consider "the strange multitude of little Things necessary in the Providing, Producing, Curing, Dressing, Making, and Finishing this one Article of Bread" (86). He must, in effect, become a guild or several, an entire Worshipful Company of Bakers, along with those who make their ovens, who mill and bolt their wheat.[68]

He begins by salvaging what grains he can from the ship's rats (57) but throws the lot away by mistake; finds himself astonished and confused when "twelve Ears" of "green barley" sprout (58) anyway; feels the benevolence of the divine in this unexpected opportunity to begin again, in "these pure Productions of Providence," the opening of a possibility. He clears some ground, plows and sows his corn. Four years later, he will have managed to reconstruct this infrastructure of daily bread. But already, even the prospect of bread, the *thing* for which he was taught to pray, ministers to what he describes as his "daily Discouragement" (86).

His decision to grow corn coincides, then, with his decision to pray again, and he enters into a self-administered pedagogy of bready spirituality. Crusoe's spiritual well-being coincides with his attempts to grow corn for bread, and just like him, the crop requires constant management and surveillance. Crusoe befriends a dog who stands guard to keep at bay the goats who, "tasting the Sweetness of the Blade, lay in it Night and Day" (84–85); he encloses the crop, shoots up the "little Cloud of Fowls" (85) that descends. He also realizes that he "wants [lacks]" pretty much everything he needs to "Fence it, Secure it, Mow or Reap it, cure and Carry it Home, Thrash, Part it from the Chaff, Save it." He also lacks "a mill to Grind it, Sieves to Dress it, Yeast and Salt to make it into Bread, and an Oven to bake it" (86–87). But he manages; he has six months of growing time to supply his wants.

Crusoe becomes a potter, turns ceramist, bakes his earthenware pots in the sun (88). Having "no Notion of Kiln, such as the Potters burn in," he learns how to bake his pots in the heat from the embers for five to six hours. Making do with "one Pair of Hands" (89), he manages to find stones that he can cut to make a mortar and pestle to grind the grains. He fashions "three small sieves" from pieces of "Seaman's Cloaths" so that he can "Search" and "dress" the meal (89–90). He forgoes the yeast—having no way (so he thinks) of "supplying the Want" (90)—and so just mixes up his "paste" and bakes it. We never learn if his dough rises, but he describes himself as becoming quite the "Pastry-Cook," making "barley Loaves" as well as "several Cakes of the rice and Puddings." In fact, he ends up with such a store of grain, so much more than enough, that the rats come back, or rather he finds himself beset by a new polity of rodents, shipborne and indigenous, that infest his stores. But his supply is such that he has no problem offering Friday "Bread and a Bunch of Raisins as a restorative" (148); makes sure that Friday gives his own rescued father the "Cake of Bread" that Crusoe carries on him for emergencies (172); and gets to explain to new arrivals the whole "Story of . . . [his] living there" along with "the way [he] . . . makes his Bread, planted [his] . . . Corn" (199). Crusoe ends up quite the little sovereign, then, or sovereign in small—his island "peopled" (174) and he, chief magistrate, Master of the Company of Bakers and shepherd to his multispecies flock of goats, sheep, cats and dogs, parrots, and assorted variously human arrivals, all of whom enjoy the fruits of his agricultural labors.

Marooned on this island, Crusoe reboots the world he has left, produces the infrastructure of daily bread that anchors the Eucharistic hieroglyph or zoologistics that the human–ovine incarnation sets in motion. We watch as he plays catch-up; time travels forward from the "state of nature" on the island to the "present" he has left. Or, more correctly, we watch as he projects a fantasy of insular self-sufficiency that comprehends the so very many entities that might otherwise intrude. His pots, for example, as Lydia Liu has shown, metaphorically see off the exteriority, the technological superiority, of Chinese porcelain making, absorbing their techniques into Crusoe's fictional ceramic turn.[69] Likewise, his bread making comprehends an entire agricultural axiomatic that rationalizes the lack of yeast from commercial brewing with what amounts to a metaphysical shrug. The loaves rise (most probably) anyway, after a while, via the chance ripening of a fungus that loves the damp and the heat. Crusoe's industry amounts to a phantasmatic, solo reprise or technological advent that not only redoubles and repeats so as to catch up, to preserve and maintain the rationality and ratio he was, but also extends his reach.[70] The island becomes, in effect, a staging ground for future exploration, future colonial projections. The island sojourn prepares readers to accompany Crusoe on his *Farther Adventures* (1719) in the South Seas and his more programmatic *Serious Reflections* (1720), which, as Robert Markley offers, "yoke . . . capitalist expansion and Protestant evangelism" to an unmoored homosocial self equipped for colonial adventure.[71] Crusoe's buffeted, shipwreck rationality offers a technology of self allied to an imaginative project that fills in faraway places, positing them as empty only so that they may be filled in in advance of the fact. Shipwreck, emergency, stands as his milieu, the true habitat against which he deploys his island self.

Here bread making stands as the grounding reconstructed cycle or relay that regularizes and maintains all the routines or concentric circles that Crusoe builds—a circling or circular motion that begins with the hasty erection of a protective barrier made out of salvaged barrels and casks of different sizes (41) but whose true beginning lies, perhaps, in the regularization of his breath, the in and out of lungs that no longer have to hold their breath against the waves that would engulf him. Crusoe inflates successive technological bubbles, ex-corporates his being into all manner of enterprises. But the air he breathes is not exactly his own. For the world Defoe imagines is not quite the "world without Others" that Gilles Deleuze posits as the true, perverse, psychotic, answer to the question: "What is a Robinsonade?"[72] Though Crusoe may make the island seem so, incorporating as he does, pressing the great many entities he finds on the island to use, the island, its plant and animal actors (whose skins he wears, whose flesh he eats, whose labor he deploys), all part of some elaborate respirator or scuba gear cum flotation device that buoys him during this wreck.

Crusoe's drive to reboot a particular history of technology, a history coterminous with Sloterdijk's sense of human history as a form of inspired catch-up to the divine, amounts to more than a story of his "hominization," then, or anthropogenesis, or even his remaining, merely, "British." It anchors one particular mode of sovereignty or ipseity that, in Jacques Derrida's words, requires the "wheel [to] describe . . . a sort of incorporated figural possibility, a *metaphora* (*metaphora* in Greek means vehicle, even auto-mobile, autobus) for all bodily movements as physical movements of return to self, auto-deictics, autonomous but physical and corporeal movements of auto-reference."[73] Here I am (again). Here Crusoe is (again), still here, working to free his hands to other labors or other devotions. To do so, he must build his own "Foundation," build and so preserve the possibility of a grounding reference, an infrastructure that inspires or enables belief. It is no coincidence, then, that this rebooting of routines, as Derrida offers, runs in tandem with the "experience of learning how to pray," with the invention, which is to say "always . . . a repetition, a reinvention, on the island, a second origin, a second genesis of the world itself, and of technology," that salvages what remains of the old and, by the salvaging, extends its reach.[74]

But if the island, itself, figures the grounding auto-reference of an "I," of a pronoun or ego that turns on itself, as it autoimmunizes, holds its breath, refuses to give air to others, then, as Derrida observes, things do not come easily or go well for Crusoe. He remains caught up in and out by fears of being buried alive, of drowning, of earthquakes, of being eaten up, and by thoughts of self-destruction.[75] Famously, these fears are keyed to the appearance of "the Print of a Man's naked Foot" (112), a single, bootless print in the sand one day that causes Crusoe to feel "himself followed by a trace, basically, hunted or tracked by a trace," by forms of writing, technical extensions of presence, other than his own.[76] The scene unfolds like some panicked inversion of a "god trick," "about Noon," the time of no shadows—a geometry lesson gone badly wrong. Great solar eye in the sky that it seemed he was becoming, primed with a synoptic power over the island that he has worked so hard for, Crusoe finds himself tripped up by this unwanted impression, his knowledge suddenly situated. "It happen'd one Day about Noon, going towards my Boat," he tells us,

> I was exceedingly surpriz'd with the print of a Man's naked Foot on the Shore, which was very plain to be seen in the Sand: I stood like one Thunder-struck, or as if I had seen an Apparition; I listen'd, I look'd round me, I could hear nothing, nor see any thing; I went up to rising Ground to look farther; I went up to the Shore and down the Shore, but it was all one, I could see no other impression but that one, I went to it again to see if there were any more, and to observe if it might not be my Fancy; but there was no room for that, for there was exactly the very Print of a Foot, Toes, Heel, and every Part of a Foot. (112)

The print's emphatic presence, its confounding facticity, petrifies. Crusoe stands "like one Thunder-struck, or as if I had seen an Apparition." He tries to listen, looks all around, but can "hear nothing, nor see any thing." He attempts to get some height on things, "went up to rising Ground to look farther," retraces his steps "up to the Shore and down the Shore." But nothing helps. He backtracks, tries to follow a track that fails by and in its singularity to constitute a trail. He returns to the scene of the crime—but it's not the work of "Fancy" or the imagination, for there remains "exactly the very Print of a Foot, Toes, Heel, and every Part of a Foot," the very "print of a Man's naked Foot on the Shore."[77]

This print will not last, of course; eventually the sea must wash it away; the wind shall blur its outline. But already Crusoe's description doubles the print, renders it mobile as it ambulates through his mind. Like it or not, he serves as its substrate, his consciousness remediating the print not once but twice, given the relay that forms between the novel in all its iterations and its readers. This island was not a world without others—certainly not without invisible actors. And this revelation or haunting takes Crusoe's legs out from under him. He floats home, "not feeling, as we say, the Ground I went on," shuts himself up in his "Castle, for so I think I call'd it ever after this" (112–13). But he cannot sleep; he fancies that the print is the work of the "Devil" or "Savages," talks himself in and out of various explanations, wonders "that all this might be a Chimera of my own; and that this Foot might be the Print of my own Foot" (115). Days later, he "peeps" out and returns to the print a third time to "measure it by my own" and finds it smaller. Fear seizes him; he wonders if he should "dig [up his] . . . two Corn Fields . . . demolish . . . [his] Bower, and Tent," and so kick over all the traces he has made (115–16), erasing everything.

In the end, he shores things up with an ever more elaborate set of defenses and with more and more routines or circles. Two years later, on discovering the arrival of "Canibals" on the island, he replays the print in his head and decides that "seeing the Print of a Man's Foot . . . was a special Providence," for he "was cast on the Side of the Island where the Savages never came" (117). The discovery of this "Print of a Man's naked Foot" reveals, as it were, a rival structure to his world that thus far had remained invisible or that Crusoe's routines had sought to comprehend, the presence of another or the possibility that his own acts of inscription rest on still others, other forms of writing—writing that he may have to acknowledge or that might overwrite his own acts. But as Crusoe's providential revisiting of the print makes plain, this haunting or revelation of other traces proves constitutive. The precariousness it installs funds his redoubling of effort, all those circles that seek to incorporate the island.

Deleuze (and the rest of us) will have to wait until Michel Tournier's *Friday* (1967) for Defoe's "well-intentioned" Robinsonade to bring the "asexual Robinson to *ends quite different from ours,*" ends that make good on the world without others that Deleuze seeks to revalue, taking perversion as a structure of possibility. Tournier manages this, in part, by rezoning the activity of bread making as the entrance not just to an elemental sexuality, a sexuality that embraces the island, but also to forms of writing otherwise than human. If Tournier's Robinson does not "deny himself the pleasure of breadmaking," his participation in "that most material and most spiritual of human activities" does not figure a second genesis but returns him to "the shameful secrets of his early childhood," which "foreshadow . . . [the] unseen flowering of his solitary state."[78] Robinson kneads the dough and recalls his childhood desire to become a baker when he grows up, a desire marked by "*the thought of a strange marriage between the dough and the baker,*" by the motions that Serres described at the beginning of this chapter. He even dreams of "*a new kind of yeast which could give the bread a musky savor, like a breath of spring*" (79).[79] His pursuit of this differently yeasty end or discourse of ends takes the smell of yeast proofing as a material and sensory enmeshment with matter. It leads him to seek after other versions of this experience, to explore interspecies or interkingdom encounters.

Envying the "elegance" of plant sex, "the process of insemination at a distance" (114), he becomes amorous with a quillai tree with which he has exchanged "sidelong glances" but gets bitten by a spider that makes its home in the fallen tree. He regards this venereal complication as a sign that his "'vegetable way' may be no more than a dangerous blind alley" (116).[80] The would-be telesexual Robinson proves naively medium specific, then, overly literal in his understanding of the flesh. Penetrating a tree stump merely intrudes into an arachnid's shelter. Understandably, the arachnid asserts its own right to shelter. In the end, the yeastiness of bread making comes to saturate the island of Speranza (herself), which begins to grow or rise like the dough of which Robinson dreamed and that he now kneads. Robinson becomes part of this milieu, twinned with it, as much substrate to it as it is to him. The two form a closed circuit, and so Friday leaves him to it, leaves the island, replaced by the newly arrived Jaan Neljapäev, "Sunday's child." Crusoe stands tall, "his feet . . . solidly planted on the rock," a living statue, "legs as sturdy and unshakable as stone" (234–35).[81]

Deleuze regards Tournier's Robinson as having discovered what he calls a "great Health," perverse and psychotic, a structure that knows no outside, that requires no foundation other than itself. He calls it a "strange Spinozism, from which the 'oxygen' is lacking, to the benefit of a more rarified air."[82] Only Robinson breathes and bubbles, and that bubbling comprehends all the *flora* and *fauna* of the island as well as the island itself. Tournier's Robinson radicalizes Defoe's Crusoe, then, even as the two remain joined at the hip. In the process he animates the zone of creaturely indistinction that both underwrites and haunts the novel, the loss of certainty as to the difference between plants, animals, and earth, between bread and stone, between trace and substrate.

A decade or so before Tournier's novel, in a hallucinatory chapter of *Tristes Tropiques,* Claude Lévi-Strauss titled his recollection of a brief encounter with the Mupé people of South America "Robinson Crusoe." He did so because the scene of the encounter is a world in which it proves impossible for him to tell land from water, plant from animal, "savage" from so-called civilized man. Upon arrival, he notices that the relationship between land and water had become unmoored and that "it was impossible to say whether the river served to irrigate this fantastic garden, or whether, it was about to be choked by a proliferation of plants and creepers."[83] This confusion or interpenetration of media extends to the animate world, for the "trees were even more a-quiver with monkeys than with leaves, and it was as if living fruits were dancing on their branches." Lévi-Strauss is put in mind of "those pictures by Brueghel in which Paradise is marked by a tender intimacy between plants, beasts, and men . . . tak[ing] . . . us back to the time when there was as yet no division among God's creatures." He wishes now that he had had more time to spend in this place, to learn the Mupé's language and their ways. But he does not, and so now, as he recalls the encounter, he is faced with a seemingly impossible task, for the codes that would enable him to name what he had experienced do not apply. "I can pick out certain scenes and separate them from the rest," he writes. "Is it this tree, this flower"? But while the "whole fills me with rapture," the nuance escapes. Like Walter Benjamin, with whom I began this chapter, high on hashish or the rhetoric of drugs, Lévi-Strauss could be anywhere: in the tropics, back home in Paris, in the rainforest or the "Bois de Meudon."[84] This zone of indistinction, or nuanced sameness, to borrow Benjamin's formulation, refuses to take the impression of Crusoe's spectral footprint. "Man Friday's footprint is missing," writes Lévi-Strauss, in what amounts, for him, to a breakdown of anthropology as a mode of inscription, a disciplinary configuration, an epistemology.

For what exactly may he now claim to know? The earth refuses to take an impression, to agree to act as substrate. All Lévi-Strauss can do, then, is tell the story of his onward movement and piece together what he can from memory. As for Tournier's Robinson and Defoe's Crusoe, whose journal comes to us impossibly, written in disappearing or fading ink, ink they both mix ever more thinly, and so which must, at some point, simply disappear, Lévi-Strauss must narrate a sensory impression of an encounter for which only he may serve as substrate. No other ground remains.[85] All they can do is keep on breathing and exhale their impressions into prose.

IDIOT

Pervert. Idiot. Fool. If there were ever an iconography of such fugitive moments of dissolution or groundlessness that I have sought to pry loose from the yeasty anthropo-zoo-genesis I have tracked in this chapter, then it might look a lot like this image of the god-denying fool that, for a certain time and place, illustrated the moral and the theme of Psalms 13 and 52, "the fool who says in his heart, There is no God [*Dixit insipiens in corde suo: non est Deus*]" (Figure 18).[86] As V. A. Kolve shows in his survey of such illustrations of the psalm throughout the Middle Ages, the figure of the fool, with torn clothes, sometimes tonsured head or white cap, is shown wandering at large, a "scarecrow of the mind" but also a "bearer of surplus." His deprivation serves to embody the tortured, exteriorized form that such a bearer of an unthinkable thought must take, a thought that, even when allowed to surface, as a logical possibility, does so only as part of a proof of the existence of the divine—even as belief still may prove difficult.[87] The image inoculates. But it also preserves and suggests the possibility of unbelief. In his reading and sifting, Kolve works hard to pry this figure loose from a tradition that seeks to discipline it but that also cannot let the possibility or facticity of nonbelief, of a denial of grounds and grounding, go unimaged or unimagined. Instead of regarding this figure pejoratively, he asks us to perceive the beauty to this figure—his clothes, his posture. He asks us to dwell upon his frailty, the precariousness and exposure that the figure embodies and endures.

Kolve also asks what exactly it might be that this fool is pictured

Figure 18. Psalm 52: Fool pictured with fool's food and a jester's "marotte." *Bible historiale,* Paris, 1357. London British Library MS Royal 17 E. vii, fol. 241. Image copyright the British Library Board. All rights reserved.

as eating? In almost every image, the fool gnaws away at a spherical iconographic constant. He seems to enter into some kind of closed circuit with this object become food item, an object on which he makes so little impression. "Art historians mostly take it to be a cake, a cheese, a loaf," observes Kolve, "but my first guess is that it was a stone—an intuition I can now verify."[88] Mistaking stone for bread, the fool's botched object choice stands as a token of his extremity, idiocy, or perversion, and perhaps also as an invitation to pity. His dog waits upon him, regarding his folly with an indefatigable but misplaced fidelity.

Yes, this fool stages the Eucharistic feast become famine. He takes stone for bread and so stands, on one hand, for the Satanic temptation to an incarnated divinity to transform that which appears not to change (stone) into bread (Matthew 4:3) and, on the other, for the unconvertible or stony-hearted heretical other—he or she who honors a different archive and a different text, the differently leavened and differently yeasty Hebrew Bible. This god-denying, stone-eating fool stands precisely as the figure I have tried to follow through this chapter: he or she for whom stone has become bread to the imagination, he or she who takes stones for bread and, by that apparent confusion, no longer knows the difference between beings, or what exactly now counts as writing. Perverts, idiots, fools, they slow down routines, arrest the action and so enable us to inquire into the conditions by which a surface that is all process, that conjoins beings, turns into something we name a ground.[89] These figures who live the nuance populate those "not-very-eventful spots" or interstices to the yeasty story I have told. They keep alive the possibility of other forms of writing, other possible inspiration communities, or, as Donna Haraway might have it, "another kind of open. Pay attention," she instructs, "it's about time."[90] As I hope my idiosyncratic or, perhaps, idiotic stream of yeasty, bready, stony bubbles demonstrates, it's been and done time. The possibility of another kind of open remains accessible always by and through the actions of a trace that these writers understand to come from without and for which they serve as substrate, whose impression they agree to take.

This foolish icon, of course, lives on into our various todays, for baker's yeast, it is said, "screams" when exposed to alcohol in a laboratory equipped to record the "vibrational movement of cell walls and amplify . . .

these vibrations so that humans can hear them."[91] As anthropologist Sophia Roosth describes, when scientist Jim Gimzeski "doused the yeast [he was working with] in alcohol, the pitch of the vibration increased. In an interview, he claimed that: 'it screams. It doesn't like it.'"[92] He "speculates that this 'screaming' is the sound of molecular pumps working overtime to expel the alcohol," the sound, as it were, of yeast attempting still to breathe even as it may drown. Quite correctly, Roosth wonders what the status of such a scream might be. Does yeast now manifest as if a subject? Or is "to say that a cell is speaking . . . to project cultural notions of what it means to be human, to be subjective and have agency," to embark on a microbiopolitical colonization?[93] Is to say that yeast "screams" to extend the reach of Crusoe's shipwreck rationality, to conscript and perhaps exhaust the animal–vegetable efficacy of yeast?

Serving now as substrate to his yeast, Gimzeski risks, like many of the writers and thinkers I have assembled in this book, being named a fool, an idiot, or being told that this scream is an artifact of the relay his experimental protocol forms between himself and his experimental object. The wrong impression, a bad or hallucinatory archive; all witness but no evidence. But therein also lies the way forward, in acknowledging and operating on the basis that our acts of writing, of inscription, necessarily coincide with still other forms of writing and coding, which they take as if a substrate, sponsoring, obliterating, ignoring them in the process. There remains always the possibility of creating scenes of inscription that enable other beings to write with us or to rewrite both of us. To write, merely to be, designates these scenes of risk and possibility, of hospitality and violence, as we do and do not take cognizance of the multispecies foundation to our own acts of inscription.

What might happen, for example, if we were to decouple our built structures (our notions of structure) from the fungal plant–animal actor we name yeast and the regimes of bread and beer it anchors and inquire into the differently fungal worlds that spring up from the ruins of our infrastructures? How, for example, as Anna Lowenhaupt Tsing puts it, might the world look if we took our cue from the mushroom that grows "at the end," not of the "world," but of a concept of "world" as something we take as if a foundation?[94] The matsutake mushroom, that prized fungus, grows in the waste spaces and lost forests of the world. You have to hunt for it. To grow these mushrooms requires a practice of care and attention much like that cultivated by the writers in this chapter, who live the fungal nuance, who find themselves convoked by bread and stone. Or, perhaps, yet more constructively, what would happen if we were to begin to understand our yeasty alliance as one instance merely of the larger "world-building work of fungi" that might lead us to imagine different realms of belonging and modes of association?[95] A whole other, differently fungal world remains to be thought, and that thinking might lead us to build on an altered set of foundations, an altered sense of foundation itself, as necessarily precarious, necessarily subject to continuous renegotiation and tuned to the ecological exigency of all we seek to come into being with.

Erasure

> Is not this a lamentable thing, that of the skin of an innocent lamb should be made parchment; that parchment, being scribbled o'er, should undo a man? Some say the bee stings, but I say, 'tis the bee's wax; for I did but seal once to a thing, and I was never mine own man since. How now? Who's there?
>
> —WILLIAM SHAKESPEARE, *Henry VI, Part 2*

I began this book with a memory of skin. Not my memory. Not your memory. Not any one person's memory. It belonged to the thing we call a "character" in a play. Rebel leader and self-proclaimed "parliament" of the Commons Jack Cade touched a parchment—writing material manufactured from the skin of a sheep, a goat, or a cow—and found himself "touched" in return. Worse still, this parchment hurt. It forced him to recall his encounters with the law. He registered this pain as a "sting." And this sting led him to bemoan the fate of the lamb or lambs from which this parchment was made.

In returning to where I began, to a memory that sympathetically transfers the pain of the knife that flays the lamb to human skin that is stung by a seal, I wish to remark on the way, throughout this book, Jack's question has proved both real and rhetorical. Idiot that he is, he introduces a retarding *praeteritio* into the action that accelerates around him, a finite statement of nonknowledge, of not knowing, that begs the question as to what is happening, what has happened, and if there might not yet still be a better way. Dick the Butcher's "cut," we know, shall come, eventuating the biopolitical field and deciding on the forms of life that shall result, but that "cut" remains hostage always to Jack's question—a question I have sought to retrieve and inhabit as it migrates through the world of animals (sheep and humans) and plants (oranges and lemons), the microscopic world of fungi (yeast), and the seemingly solid world of stone.

If it seems to you that the concept of the "multispecies" has become too capacious in the process, or too variable in its crosscutting of differing, vying polities of fungal, plant, and animal actors, then that seems right to me also. But that is because the word does not designate an achieved or achievable community or polity of beings so much as it names an orientation that understands the way our concepts for being make possible certain worlds even as they unmake others. The word remains a proposition, orientation, or invitation, which proceeds on the basis that it does not quite manage to refer and so to ground itself as a valid reference. A multispecies impression offers no answers or necessarily positive outcomes, but the awareness it cultivates, an awareness that understands that to be means to become with many, fundamentally alters what it means to be now. It asks that you open the question of what grounds your world, what serves as its foundation and to consider other modes of organization.

And now that I am at an end, my own ink fading on me, even as I keep on writing, it occurs to me that, like Lévi-Strauss, I too have this image in my mind of a painting. It's by Titian or perhaps by Leonardo da Vinci. The scene it depicts has become something of a blur. It's of a feast, a meal, a *cena*.[1] Once upon a time, there was this fashion for placing oranges or lemons or sometimes citrons, whole or cut, most often cut, on a table at which a man who was also a lamb, and bread, and wine, and blood, sat along with twelve other men. Together they feasted on a lamb that had been killed and cooked, along with some quotidian bread that they washed down with the wine. The oranges served as their sauce or possibly as palate cleanser, testimony, at the time, to a particular level of Italian sophistication or fashion.[2] It seems to me that throughout this book, all the while I have been party to this table, this Seder, and this Supper, though like the god-denying fool and the cast of idiots I have assembled, I have sought not to understand the arrangement of beings it presents but rather allowed the lamb, the orange, and the bread to dislocate the table to their own ends. Never exactly "last," this supper, this table, the figure of a counting or logistics, a festive meal and a tabulation, proceeds always as a repetition and redoubling, a troping of the feasts that precede it.

And so it seems to me that in siding with the lamb, the oranges, and the bread, or in not knowing, any longer, their role exactly, I have tried to dislocate this table's shape, prorogue its routines so as to make possible other, more hospitable modes of organization. I have attempted to undo one organizing technotheological structure that serves to keep beings separate, or, in truth, confuses them strategically, so as to make possible one type of world. In its place I have tried to imagine another, altered and altering regime of description in which it remains hard to know the difference between a person and a sheep, between a prison warder and the oranges for which he has such a taste, between bread and stone.

It shall be up to you if you want to search out this painting. I shall not show it to you. For it is merely one icon among many, one image of a putative future rooted in an infinitely receding past that promises a cancellation of present antagonisms. Come the end, all I shall offer you is the looming blankness of an empty page or the flicker of an empty screen. The future, if there is to be one, may not be imaged even as it requires that we imagine another order of table or world in common, a world that owns its existence as a series of competing, sometimes complementary, sometimes violent, sometimes sustaining multispecies impressions.

Acknowledgments

This book largely emerged from a series of papers written for colloquia and conferences. Were it not for these invitations to present work in progress, the persons who extended them, and those who agreed to be an audience, it might not have come to be. I fear that the list of individuals and venues that follows is incomplete. I know that it fails to acknowledge the generosity of audience members who asked me questions I could not answer, smiled at the answers I managed, and were kind enough to send me poems, essays, images, and references that became part of the project. Thank you all. This list fails also to capture the laughter and sometimes giddy delight of many of those occasions. But I like to think that some of that laughter and delight made it into the book, even if all the requisite thanks did not. For all the omissions that follow, I am sorry. But know that this regret is nothing in comparison to the gratitude I feel for the opportunities given me to speak to you and the care you showed me (and these ideas) by your responses.

Thank you to Jacques Lezra, Henry Turner, Susanne Wofford, and the Department of English at University of Wisconsin, Madison; to Tom Conley and the Renaissance Seminar at the Harvard Humanities Center; to Matt Cohen and Duke University's English department; to Sharon O'Dair and the Hudson Strode Program at University of Alabama, Tuscaloosa; to Rebecca Zorach and the Department of Art History at University of Chicago; to the Medieval and Renaissance Workshop at University of Chicago, convened by Carla Mazzio; to Gary Taylor and the Department of English at Florida State University; to Matthew Kinservik and the Mellon-funded Colloquium on Material Culture Studies and Public Engagement at University of Delaware; to Alan Stewart, Molly Murray, and the Department of English at Columbia University; to Jeffrey Jerome

Cohen for frequent visits to the Medieval and Early Modern Studies Institute at George Washington University; to Robert N. Watson and the Mellon Colloquium on the "Pre-history of Environmentalism" at UCLA; to Christopher Wood and the New England Renaissance Conference held at Yale University; to Liza Blake, Kathryn Vomero-Santos, and the Renaissance Workshop at New York University; to Garrett A. Sullivan and the Department of English at Pennsylvania State University; to Julia Reinhard Lupton and the Department of English at University of California, Irvine; to Joseph Campana and Scott Maisano for "Renaissance Posthumanisms" at the Humanities Center at Rice University; to Joe again, along with Timothy Morton, Judith Roof, Cary Wolfe, and the Morse Lecture Series, also at Rice; to Alexandra Halasz and the Department of English's Faculty Seminar at Dartmouth College; to Lowell Gallagher and the Center for Medieval and Renaissance Studies at UCLA; to Laurie Shannon, Jeff Masten, Wendy Wall, and the Early Modern Colloquium at Northwestern University; and to Amanda Bailey and the Marshall Grossman Lecture Series at University of Maryland.

Long before the ideas in this book were aired, their earliest glimmers and their style of writing were nurtured by a Philadelphia-area work-in-progress group that, at the time, included Scott Black, Edmund Campos, Jane Hedley, Nora Johnson, Steve Newman, Kristen Poole, Katherine A. Rowe, Lauren Shohet, Jeffrey Shoulson, and Lyn Tribble. Many of us have since moved on, but I remember our conversations very fondly. Since then, I have found myself happily engaged (sometimes daily but sometimes also at an interval of years) with friends and fellow travelers: A. R. Braunmuller, Jim Brophy, Richard Burt, Joseph Campana, Jeffrey Jerome Cohen, Lowell Duckert, Mary Floyd-Wilson, Lowell Gallagher, David Glimp, Jonathan H. Grossman, Eileen Joy, Tom Leitch, Julia Reinhard Lupton, Steve Mentz, Nova Myhill, Vin Nardizzi, Gail Kern Paster, Lois Potter, Kellie Robertson, Brad Ryner, Bruce R. Smith, Karl Steel, Garrett A. Sullivan, Ayanna Thompson, Rob Wakeman, Lance Winn, and Michael Witmore. I am especially thankful to friends and welcome acquaintances who demanded more of me than I had at the time: Carla Mazzio once felt sure that I must have something to say about pastoral; Jonathan Gil Harris insisted that sheep, oranges, and yeast belonged together in one book when I was not so sure; Carolyn Dinshaw had a hunch that David Halperin's first book might have something for me; Stephen J. Campbell told me I was mad (in a good way) and has never failed since to offer his friendship and art historical expertise; Madhavi Menon inspired me to try to keep my argument turning; Laurie Shannon suggested that I think more about *otium*; and Bryan Reynolds offered jolts of energy by way of his example and encouragement.

Seed money for a summer's research in the United Kingdom came from the University of Delaware in the form of a General University Grant. More sustained research was made possible under the auspices of a long-term National Endowment for the Humanities fellowship at the Folger Shakespeare Library and a Franklin Award from the American Philosophical Society. I am grateful for the intellectual camaraderie of the other long-term fellows then at the Folger, Jim Bono, Paul Hammer, Tim Harris, and Carole Levin, and to the library's staff for making us feel so welcome. I found similarly generous intellectual friendship from the librarian and archivist at Stonyhurst College in Lancashire and also from the staffs of the University of Delaware's Morris Library and Swarthmore College Library.

I have had the privilege of learning from a cohort of talented graduate students at Delaware and beyond. Congratulations to Sabina Amanabayeva, Lizz Angello, Jim Beaver, Kevin Burke, Joshua Calhoun, Mike Clody, Hannah Eagleson, Darlene Farabee, Tom Hamill, Kyle Meikle, Kelly Nutter, Brie Parkin, Brad Ryner, David Satran, Joe Turner, Kyle Vitale, and Jane Wessel, all of whom wrote dissertations while this book was cooking.

I am thrilled that the book appears with University of Minnesota Press. I have been incredibly fortunate in my editors: Richard Morrison, who made publishing my first book a joy, and now the immensely supportive Doug Armato and the inspiring Cary Wolfe. Erin Warholm-Wohlenhaus was an amazingly attentive and helpful presence in shepherding the final manuscript when I was tempted to stray. I was also fortunate in my copy editor, Holly T. Monteith, whose gentle interventions surprised and delighted. Karl Steel read the manuscript with his inimitable rigor, eye for detail, and list of undeniable improvements. I am grateful for readings and challenges along the way from Jeffrey Jerome Cohen, Mary Floyd-Wilson, David Glimp, Julia Reinhard Lupton, Garrett A. Sullivan, Rebecca Winer, Michael Witmore, and Cary Wolfe.

This book is dedicated to Noah and Naomi Yates, who grew a lot while I was writing, and to all the imaginary creatures that inhabit our songs, stories, and conversations (the Juvenile Delinquents, Pig Slop, Muddy Spider, Funky Llama, Mr. Snake, and Gobble). Your imagination, love, laughter, and hard questions mark these pages at every point.

My partner in all things remains Rebecca Lynn Winer. I count myself luckier than I have any right to be for her love, support, and conversation.

—Swarthmore, Pennsylvania
Thanksgiving 2015

Notes

IMPRESSION

1 William Shakespeare, *Henry VI, Part 2,* ed. Roger Warren (Oxford: Oxford University Press, 2002), 4. 8. 56–57. Unless otherwise indicated, subsequent references appear parenthetically in the text.

2 For medieval analogues to this moment, see Sarah Kay, "Legible Skins: Animals and the Ethics of Medieval Reading," *postmedieval* 2, no. 1 (2011): 13–32; Kay, "Original Skin: Flaying, Reading and Thinking in the Legend of Saint Bartholomew and Other Works," *Journal of Medieval and Early Modern Studies* 36, no. 1 (2006): 35–74; and Bruce Holsinger's insightful "Of Pigs and Parchment: Medieval Studies and the Coming of the Animal," *PMLA* 124, no. 2 (2009): 616–23. For a useful survey of instances in which leather (the skin of a sheep, goat, or cow) appears in Shakespeare's plays, see Anston Bosman, "Shakespeare in Leather," in *The Forms of Renaissance Thought: New Essays in Literature and Culture,* ed. Leonard Barkan, Bradin Cormack, and Sean Keilen, 225–45 (New York: Palgrave Macmillan, 2009).

Jack Cade's skin memory stands as a powerful example also of what Erica Fudge has called the "persistent presence of the animal in the animal-made-object," which has the effect of "actually produc[ing] new agents." Fudge, "Renaissance Animal Things," *New Formations* 76 (2012): 94. On touch as producing this order of destabilizing sensation, among others, see Eve Kosofsky Sedgwick, *Touching Feeling: Affect, Pedagogy, Performativity* (Durham, N.C.: Duke University Press, 2003), and in Renaissance studies, Elizabeth D. Harvey, *Sensible Flesh: Touch in Early Modern Culture* (Philadelphia: University of Pennsylvania Press, 2002).

3 On the manufacture of parchment as indexed to numbers of calves or sheep, see Lucien Febvre and Henri Jean Martin, *The Coming of the Book: The Impact of Printing 1450–1800,* trans. David Gerrard (London: Verso, 1976), 18, and, more generally, Ronald Reed, *The Nature and Making of*

Parchment (Leeds: Elemete Press, 1975). For an excellent overview of the materials of book production, see Alexandra Gillespie and David Wakelin, eds., *The Production of Books in England 1350–1500* (Cambridge: Cambridge University Press, 2014). For a programmatic sense of how zooarchaeology might use a biomolecular analysis of parchment as a way of recovering the history of breeds of sheep and so motivate this archive as a way of approaching agricultural history, see M. D. Teasdale, N. L. Van Doorn, S. Fiddyment, C. C. Webb, T. O'Connor, M. Hofreiter, M. J. Collins, and D. G. Bradley, "Paging through History: Parchment as a Reservoir of Ancient DNA for Next Generation Sequencing," *Philosophical Transactions of the Royal Society: Biological Sciences* (December 8, 2014), doi:10.1098/rstb.2013.0379. This mode of analysis potentially enables a history of the book that would attend to individual animal and plant actors pressed to use in the making of individual books and manuscripts. For pioneering work in this direction, see Joshua Calhoun, "The Word Made Flax: Cheap Bibles, Textual Corruption, and the Poetics of Paper," *PMLA* 126, no. 2 (2011): 327–44.

4 On the "quasi-machine-like" operation of writing, see, e.g., Jacques Derrida, "Typewriter Ribbon: Limited Ink (2) (within Such Limits)," in *Material Events: Paul de Man and the Afterlife of Theory,* ed. Tom Cohen, Barbara Cohen, J. Hillis Miller, and Andrzej Warminski, 277–334 (Minneapolis: University of Minnesota Press, 2001). For a more wide-ranging study beyond the early modern period, see Jeffrey Masten, Peter Stallybrass, and Nancy J. Vickers, eds., *Language Machines: Technologies of Literary and Cultural Production* (New York: Routledge, 1997).

5 As Michael L. Ryder offers, parchment was stretched as opposed to tanned (as in leather working), producing a smooth and highly durable, almost laminated surface, but "despite all the smoothing during manufacture the flesh side can usually be distinguished from the grain side (skin surface) by its rougher texture and often darker colour." Ryder, "The Biology and History of Parchment," in *Pergament: Geschichte, Struktur, Restaurierung, Herstellung,* ed. Peter Rück (Sigmaringen: Thorbecke, 1991), 26.

6 Ronald Reed, "Some Thoughts on Parchment for Bookbinding," in *Pergament,* 218. Beyond such routinized biosemiotic transfers, Katherine M. Rudy documents the range of human bodily substances that leave their marks on medieval parchments—oils from fingerprints, food stains, tears, blood stains. Rudy, "Dirty Books: Quantifying Patterns of Use in Medieval Manuscripts Using a Densitometer," *Journal of Historians of Netherlandish Arts* 2, no. 1–2 (2010), doi:10.5092/jhna.2010.2.1.1.

7 Paul de Man, *The Resistance to Theory* (Minneapolis: University of Minnesota Press, 1986), 44.

8 On *ethos* and impersonation as a schoolboy exercise in Tudor England, see Lynn Enterline, *Shakespeare's Schoolroom: Rhetoric, Discipline, Emotion* (Philadelphia: University of Pennsylvania Press, 2011), 132–33. For a modeling of human users as "wetware," see Richard Doyle, *Wetwares: Experiments in Postvital Living* (Minneapolis: University of Minnesota Press, 2003).

9 I take the term "insurgent writing" from Steven Justice's account of the Peasant's Revolt of 1381 as anything but an all-out attack on writing and instead on a carefully orchestrated and so tactically significant set of attacks on specific forms of legal documents as opposed to on writing per se; see Justice, *Writing and Rebellion: England in 1381* (Berkeley: University of California Press, 1994), 41 and, more generally, 13–60. For a reading of the Cade sequence as an apparent valorization of speech over writing, see Roger Chartier, "Jack Cade, the Skin of a Dead Lamb, and the Hatred for Writing," *Shakespeare Studies* 34 (2006): 77–89. For a longer treatment of the play in relation to its sources and the image of the populace it offers, and as a form of archival politics, see Julian Yates, "Skin Merchants: Jack Cade's Futures and the Figural Politics of Shakespeare's *Henry VI Part II,*" in *Go Figure: Forms, Energy, Matter in Early Modern England,* ed. Judith Anderson and Joan Pong Linton (Bronx, N.Y.: Fordham University Press, 2011), 149–69 and 199–203.

10 *Oxford English Dictionary,* s.v. "nap." http://www.oed.com/view/Entry/124991?rskey=TmLZkF&result=2&isAdvanced=false.

11 Porcupines were understood to discharge their spines when hunted. As Edward Topsell describes, "when they are hunted the beast stretcheth his skin and casteth [quills] off, one or two at a time." Such porcupines that might have been seen (barring hedgehogs) were "brought up and downe" from Africa to "Europe to be seene for mony." Topsell, *The History of Foure-Footed Beasts* (London, 1607), 588.

12 Chartier, "Jack Cade," 85. For a reconstruction of the Jack Cade rebellion, see I. M. H. Harvey, *The Jack Cade Rebellion* (Oxford: Clarendon Press, 1991). The list of possible source rebellions derives from Annabel Patterson's *Shakespeare and the Popular Voice* (Oxford: Blackwell, 1989), which is scrupulous in its treatment of the rebels as capable of political thought and action. The best reconstruction of how the play uses the chronicle histories of the rebellion remains Paola Pugliatti, "'More Than History Can Pattern': The Jack Cade Rebellion in Shakespeare's *Henry VI, 2,*" *Journal of Medieval and Renaissance Studies* 22 (1992): 451–78.

13 I have used Roger Warren's Oxford edition of the play because he renders this scene as a flat text without the usual editorial insertions that mark the rebels' voices as "asides." The issue for Warren is one of prescription. Mark their speeches as asides and you reduce the potential for reading the Cade sequence as disruptive or self-ironizing carnival and so sponsor a necessarily negative appraisal of the rebels. For Warren's rationale, see Shakespeare, *Henry VI, Part 2,* 52. On Shakespeare's use of metaplasm in his early histories as a tool for maintaining metrical regularity while incorporating allusions to or quotations from other texts, see Russ Macdonald, *Shakespeare's Late Style* (Cambridge: Cambridge University Press, 2006), 81. On Falstaff's use of a horizontal quasi-democratic "we," see Harry Berger Jr., "The Prince's Dog: Falstaff and the Perils of Speech-Prefixity," *Shakespeare Quarterly* 49, no. 1 (1998): 40–73.

14 On idiocy or the appearance of stupidity (of not knowing) as a way of slowing routines down and so registering the cosmopolitical wager or effect of a phenomenon, see Isabelle Stengers, "The Cosmopolitical Proposal," in *Making Things Public: Atmospheres of Democracy*, ed. Bruno Latour and Peter Weibel, 994–1003 (Cambridge, Mass.: MIT Press, 2005). Implicitly, I am reading Jack and the Jack Cade sequence as one such "idiocy" overtaken by the routines it seeks to interrupt. On the knife as a tool for erasing or altering writing on parchment or vellum, see Jonathan Goldberg, *Writing Matter: From the Hands of the English Renaissance* (Stanford, Calif.: Stanford University Press, 1990), 59–107.

15 The metaphor of the lamb and the flock reappears within what John Michael Archer names the play's "civil butchery." See Archer, *Citizen Shakespeare: Freemen and Aliens in the Language of the Plays* (New York: Palgrave, 2005), 19–20 and, esp., 80–81.

16 On Jack's self-narrating demise, see Thomas Cartelli, "Jack Cade in the Garden: Class Consciousness and Class Conflict in *2 Henry VI,*" in *Enclosure Acts: Sexuality, Property, and Culture in Early Modern England,* ed. Richard Burt and John Michael Archer, 48–67 (Ithaca, N.Y.: Cornell University Press, 1994). On his becoming earth, see Jean E. Feerick, "Groveling with Earth in Kyd and Shakespeare's Historical Tragedies," in *The Indistinct Human in Renaissance Literature,* ed. Jean E. Feerick and Vin Nardizzi, 231–52 (London: Palgrave Macmillan, 2012). For a discussion of this scene in terms of what it tells us about the experience of hunger in the period as a topic of political debate, see Hillary Eklund, "Revolting Diets: Jack Cade and the Politics of Hunger in *2 Henry VI,*" *Shakespeare Studies* 42 (2014): 51–63.

17 On "ontological choreography," see Charis Thompson, *Making Parents: The Ontological Choreography of Reproductive Technologies* (Boston: MIT Press, 2007).

18 Donna Haraway, *When Species Meet* (Minneapolis: University of Minnesota Press, 2008), 11. See also the movement within cultural anthropology to craft a descriptive regime that includes nonhuman actors within its purview and that assumes that "culture" is in no way a human preserve. Most influential here is the work of Philippe Descola. In *Beyond Nature and Culture,* trans. Janet Lloyd (2005; repr., Chicago: University of Chicago Press, 2013), Descola offers a program for re-imagining an anthropology that faces "the daunting challenge: either to disappear as an exhausted form of humanism or else to transform itself by rethinking its domain and its tools in such a way as to include in its object far more than the *anthropos*: that is to say, the entire collective of beings that is linked to him but is at present relegated to the position of a merely peripheral role" (xx). Among the burgeoning studies with an explicitly multispecies orientation, see Eben S. Kirksey and Stefan Helmreich, "The Emergence of Multispecies Ethnography," *Cultural Anthropology* 25, no. 4 (2010): 545–76, and, less programmatically, Stefan Helmreich, *Alien Ocean: Anthropological Voyages in Microbial Seas* (Berkeley: University of California Press, 2009); perhaps most expansively, Eduardo Kohn, *How Forests Think: Toward an Anthropology beyond the Human* (Berkeley: University of California Press, 2013); John Hartington Jr., *Aesop's Anthropology: A Multispecies Approach* (Minneapolis: University of Minnesota Press, 2014); the collection of essays in cultural anthropology *The Multispecies Salon,* ed. Eben Kirksey (Durham, N.C.: Duke University Press, 2014); Anna Lowenhaupt Tsing, *The Mushroom at the End of the World: On the Possibility of Life in Capitalist Ruins* (Princeton, N.J.: Princeton University Press, 2015); and Eben Kirksey, *Emergent Ecologies* (Durham, N.C.: Duke University Press, 2016). For a consideration of the ties between (environmental) history and biology that thinks through and with questions of co-evolution, scale, neuroscience, and the anthropocene, see "AHA Roundtable: History Meets Biology," *American Historical Review* 119, no. 5 (2014): 1492–629.

19 Haraway, *When Species Meet,* 3–4.

20 On the representational possibilities to be found in rock formations by human observers that exceed human agency in their making, see Roger Callois, *The Writing of Stones,* trans. Barbara Bray (Charlottesville: University Press of Virginia, 1985). See esp. his discussions of a series of limestones from Tuscany whose form evokes from him the words

"Castle" and "Portrait" (85–104). For a recent intriguing use of Callois, see Jeffrey Jerome Cohen, *Stone: An Ecology of the Inhuman* (Minneapolis: University of Minnesota Press, 2015), 86–87 and 169–71.

21 Michel Serres, *Biogea,* trans. Randolph Burks (Minneapolis, Minn.: Univocal, 2012), 171–72.

22 Haraway, *When Species Meet,* 4, 16–17.

23 Donna Haraway, *The Companion Species Manifesto* (Minneapolis: University of Minnesota Press, 2008), 20.

24 Erich Auerbach, *Mimesis: The Representation of Reality in Western Literature,* trans. Willard R. Trask (Princeton, N.J.: Princeton University Press, 1953). My use of the phrase "nonconventional entity" is indebted to Brian Massumi's *What Animals Teach Us about Politics* (Durham, N.C.: Duke University Press, 2014), 38. Massumi explains that the phrase derives from a desire, in the wake of an expanded sense of nonhuman agents, to "avoid the implicit anthropomorphism of designating the other only as the negative of the human." Less traumatized by the apparent threat of the reciprocal relation between anthropomorphism and zoomorphism (the co-making of beings), he adds that "it is necessary to recognize that *we are our own nonconventional entities.*"

25 On inquiry as a mode of hunting that subordinates the investigator to the phenomenon in the mode of what she calls "reciprocal capture," see Isabelle Stengers, *Cosmopolitics I,* trans. Robert Bononno (Minneapolis: University of Minnesota Press, 2010), 35–37. For an allied model of inquiry that derives from a lifetime's work in archives, metaphorized in terms of diving, currents, and the threat of drowning, see Arlette Farge, *The Allure of the Archives,* trans. Thomas Scott-Railton (New Haven, Conn.: Yale University Press, 2013). On the inventory as a genre tuned to the phenomenology of everyday objects or what he termed "l'infra-ordinaire," see Georges Perec, *Species of Spaces and Other Pieces,* ed. and trans. John Sturrock (London: Penguin Books, 1999), 207–44. For an allied sense of storytelling, tracking, and sticking close to an entity as not just story but method, see Tsing, *Mushroom at the End of the World,* 111.

26 The phrase "flat ontology" appears first in Manuel De Landa's *A Thousand Years of Nonlinear History* (New York: Swerve, 2000) and derives from his elaboration of Gilles Deleuze and Félix Guattari's figures of the "rhizome," the "assemblage," and the "axiomatic" as elaborated in *A Thousand Plateaus: Capitalism and Schizophrenia,* 2 vols., trans. Brian Massumi (Minneapolis: University of Minnesota Press, 1987).

27 Cary Wolfe, "Human, All Too Human: 'Animal Studies' and the Humanities," *PMLA* 124, no. 2 (2009): 571. Like Kohn in *How Forests Think,* I am

interested in what it means to "provincialize" human language systems in Dipesh Chakrabarty's sense of the term, but I think that the lesson of Kohn's always mesmerizing attempts to go beyond language lie in the way he is returned, like it or not, to inventorying the way other forms of expression or semiosis inundate human language systems in the form of trans-species pidgins and modes of address. He is forced, in other words, to consider the impression such acts of marking make on our language systems become substrate to otherwise than human texts. See esp. chapter 4, "Trans-Species Pidgins," in Kohn, *How Forests Think,* 131–52. On the rhetoric of "provincializing," see Dipesh Chakrabarty, *Provincializing Europe: Postcolonial Thought and Historical Difference* (Princeton, N.J.: Princeton University Press, 2000).

28 Jacques Derrida, *Of Grammatology,* trans. Gayatri Chakravorty Spivak (Baltimore: Johns Hopkins University Press, 1974), 84.

29 "'Eating Well,' or the Calculation of the Subject: An Interview with Jacques Derrida," in *Who Comes after the Subject?,* ed. Eduardo Cadava, Peter Connor, and Jean-Luc Nancy (New York: Routledge, 1991), 116.

30 Ibid., 106. Here I rely also on David Wills's commentary in "Techn*e*ology or the Discourse of Speed," in *The Prosthetic Impulse: From a Posthuman Present to a Biocultural Future,* ed. Marquard Smith and Joanne Morra (Cambridge, Mass.: MIT Press, 2006), 237–64.

31 Cary Wolfe, *Before the Law: Humans and Other Animals in a Biopolitical Frame* (Chicago: University of Chicago Press, 2013), 74.

32 Timothy Morton articulates the issue a little differently, drawing on Graham Harman's equipmental reading of Martin Heidegger as elaborated in *Tool-Being: Heidegger and the Metaphysics of Objects* (Chicago: Open Court Press, 2002), *Guerrilla Metaphysics* (Chicago: Open Court Press, 2005), and *The Quadruple Object* (New York: Zero Books, 2011). "We are not going to try to bust through human finitude," he writes, "but to place that finitude in a universe of trillions of finitudes, as many as there are things—because a thing just is a rift between what is and how it appears, for any entity whatsoever, not simply for that special entity called the (human) subject)." Morton, *Hyperobjects: Philosophy and Ecology after the End of the World* (Minneapolis: University of Minnesota Press, 2013), 18. At issue here is Morton and Harman's divergence from Quentin Meillassoux's refusal of finitude, suspension of rhetoric, and assertion of a purified, witnessless technology of observation in Meillassoux, *After Finitude: An Essay on the Necessity of Contingency,* trans. Ray Brassier (London: Continuum Books, 2008).

33 Wolfe, *Before the Law,* 76.

34 Jacques Derrida, *Archive Fever,* trans. Eric Prenowitz (Chicago: University of Chicago Press, 1996), 11. On the need to think biopolitics with bibliography or media specificity, see Jacques Derrida, *Paper Machine,* trans. Rachel Bowlby (Stanford, Calif.: Stanford University Press, 2005), esp. the interview "Paper or Me, You Know . . . (New Speculations on a Luxury of the Poor)," 41–60, in which paper/s and person/s prove inseparable. On the need to think through the specific forms of "bare life" or "denuded life" *(vita nuda)* as they are "backed" or mediated and the difficulty for Giorgio Agamben in doing so, see Jacques Derrida, *The Beast and the Sovereign,* vol. 1, trans. Geoffrey Bennington, ed. Michel Lisse, Marie-Louise Mallet, and Ginette Michaud, 305–34 (Chicago: University of Chicago Press, 2009), and Giorgio Agamben, *Homo Sacer: Sovereign Power and Bare Life,* trans. Daniel Heller-Roazen (Stanford, Calif.: Stanford University Press, 1998).

35 Wolfe, *Before the Law,* 54–55, 27.

36 Ibid., 50. Here Wolfe parses the concept of "flesh" Roberto Esposito adapts from the phenomenological "flesh" of Maurice Merleau Ponty. See Esposito, *Immunitas: The Protection and Negation of Life* (London: Polity Press, 2011), 118–21, 140–41, and also Merleau Ponty, *The Visible and the Invisible,* trans. Alphonso Lingis (Evanston, Ill.: Northwestern University Press, 1968), 143–47. For Merleau Ponty's phenomenology, "flesh" is in no way reducible to matter but refers to the "coiling over of the visible upon the seeing body" (146). It takes on an archival relation to the human, a folding in and over and on itself that creates intruded exteriorities, or what Merleau Ponty terms "the intertwining—the chiasm."

37 Wolfe, *Before the Law,* 54.

38 Nicole Shukin, *Animal Capital: Rendering Life in Biopolitical Times* (Minneapolis: University of Minnesota Press, 2009), 11. Like Shukin, I am interested in "a different trajectory of biopolitical—or, we might say, zoopolitical—critique, one beginning with a challenge to the assumption that the social flesh and 'species body' at stake in the logic of biopower is predominantly human" (9). That said, I am not sure that the "tautology" of our mimetic relation to other beings (animal, plant, fungal, mineral, and so on) is necessarily always compromised. Hence my fugitive acts of theft, errant beachcombing, and examination of the technically reciprocal relations of anthropo-zoo-morphism. For an allied approach that inquires into the way different hierarchies of animation stack animal, plant, and mineral actors in ways crosscut by discourses of race, gender, and sexuality, see Mel Y. Chen, *Animacies: Biopolitics, Racial Mattering, and Queer Affect* (Durham, N.C.: Duke University Press, 2012).

39 Jacques Derrida, *The Animal That Therefore I Am,* trans. David Wills (New York: Fordham University Press, 2008), 29.

40 Jakob von Uexküll, *A Foray into the Worlds of Animals and Humans* with *A Theory of Meaning,* trans. Joseph D. O'Neil (Minneapolis: University of Minnesota Press, 2010), 44–52, 219–21.

41 On Despret's model for a mutually constitutive process rooted in trust, see Viciane Despret, "The Body We Care For: Figures of Anthropo-zoo-genesis," *Body and Society* 10, no. 2–3 (2004): 120–21. For Michel Serres's attempts to think beyond the neutrality of a parasitic chain with its excluded middles ("the third man") toward successive figures of symbiosis along with what frequently sounds like despair at what he takes to be "appropriation through pollution," writing as a form of excremental marking or re-marking, see Serres, *The Parasite,* trans. Lawrence R. Schehr (Minneapolis: University of Minnesota Press, 2007); Serres, *Angels: A Modern Myth,* trans. Francis Cowper (Paris: Flammarion, 1993); Serres, *The Natural Contract,* trans. Elizabeth MacArthur and William Paulson (Ann Arbor: University of Michigan Press, 1995); and Serres, *Malfeasance: Appropriation through Pollution,* trans. Anne-Marie Feenberg-Dibon (Stanford, Calif.: Stanford University Press, 2011).

42 "'*Lupus est homo homini, non homo, quom qualis sit non novit*' ('When one does not know him, man is not a man but a wolf for man')." Plautus, *Asinaria* (495), quoted in Jacques Derrida, *The Beast and the Sovereign,* 1:11. See also Serres's elaboration of this formula to include other animal figures: "Man is a wolf for men, an eagle for sheep, a rat for rats. In truth a *rara avis.*" Serres, *Parasite,* 7.

43 Haraway, *When Species Meet,* 17.

44 Ibid., 17–18. Haraway's charting of the word's use and etymology replays, in small, Sigmund Freud's strategy in approaching the uncanny. See Sigmund Freud, "The Uncanny," *Standard Edition* 17 (1917–19): 217–56.

45 Strikingly, this urge to document proof of life in the wake of a looming ecological disaster corresponds to the polarized predicament Derrida earlier called "nuclear criticism." As the soon to be over Cold War raised the threat of nuclear catastrophe, of "remainderless destruction," a total loss of the archive and the "human," to delirious heights, Derrida detected a set of moves that taxed the present with the fabulation of a referent: on one hand, the mutually assured destruction of nuclear oblivion, and on the other, the postivizing of textual traces in the name of "life"—an immanent ideology. See Jacques Derrida, "No Apocalypse, Not Now. (Full Speed Ahead, Seven Missiles, Seven Missives)," *Diacritics* 14, no. 2 (1984): 20–31. For a rethinking of this matrix in regard to global

warming, see the special issue of *Diacritics* devoted to climate change criticism and especially Richard Klein's "Climate Change through the Lens of Nuclear Criticism," *Diacritics* 41, no. 3 (2013): 82–87. For critiques positing "life" as some retrievable "good," see Tom Cohen, Claire Colebrook, and J. Hillis Miller, *Theory and the Disappearing Future: On De Man, on Benjamin* (New York: Routledge, 2012), 9–10. For a sense of the stakes to what to name our present moment, see Donna Haraway, "Anthropocene, Capitalocene, Plantationocene, Chthulucene: Making Kin," *Environmental Humanities* 6 (2015): 159–65, and Clive Hamilton, Christophe Bonneuil, and François Gemenne, eds., *The Anthropocene and the Global Environmental Crisis: Rethinking Modernity in a New Epoch* (London: Routledge, 2015).

46 In essence, I am suggesting that the likes of Bruno Latour's modeling of techniques as always also inquiries into their own propositional content attempts to craft a necessary link between *poiesis* and critique, between construction and deconstruction conceived, following Niklas Luhmann, as "second order observing." See Luhmann, "Deconstruction as Second-Order Observing," in *Distinction: Re-describing the Descriptions of Modernity,* trans. Joseph O'Neil, Elliott Schreiber, Kerstin Behnke, and William Whobrey, 94–112 (Stanford, Calif.: Stanford University Press, 2002), and Bruno Latour, "An Attempt at a Compositionist Manifesto," *New Literary History* 41, no. 3 (2010): 471–90.

47 Or, in Judith Butler's terms, "grievable life." Butler, *Precarious Life: The Powers of Mourning and Violence* (London: Verso, 2004). See also Wolfe's commentary and critique in *Before the Law,* 18–21.

48 Wolfe, *Before the Law,* 103.

49 Ibid.

50 On the violence of decision as cutting or the creation of an edge, see, in different registers, Serres, *Natural Contract,* 55, and Jacques Derrida, *The Gift of Death,* trans. David Wills (Chicago: University of Chicago Press, 1992), 53–82.

51 For these terms, see, among others, Bruno Latour, *Science in Action,* trans. Catherine Porter (Cambridge, Mass.: Harvard University Press, 1987), to his most recent *An Inquiry into Modes of Existence: An Anthropology of the Moderns,* trans. Catherine Porter (Cambridge, Mass.: Harvard University Press, 2013); Haraway, *Companion Species Manifesto*; Haraway, *When Species Meet*; Haraway, *Staying with the Trouble: Making Kin in the Chthulucene* (Durham, N.C.: Duke University Press, 2016); Jane Bennett, *Vibrant Matter: A Political Ecology of Things* (Durham, N.C.: Duke University Press, 2010); Stacy Alaimo, *Bodily Natures: Science, Environment,*

and the Material Self (Bloomington: University of Indiana Press, 2010); Timothy Morton, *Ecology without Nature: Rethinking Environmental Aesthetics* (Cambridge, Mass.: Harvard University Press, 2007); Morton, *The Ecological Thought* (Cambridge, Mass.: Harvard University Press, 2010); Morton, *Hyperobjects*; Rosi Braidotti, *The Posthuman* (London: Polity Press, 2013); Karen Barad, *Meeting the Universe Halfway: Quantum Physics and the Entanglement of Matter and Meaning* (Durham, N.C.: Duke University Press, 2007). For Stengers's modeling of ethics and technics, see Isabelle Stengers, *Power and Invention: Situating Science* (Minneapolis: University of Minnesota Press, 1997), 216.

52 Bruno Latour, *The Politics of Nature: How to Bring the Sciences into Democracy,* trans. Catherine Porter (Cambridge, Mass.: Harvard University Press, 2004), 62–64; Haraway, *Companion Species Manifesto*; Haraway, *When Species Meet.*

53 Timothy Morton makes a case for such a retooled notion of aesthetics in "An Object-Oriented Defense of Poetry," *New Literary History* 43, no. 2 (2012): 205–24. For an allied argument for the "human" refigured as telephone or screen, see Julian Yates, "It's (for) You: The Tele-T/r/opical Post-human," *Postmedieval* 1, nos. 1/2 (2010): 223–34, and Yates, "'Hello Everything': Renaissance/Post/Human," in *The Return of Theory in Early Modern English Studies 2: From Metaphysics to Biophysics,* ed. Paul Cefalu, Gary Kuchar, and Bryan Reynolds, 13–33 (New York: Palgrave Macmillan, 2014). For an allied model, see Ian Bogost, *Alien Phenomenology, or What It's Like to Be a Thing* (Minneapolis: University of Minnesota Press, 2012).

54 On interrupting or "freeze-framing" chains of reference, see Bruno Latour, *On the Modern Cult of the Factish Gods* (Durham, N.C.: Duke University Press, 2010), 99–123. On causation, see Graham Harman, "On Vicarious Causation," in *Collapse II,* ed. R. Mackay (Falmouth, U.K.: Urbanomic, 2007), 171–205.

55 Such a redefined humanities corresponds, I think, to the caution Barbara Johnson takes in thinking through the status of *prosopopoeia* as one of those relays between lyric poetry and the legal apparatus in her *Persons and Things* (Cambridge, Mass.: Harvard University Press, 2008). Working through Paul de Man's foundational essay "Anthropomorphism and Trope in the Lyric," in *The Rhetoric of Romanticism* (New York: Columbia University Press, 1984), she worries as to the way lyric finds itself conjoined with the law such that "legislators count on lyric poetry to provide" a "fallacious lyrical reading of the unintelligible," enabling them therefore to operate on the "assumption that the human *has been*

or *can be* defined so that it can then be presupposed without the question of its definition being raised as a question—legal or otherwise" (206–7). She regards de Man's decision to end his essay by giving the last words to an impersonal construction that personifies mourning itself as instructive in this regard. "True 'mourning' is less deluded," observes de Man, "the most *it* can do is to allow for non-comprehension and enumerate non-anthropomorphic, non-elegiac, non-celebratory, non-lyrical, non-poetic, that is to say prosaic, or, better *historical* modes of language power" (262). De Man's "true mourning," in this regard, comes to function as one instance, instead, in which we encounter "the loss of unconsciousness about the lack of humanness" to the "human." Johnson, *Persons and Things,* 207.

56 The question of sites of enunciation is what I take to be the material problem elaborated in Gayatri Chakravorty Spivak's classic "Can the Subaltern Speak?," in *Marxism and the Interpretation of Culture,* ed. Cary Nelson, 271–313 (Urbana: University of Illinois Press, 1988).

57 For this model of scientific inquiry as necessarily linking sets of conditions of knowledge production and existence, see Stengers, *Power and Invention,* 171.

58 For a passionate attempt to conserve the affective and political power of the "creaturely voice" as a privileged site for nonanthropic forms of justice, see Tobias Menely, *The Animal Claim: Sensibility and Creaturely Voice* (Chicago: University of Chicago Press, 2015). My treatment of Jack's skin memory puts me in sympathy with Menely's arguments for the purchase still to be had from lyric investments in voice, investments that, quite rightly, he argues mean that "any posthumanist theory of justice, any 'affirmative biopolitics' . . . will return . . . to the necessity of accounting for the communicative conditions in which we find ourselves answerable to the clamor of other beings who are like ourselves passionate and finite" (204). It is not possible to treat Menely's careful argument fully here, but I read him as offering an optimistic rejoinder to Barbara Johnson on the links between law and lyric in *Persons and Things* (see note 55).

59 On theft as a political technique, see Roland Barthes, *Mythologies,* trans. Annette Lavers (New York: Hill and Wang, 1972), 135.

60 Timothy C. Campbell, *Improper Life: Technology and Biopolitics from Heidegger to Agamben* (Minneapolis: University of Minnesota Press, 2011), 142.

1. COUNTING SHEEP IN THE BELLY OF THE WOLF

1 Michel Serres, with Bruno Latour, *Conversations on Science, Culture, and Time,* trans. Roxanne Lapidus (Ann Arbor: University of Michigan Press, 1995), 57.

2 Shakespeare, *Henry VI, Part 2,* 4.2.72–76.

3 Karl Marx, *Capital,* 3 vols., trans. Ben Fowkes (Harmondsworth, U.K.: Penguin, 1990), 143. In a similar instance of ovine prejudice in *The New Atlantis,* Francis Bacon refers to the audience that witnesses the miraculous arrival of the shipborne ark of biblical texts that cements the House of Solomon's control of the island world of Bensalem, "Renfusa," a combination of the Greek *rhen* (sheep) and *phusis* (nature/being). It remains unclear whether this founding event presents as divine fact or human rhetoric—the House of Solomon (science) having within its power the ability to produce optical illusions. See Bacon, *The New Atlantis and the Great Instauration,* ed. Jerry Weinberger (Wheeling, Ill.: Harlan Davidson, 1989), 47–48.

4 After writing this chapter, I became aware of Nicole Shukin's highly insightful opening of the problem I am tackling: the presence of an undergirding sheepy stratum to biopolitical thought. As she puts it with customary precision, "sheep are metaphorically omnipresent yet materially missing from the study of a technology of power that, according to Foucault, enfolds *human* individual populations who become subject to forms of pastoral care first institutionalized by the Christian Church and subsequently secularized by the modern state." See Shukin, "Tense Animals: On Other Species of Pastoral Power," *New Centennial Review* 11, no. 2 (2011): 145. As Shukin puts it, the question as to "how the government of human life might be biopolitically imbricated with that of other species is potentially opened up—yet actually foreclosed—by Foucault" (145). As I argue in this chapter, the archives of biopolitics are littered with ovine figures that thematize this foreclosure within the discourses of pastoral and pastoral care. Moreover, to think the biopolitics of pastoral across the lines of species requires a concerted thinking across or between the lines between animal and plant. Foucault's "foreclosure," as it were, might be understood as a further dislocation of sheep who return, I argue, by and through figures of plant being.

5 On sheep as stock and the origins of property, see Marc Shell, *Money, Language, Thought* (Baltimore: The Johns Hopkins University Press, 1982), esp. his chapter "The Whether and the Ewe: Verbal Usury and *The Merchant of Venice,*" 47–83. On the link between "cattle" and "capital"

and the Latin *pecus* (cattle) and *pecunia* (money), see Rhoda M. Wilkes, *Livestock/Deadstock: Working with Farm Animals from Birth to Slaughter* (Philadelphia: Temple University Press, 2010), 25–26.

6 Michel Foucault, *Security, Territory, Population: Lectures at the Collège de France, 1977–1978,* trans. Graham Burchell (New York: Picador, 2007), 169.

7 Michel Foucault, *The Archaeology of Knowledge,* trans. A. M. Sheridan Smith (New York: Pantheon Books, 1972), 125–31. For a brilliant speculation into the potential origins of pastoral power, or what I would call its "a priori," that posits the comanagement of human and sheepy flocks in such practices as "grazing," see Anand Pandian, "Pastoral Power in the Postcolony: On the Biopolitics of the Criminal Animal in South India," *Cultural Anthropology* 23, no. 1 (2008): 85–117.

8 In his catalog of oddities or bodily wonders, Dutch physician Nicolaes Tulp tells of an Irish boy, raised by sheep, who was "magis ferae, quam hominis speciem" (more a beast than a type of human) and who "manducabat solum gramen, ac foenum, et quidem eo delectu, quo curiosissimae oves" (ate only grass or hay, with the same choice as the fussiest of sheep). Tulp, *Observationes medicae* (Amsterdam: Elzevir, 1641), 312–13. Quoted and translated by Karl Steel in his "With the World, or Bound to Face the Sky: The Postures of the Wolf-Child of Hesse," in *Animal, Vegetable, Mineral: Ethics and Objects,* ed. Jeffrey Jerome Cohen (Washington, D.C.: Oliphaunt Books, 2012), 32.

9 Foucault, *Security, Territory, Population,* 198–200.

10 Michel Foucault, *"Society Must Be Defended": Lectures at the Collège de France, 1975–76,* trans. David Macey (New York: Picador, 2003), 254–55.

11 Haraway, *When Species Meet,* 73–74.

12 Beyond factory farming, these abuses would include the depredations of the Middle Passage and chattel (hear also cattle/capital) slavery that articulated their victims as if cattle, and so as a fungible biomass, the early modern circum-Atlantic circuit a forerunner of today's network of container ships and transport links. They would include also and obviously the systematic genocide of the Shoah.

13 Foucault, *Security, Territory, Population,* 169.

14 Ibid., 128–29. The French text uses the verb *dénombrer* as opposed to *compter* for "count," with undertones of enumeration and marking as opposed to a simple counting that treats sheep interchangeably as in the more colloquial expression *compter les moutons* (to count sheep). Michel Foucault, *Sécurité, Territoire, Population, Cours au Collège de France, 1977–1978,* ed. Alessandro Fontana and Michael Senellart (Paris: Gallimard Seuil, 2004), 132.

15 Ibid., 129.
16 Ibid., 194.
17 William Shakespeare, *Hamlet,* ed. Ann Thompson and Neil Taylor (London: Arden Shakespeare, 2006), 5. 1, 107–10.
18 Antonio Gramsci, *Prison Letters,* trans. Hamish Henderson (London: Pluto Press, 1988), 56. Subsequent references appear parenthetically in the text.
19 The Italian original of the epigraph to this chapter reads as follows: "In ogni momento sarò capace con una scossa di buttar via la pellaccia mezzo di asino e mezzo di pecora che l'ambiente sviluppa sulla vera propria naturale pelle . . . questo inverno, quasi tre mesi senza vedere il sole, altro che in qualche lontano riflesso. La cella riceve una luce che sta di mezzo tra la luce di una cantina e la luce di un acquario." Antonio Gramsci, *Lettere dal carcere,* trans. and ed. Sergio Caprioglio and Fubini Caprioglio (Torino: Einaudi, 1965), 110.
20 Agamben, *Homo Sacer.*
21 On this formulation of mimesis, see Tom Cohen, *Ideology and Inscription: "Cultural Studies" after Benjamin, De Man, and Bakhtin* (Cambridge: Cambridge University Press, 1998), 136, and Michael Taussig, *Mimesis and Alterity: A Particular History of the Senses* (London: Routledge, 1993).
22 John Whitfield, "Celebrity Clone Dies of Overdose," *Nature,* February 18, 2003, doi:10.1038/news030217-6.
23 As researchers at the Roslin Institute confirm, Dolly was euthanized because of an incurable viral disease, "sheep pulmonary adenomatosis (SPA)." http://www.roslin.ed.ac.uk/public-interest/dolly-the-sheep/a-life-of-dolly/. The prospect of "short-lived clones" arose from questions concerning "a cellular 'timer'" called "a telomere, which consists of a strand of genetic material DNA on the end of all chromosomes, rather like the plastic at the end of a shoelace. Every time a cell divides, the telomeres become shorter . . . we found, in the case of Dolly, that, although we could turn back time in one sense, converting an adult cell into an embryonic cell, we did not reset the cellular aging clock, so that her telomeres were 40 percent shorter (older) than those of a typical sheep of her age." Ian Wilmut and Roger Highfield, *After Dolly: The Uses and Misuses of Cloning* (New York: W. W. Norton, 2006), 233, 232–34 more generally. As Wilmut notes, this problem is not true of all animals when cloned, nor is it true of all sheep. The key cultural study of Dolly, thus far, remains Sarah Franklin, *Dolly Mixtures: The Remaking of Genealogy* (Durham, N.C.: Duke University Press, 2007), who reminds us that the word "clone is a term from botany, derived from the Greek *klon,* for twig" (19).

On the refiguring of human reproductive technologies by way or through Dolly, see also the collection of essays edited by E. Ann Kaplan and Susan Squier, *Playing Dolly: Technocultural Formations, Fantasies, and Fictions of Assisted Reproduction* (New Brunswick, N.J.: Rutgers University Press, 1999), and Susan Squier, *Liminal Lives: Imagining the Human at the Frontiers of Biomedicine* (Durham, N.C.: Duke University Press, 2004).

24 Franklin, *Dolly Mixtures,* 21, quoting Ian Wilmut, Keith Campbell, and Colin Tudge, *The Second Creation: The Age of Biological Control by the Scientists Who Cloned Dolly* (London: Headline, 2000), 24.

25 Thomas Shadwell, *The Virtuoso* (1676), ed. David Stuart Rhodes and Marjorie Hope Nicolson (Lincoln: University of Nebraska Press, 1966), 2.2.190–94. All subsequent references appear parenthetically in the text.

26 Marjorie Hope Nicolson, *Pepys' Diary and the New Science* (Charlottesville: University of Virginia Press, 1965), 68, 75.

27 The most comprehensive account of the story of these transfusions remains ibid., 55–99.

28 Ibid., 55–99, 166–69.

29 Thomas Birch, *The History of the Royal Society* (London, 1756), 2:212–14. A full list of all the experiments with descriptions appears in volumes 1–3 of the *Philosophical Transactions* (1665–68).

30 Edmund King to Boyle, November 25, 1667, in Boyle, *The Correspondence of Robert Boyle,* ed. Michael Hunter, Antonio Clericuzio, and Lawrence M. Principe (London: Pickering and Chatto, 2001), 3:366.

31 On Coga's unreliability, see Nicolson, *Pepys' Diary and the New Science,* 79–82. On the issue of witnessing and the status of scientific reliability in the period, see Steven Shapin, *A Social History of Truth* (Chicago: University of Chicago Press, 1992).

32 Henry Stubbe, *The Plus Ultra Reduced to a Non Plus* (London, 1670), 133.

33 King to Boyle, November 25, 1667, 3:367.

34 Louis Marin coins the term "trans*signifiance*" in his study of the seventeenth century, Jansenist, *Port-Royal Logic,* Eucharistic poetics, and beast fables in *Food for Thought,* trans. Mette Hjort (Baltimore: The Johns Hopkins University Press, 1989), 121–22. He understands thereby the rhetorical procedure by which "as in the *Logic of Port-Royal,* the miraculous Eucharistic sign tampered with the taxonomic boundaries separating the various kinds of signs from each other, the marvelous cookery of the tales[he reads] tampers with a full range of practices of *significance* . . . to permit . . . all manner of slippages, displacements, and transformations" as in the logic of Coga's theory.

35 For a provocative analysis of the grounding function of blood as the core term that operates in Christianity, see Gil Anidjar, *Blood: A Critique of Christianity* (New York: Columbia University Press, 2014).

36 Henry Stubbe, *A Specimen of Some Animadversions upon a Book Entituled, Plus Ultra* (London 1670), 179. Quoted also in Nicolson, *Pepys' Diary and the New Science,* 169.

37 Haraway, *When Species Meet,* 17. Throughout *Companion Species Manifesto* and *When Species Meet,* Haraway nods to what remains of the Catholicism of her childhood. In one instance, she writes, "I grew up in the bosom of two major institutions that counter the modernist belief in the no-fault divorce, based on irreconcilable differences, of story and fact. Both of these institutions—the Church and the Press [her father was a sports reporter] . . . Sign and flesh, story and fact. In my natal house, the generative partners could not separate. . . . No wonder culture and nature imploded for me as an adult." Haraway, *Companion Species Manifesto,* 18. In another, Haraway writes, "Raised a Roman Catholic, I grew up knowing that the Real Presence was present under both 'species,' the visible form of the bread and the wine. Sign and flesh, sight and food, never came apart for me again after seeing and eating that hearty meal." Haraway, *When Species Meet,* 18. Accordingly, she remains profoundly and productively invested in the convoking of matter and discourse in transubstantiation.

38 For a sociological investigation of the attitudes of workers in industrialized food webs to cows and sheep, see Wilkie, *Livestock/Deadstock,* esp. 115–86.

39 Sue Coe and Judith Brody, *Sheep of Fools: A Song Cycle for Five Voices* (Seattle, Wash.: Fantagraphics Books, 2005). See also Steve Baker, *Artist/Animal* (Minneapolis: University of Minnesota Press, 2013), 148.

40 http://www.serta.com/counting-sheep.

41 In *Camera,* July 2001.

42 Thomas More, *The Yale Edition of the Complete Works of St. Thomas More,* ed. Edward Surtz and J. H. Hexter (New Haven, Conn.: Yale University Press, 15), 4:1. All subsequent references appear parenthetically in the text.

43 Jane Schneider, "The Anthropology of Cloth," *American Review of Anthropology* 16 (1987): 419. For a comprehensive treatment of the way cloth manufacturing haunts sixteenth- and seventeenth-century English literature and culture in the form of complaints against the destructive power of sheep farming for wool, see Roze Hentschell, *The Culture of Cloth in Early Modern England: Textual Constructions of a National Identity* (Aldershot, U.K.: Ashgate, 2008), esp. 19–50.

44 Joan Thirsk, *Tudor Enclosures,* pamphlet 41 (London: Historical Association, 1959), and Thirsk, "The Common Fields," in *The Rural Economy of England: Collected Essays* (London, 1984), 35–36.

45 Hugh Latimer, *The Sermons of Hugh Latimer, Sometime Archbishop of Worcester,* ed. Rev. George Elwes Corrie for the Parker Society (Cambridge: Cambridge University Press, 1844), 38–78.

46 For an account of the wool trade in the period, see Peter J. Bowden, *The Wool Trade in Tudor and Stuart England* (London: Macmillan, 1962). For the period-specific language of "graziers" and "wool growing," see esp. 7–8 and also Erica Fudge, "Renaissance Animal Things," *New Formations* 76 (2012): 86–87.

47 On the language of mutuality and obligation, see Craig Muldrew, *The Economy of Obligation: The Culture of Credit and Social Relations in Early Modern England* (London: Macmillan, 1998), and esp. Andrew McRae, *God Speed the Plough: The Representation of Agrarian England, 1500–1660* (Cambridge: Cambridge University Press, 1996), esp. 23–57.

48 I am indebted to David Glimp for this observation and to his excellent essay "'Utopia' and Global Risk Management," *ELH* 75, no. 2 (2008): 263–90.

49 [Sir Thomas Smith], *A Discourse of the Commonweal of This Realm of England,* ed. Mary Dewer (Charlottesville: University of Virginia Press, 1969), 21.

50 For similar proverbial statements, see "The Decay of Tudor England Only by the Great Multitude of Sheep," cited in R. H. Tawney and Eileen Power, eds., *Tudor Economic Documents* (London, 1953), 3:52. See also Thomas Becon, *The Jewell of Joy* (1550), fol. 16b, quoted also in McRae, *God Speed the Plough,* 43.

51 Topsell, *Historie of Four-Footed Beastes,* 626. On the reappearance of the ovine figure in the sixteenth century, see Cathy Shrank, *Writing and Nation in Reformation England, 1530–1580* (Oxford: Oxford University Press, 2004), 44, 145, 217–19. An invaluable study cum survey of early and mid-Tudor writing, Shrank traces the passage of the figure into a more general economic discourse that worries over unequal exchange and especially usury, as in Thomas Wilson's *A Discourse upon Vsurye* (1572), where an entire bestiary of parasites is summoned to explain predatory economic practices. Wilson trades also on the visual pun that metaplasmically derives "Jews" from "iewes" in early modern orthography.

52 M. W. Beresford, "The Poll Tax and Census of Sheep, 1549," in *Time and Place: Collected Essays* (London: Hambledon Press, 1984), 137.

53 Ibid., 137–38.

54 "Causes of Dearth," in *State Papers Domestic,* Edward VI, v, no. 20. Quoted in Beresford, "Poll Tax," 141.
55 Beresford, "Poll Tax," 145.
56 Ibid., 155.
57 Ibid., 144.
58 On Kett's Rebellion, see Alexander Nevil, *Norfolkes Furies; or A View of Ket's Campe* (London, 1615), a translation of *De Furoribus Norfolciensium Ketto Duce* (1575), D3r.
59 Ibid., K3r.
60 M. L. Ryder, *Sheep and Man* (London: Duckworth, 1983). Ryder remains the best survey of the codomestication of sheep and human animals, though he dates the mutual domestication to 9000 BCE (3). Current social zooarchaeologists put the date much earlier; see, e.g., a standard textbook by Nerissa Russell, *Social Zooarchaeology: Humans and Animals in Prehistory* (Cambridge: Cambridge University Press, 2012), 208.
61 Karen Raber, *Animal Bodies, Renaissance Culture* (Philadelphia: University of Pennsylvania Press, 2013), 164. Raber offers a very persuasive supplemental countertext to the period's adulation of sheep, assembling her own archive of what she calls sheep's "material natures" as they manifest in husbandry manuals in the period as well as in More's *Utopia.* Raber reads *Utopia*'s cancellation of private property as opening up the possibility of a more differentiated and so hospitable series of relations between humans and other animals by and through the recognition of labor as a category. I agree entirely, though, as shall become clear in my next chapter, such "positive" representations that we may find in More's text are complicated by the rhetorical exigency of the frame narrative and are overdetermined by its winking antimimetic effects and the uncertain, unstable relation between its two books, all of which renders Raphael's description of the Utopian *res* almost impossible to adjudicate. Utopia always, always wins. For Raber's reading, see *Animal Bodies, Renaissance Culture,* 151–78.

For an excellent consideration of sheep as an icon of New Zealand's bestiary that works to undo their supposed "sheepishness," see Annie Potts, Philip Armstrong, and Deidre Brown, *A New Zealand Book of Beasts: Animals in Our Culture, History, and Everyday Life* (Auckland: Auckland University Press, 2013), 33–61. Allied to this book comes Philip Armstong's *Sheep* (London: Reaktion Books, 2016), which offers a comprehensive survey of sheep attentive to undoing their passivity and returning to them what I would call their charisma. Armstrong's excellent study arrived during the copy editing of this book, but I have tried to incorporate his arguments, allied to my own, where possible.

62 See, e.g., Philip Slavin's analysis in times of plague, "The Great Bovine Pestilence and Its Economic and Environmental Consequences in England and Wales, 1318–50," *Economic Historical Review* 65, no. 4 (2012): 1239–66, and Timothy A. Newfield, "A Cattle Panzootic in Early Fourteenth-Century Europe," *Agricultural History Review* 57 (2009): 155–90. For a consideration of outbreaks of "foot and mouth" in the United Kingdom and beyond, see Franklin, *Dolly Mixtures,* 168–94. On the deleterious effects of antibiotics in industrialized farming as precisely a matter of biopolitics, see Wolfe, *Before the Law,* 48–50.

63 In Japan, for example, sheep only began to be imported from China in the nineteenth century and then not successfully. Most of the five hundred thousand or so sheep living in Japan today derive from a flock of Corriedales imported in 1929, mostly on the island of Hokkaido; see Ryder, *Sheep and Man,* 301. Accordingly, sheep occupy a very different place within Japan's cultural imaginary, identified most readily with its interwar modernization and industrialization. Still, Hokkaido figures as a scene of pastoral retreat in Haruki Murakami's magical realist novels *Wild Sheep Dance* and *Dance, Dance, Dance,* in which the hero's picaresque adventures are punctuated by visits from a hybrid sheep-man as he searches after a magical-mystical sheep he has been recruited to locate by a mysterious crime boss. At one moment our hero too finds himself counting the sheep in a photograph, counting them again and again as the number comes out sometimes as thirty-two and at others thirty-three. For Murakami, then, counting sheep translated to a photographic medium has become a magi-technical forensic exercise that peeks into the organizing basis of the novel's world. See Murakami, *A Wild Sheep Chase,* trans. Alfred Birnbaum (New York: Vintage Books, 1989), esp. 71–74, and, for further adventures, *Dance, Dance, Dance,* trans. Alfred Birnbaum (New York: Vintage Books, 1994).

64 Marx, *Capital,* 1:556. Quoted also in Morton, *Ecology without Nature,* 86.

65 On transhumance in Europe, see John A. Marino, *Pastoral Economics in the Kingdom of Naples* (Baltimore: The Johns Hopkins University Press, 1988); Carla Rahn Phillips and William D. Phillips Jr., *Spain's Golden Fleece: Wool Production and the Wool Trade from the Middle Ages to the Nineteenth Century* (Baltimore: The Johns Hopkins University Press, 1997), and the Annales School classic by Emmanuel Le Roy Ladurie, *Montaillou: The Promised Land of Error,* George 30th Anniversary ed.

(New York: Braziller, 2008). On sheep in sixteenth-century Mexico, see Elinor G. K. Melville, *A Plague of Sheep: Environmental Consequences of the Conquest of Mexico* (Cambridge: Cambridge University Press, 2001); in early America, Virginia DeJohn Anderson, *Creatures of Empire: How Domestic Animals Transformed Early America* (Oxford: Oxford University Press, 2004); in Australia, see Sarah Franklin, *Dolly Mixtures,* 118–57. On the emerging ecological impact of sheep in the Middle East, see Paul Sillitoe, Ali A. Ashawi, and Abdul Al-Amir Hassan, "Challenges to Conservation and Land Use: Change and Local Partnerships in Al Reem Biosphere Reservation, West Qatar," *Journal of Ethnobiology and Ethnomedicine* 6, no. 28 (2010): 1–31. I am grateful to cultural anthropologist Kate McCellan for this last reference and look forward to her future work on emerging ecological forms in the Middle East.

66 Marx, *Capital,* 1:163–64.

67 Jameson, "Of Islands and Trenches: Naturalization and the Production of Utopian Discourse," *Diacritics* 7, no. 2 (1977): 4. Note: "naturalization" should read "neutralization."

68 Louis Marin, *Utopics or the Semiological Play of Textual Spaces,* trans. Robert A. Vollrath (Atlantic Highlands, N.J.: Humanities Press International, 1984), 143–50.

69 Richard Halpern, *The Poetics of Primitive Accumulation: English Renaissance Culture and the Genealogy of Capital* (Ithaca, N.Y.: Cornell University Press, 1991), 154.

70 The choice is never between having a "fetish" or not but between competing fetishes, some bad, some good, some too good to be true. See Peter Stallybrass, "Marx's Coat," in *Border Fetishisms: Material Objects in Unstable Spaces,* ed. Patricia Spyer (New York: Routledge, 1998), 184. As William Pietz argues, any accusation of "fetish" refers only to an irreconcilable difference between competing systems of value, "The Problem of the Fetish, 1," *Res* 9 (1985): 5–17.

71 Topsell, *Historie of Four-Footed Beastes,* 2v and 626–27. The term *profit* indicates not merely financial gain but the target word for translating the Latin *utilitas* into early modern English. Perhaps the most formative use of the term for writers at the end of the sixteenth century is in Roger Ascham, *The Schoolmaster,* ed. Lawrence V. Ryan (Charlottesville: University of Virginia Press, 1974). Throughout this text, Ascham uses the word *profit* to refer to a very broad range of uses that may be derived from natural and conceptual resources.

72 While there has as yet been no full-scale solo treatment of sheep as a privileged actor or actant in early modern England, within the variously

inflected discourse of critical animal studies or critical animal studies–inspired historicist criticism, the following scholars have made inroads into such an account. No one has done more than Erica Fudge to advance this endeavor. In a series of books and edited collections of essays, the highlights of which remain *Perceiving Animals: Humans and Beasts in Early Modern English Culture* (Basingstoke, U.K.: Macmillan, 2000) and *Brutal Reasoning: Animals, Rationality, and Humanity in Early Modern England* (Ithaca, N.Y.: Cornell University Press, 2006), she has unmoored the default privileging of the human subject as the orienting unit of our narratives and challenged the Cartesian impulse that renders animals automatons, producing a counterhistory to this overly familiar story. Building on this work, in *The Accommodated Animal: Cosmopolity in Shakespearean Locales* (Chicago: University of Chicago Press, 2013), Laurie Shannon finds a series of latent cosmopolitan possibilities and even what she names a "zootopian constitution" in various Shakespearean locales. Shannon's work sets the tone for the salvaging or prospecting of what Michel Foucault might have called "subjugated [historical] knowledges" that have been systematically excluded or removed from view by the default emplotting of the past as a story of human emergence.

Other key arguments come from Bruce Boehrer, *Animal Characters: Nonhuman Beings in Early Modern Literature* (Philadelphia: University of Pennsylvania Press, 2010); Andreas Hofele, *Stage, Stake, and Scaffold: Humans and Animals in Shakespeare's Theatre* (Oxford: Oxford University Press, 2011); and Raber, *Animal Bodies, Renaissance Culture.* And for a very interesting stand-alone essay on sheepishness in the period, see Paul A. Yachnin, "Sheepishness in The Winter's Tale," in *How to Do Things with Shakespeare,* ed. Laurie Maguire (Oxford: Blackwell, 2008), 210–29.

For the Middle Ages, key recent contributions come from Karl Steel's *How to Make a Human: Animals and Violence in the Middle Ages* (Columbus: The Ohio State University Press, 2011), and Susan Crane, *Animal Encounters: Contacts and Concepts in Medieval Britain* (Philadelphia: University of Pennsylvania Press, 2013).

73 Leonard Mascall, *The First Booke of Cattell* (1591), O1v.

74 Gervase Markham, *Cheap and Good Husbandrie* (London, 1616), 104.

75 Topsell, *Historie of Four-Footed Beastes,* 607–8.

76 http://cinema-scope.com/cinema-scope-magazine/1107/.

77 The ethical derivative here propels the experience of shepherding into a moral philosophical register, especially when the practitioner begins as a novice or incompetent. The popularity of Tudor how-to books, such as Mascall's, derives in part from the need to restore knowledge of how to

manage an estate for persons of a middling sort made good. Today, the genre of memoir includes a similar form of husbandry/self-care manual, as former high-flying professionals opt out and retrain as something else, frequently something deemed more "grounded" or "real." See, e.g., Joan Jarvis Ellison's extremely readable *Shepherdess: Notes from the Field* (West Lafayette, Ind.: Purdue University Press, 1995), which is explicitly lauded and introduced as an example of wisdom literature. On the gendering of different forms of livestock management and the advent of the "hobby farmer" who turns back chronological time, see Wilkes, *Livestock/Deadstock,* 43–64, 89–114.

78 Thomas Fella, *Commonplace Book,* MS V. a. 311, 51r–52v, Folger Shakespeare Library, Washington, D.C.

79 Ibid.

80 On sizing, see the pioneering work of Joshua A. Calhoun, "Legible Ecologies: Animals, Vegetables, and Reading Matter in Renaissance England" (PhD diss., University of Delaware, 2011), and "The Word Made Flax: Cheap Bibles, Textual Corruption, and the Poetics of Paper," *PMLA* 126, no. 2 (2011): 327–44.

81 For a sense of devastation that the removal of sheep from the English landscape might produce, see John Taylor the water poet's *Taylor's Pastorall,* in which he considers how "infinite numbers of people rich and poore" owe their livelihood to sheep. "No Ram," he writes, "no Lambe, no Lambe no Sheepe, no Sheepe no Wooll, no Wooll no Woolman, no Woolman no Spinner, no Spinner no Weaver, no Weaver no Cloth, no Cloth no Clothier, no Clothier no Clothworker, Fuller, Tucker, Shearman, Draper, or scarcely a rich Dyer." Taylor, *All the Workes . . . Collected in One Volume* (1630), facsimile ed. (London: Spenser Society, 1868–69), 538. Quoted in McRae, *God Speed the Plough,* 226.

82 On the translation of alleged ovine passivity into figures of vegetal growth, see Armstrong, *Sheep,* 146–48. Armstrong cites the example of the "Vegetable Lamb of Tartary" from *The Travels of Sir John Mandeville* as the most famous literalization of this construction of sheep. See also *The Travels of Sir John Mandeville,* ed. M. C. Seymour (Oxford: Oxford University Press, 1967), 204.

83 Quoted in Louis A. Montrose, "Of Gentlemen and Shepherds: The Politics of Elizabethan Pastoral Form," *English Literary History* 50, no. 3 (1983): 439–42.

84 On the permutations and celebratory connotations of Psalm 23, see Bruce R. Smith, *The Key of Green: Passion and Perception in Renaissance Culture* (Chicago: University of Chicago Press, 2008), 181–90. Hatton's

letter is quoted in Agnes Strickland, *The Lives of the Queens of England* (London: George Bell, 1892), 321–22. See also Ian W. Archer et al., eds., *Religion, Politics, and Society in Sixteenth-Century England* (Cambridge: Cambridge University Press, 2003).

85 William Bradford, *Of the Plymouth Plantation, 1620–1647,* ed. Francis Murphy (New York: Random House, 1981), 355–56. One of the best commentaries on this case remains Jonathan Goldberg's in *Sodometries: Renaissance Texts, Modern Sexualities* (Stanford, Calif.: Stanford University Press), 223–49. Goldberg unpacks what it means for Granger to serve as a "transfer point for energies directed against Indians and women; his crossing of species is also a racial and gender crossing" (241). Goldberg's work presages the way for readings that chart the way certain genres of writing about animals cohabit with dissident sexualities such as Susan McHugh's *Animal Stories: Narrating across Species Lines* (Minneapolis: University of Minnesota Press, 2011) and Mel Y. Chen's *Animacies: Biopolitics, Racial Mattering, and Queer Affect* (Durham, N.C.: Duke University Press, 2012). On interspecies sex acts in early modern England, see Erica Fudge, "Monstrous Acts: Bestiality in Early Modern England," *History Today* 50, no. 8 (2000): 20–25, and Thomas Courtney, "'Not Having God before His Eyes': Bestiality in Early Modern England," *Seventeenth Century* 26, no. 1 (2011): 149–73.

86 Bradford, *Of the Plymouth Plantation,* 355–56.

87 Philip K. Dick, *Do Androids Dream of Electric Sheep?* (New York: Del Rey Books, 1968), 177. In earlier editions, the novel is set circa 1992. Subsequent references appear parenthetically in the text.

88 On "affective labor," see Michael Hardt and Antonio Negri, *Multitude: War and Democracy in the Age of Empire* (New York: Penguin Books, 2005).

89 The novel's opening scene, in which Iran and Rick use their "mood [altering] organ" dialing up successive psychological states to annoy one another, renders Descartes's mind–body dualism a humorous, impossible shuttling back and forth between states of being. Dick, *Do Androids Dream of Electric Sheep?,* 3–5.

90 On the abyssal "indistinction" between reaction and response and the difficulty of parsing the essence of either as the preserve or condition of the human, other animals, plants, and so on, see Jacques Derrida, *The Animal That Therefore I Am,* trans. David Wills (New York: Fordham University Press, 2008).

91 Ursula K. Heise, "From Extinction to Electronics: Dead Frogs, Live Dinosaurs, and Electric Sheep," in *Zoontologies,* ed. Cary Wolfe (Minneapolis: University of Minnesota Press, 2003), 59–81.

92 On this dynamic, see Nancy Armstrong, *Desire and Domestic Fiction: A Political History of the Novel* (Oxford: Oxford University Press, 1990).

93 Derrida, *The Animal That Therefore I Am,* 62–63.

94 Jacques Derrida, *The Beast and the Sovereign,* 2 vols., trans. Geoffrey Bennington, ed. Michel Lisse, Marie-Louise Mallet, and Ginette Michaud (Chicago: University of Chicago Press, 2009), 1:328 and 336.

95 http://thebigsheep.co.uk/.

96 On the naming of domestic animals in early modern England, see Keith Thomas, *Man and the Natural World: Changing Attitudes in England 1500–1800* (London: Penguin Books, 1983), 113–15. On the potential to be found by attending carefully to sheep and cows as potential members of human households in early modern England, see Erica Fudge, "Farmyard Choreographies in Early Modern England," in *Renaissance Posthumanism,* ed. Joseph Campana and Scott Maisano, 145–66 (Brooklyn, N.Y.: Fordham University Press, 2016).

97 For this sense of a potential or excess even in the context of zoos, see Brian Massumi, *What Animals Teach Us about Politics* (Durham, N.C.: Duke University Press, 2014), 65–97.

98 On the disanimating arithmetic of "livestock" as its origins find themselves routed from Noah's Ark through early modern imaginary arithmetic to the horrors of the Middle Passage and on to the interchangeable infrastructures of container ships, see Shannon, *Accommodated Animal,* 270–83.

99 Thelma Rowell, "A Few Peculiar Primates," in *Primate Encounters: Models of Science, Gender, and Society,* ed. Shirley C. Strum and Linda Fedigan (Chicago: University of Chicago Press, 2000), 65–66.

100 Ibid., 69.

101 For an attempt to take on the perspective of cows and sheep, or to offer an account of the past from that position, see the installations presented in print and photographic form in Laura Gustafsson and Haapoja Terika, eds., *History According to Cattle* (Brooklyn, N.Y.: Punctum Books, 2015).

2. WHAT WAS PASTORAL (AGAIN)?

1 On *logos* as counting, calculus, or ratio, see Derrida, *Beast and the Sovereign,* 1:337–38.

2 As Empson writes, "pastoral though 'about' is not 'by' or 'for'" those it describes. William Empson, *Some Versions of Pastoral* (1935; repr., New York: New Directions, 1974), 6. My reading of Empson is indebted to "The Dead End of Formalist Criticism," in which Paul de Man remarks that *Some Versions of Pastoral* masquerades as a genre study but deals with

the fundamental "ontology of the poetic," of Being itself, "wrapped . . . as is his [Empson's] wont, in some extraneous matter that may well conceal the essential." De Man, *Blindness and Insight: Essays on the Rhetoric of Contemporary Criticism,* 2nd rev. ed. (1977; repr., Minneapolis: University of Minnesota Press, 1983), 239.

3 As Louis A. Montrose objects, "merely to pose the question of 'what pastoral really is' is to situate oneself within an idealist aesthetics that represses the historical and material determinations of any written discourse." Montrose, "Of Gentlemen and Shepherds: The Politics of Elizabethan Pastoral Form," *English Literary History* 50, no. 3 (1983): 416. For an attempt to revalue or redefine pastoral as an active engagement with issues of land use and social critique in sixteenth- and seventeenth-century England, see Ken Hiltner, *What Else Is Pastoral? Renaissance Literature and the Environment* (Ithaca, N.Y.: Cornell University Press, 2011).

4 What is or what proletarian literature will be is the question with which *Some Versions of Pastoral* begins and which Empson attempts to adjudicate—will it, or must it, be pastoral? (1–27). Subsequent references appear parenthetically in the text.

5 On sleep as a state of being in which animals enter into a similarity with plants and the keying of this model to the vegetative soul in Aristotelian traditions and its denial by post-Cartesian thought, and return in contemporary sleep science, see Garrett A. Sullivan Jr., *Sleep, Romance, and Human Embodiment: Vitality from Spencer to Milton* (Cambridge: Cambridge University Press, 2012), esp. 99–100 and 112–14.

6 Brian Vickers, "Leisure and Idleness in the Renaissance: The Ambivalence of *Otium* (Part II)," *Renaissance Studies* 4, no. 2 (1990): 130. See also the previous essay, "Leisure and Idleness in the Renaissance: The Ambivalence of *Otium* (Part I)," *Renaissance Studies* 4, no. 1 (1990): 1–37. On the origins of leisure, see also Peter Burke, "The Invention of Leisure in Early Modern Europe," *Past and Present* 146 (1995): 136–50.

7 By far the most subtle and perceptive reading of the painting remains Stephen J. Campbell's lyrical, chapter-long treatment "Mantegna's Mythic Signatures in *Pallas and the Vices,*" in *The Cabinet of Eros: Renaissance Mythological Painting and the Studiolo of Isabella D'Este,* 146–68 (New Haven, Conn.: Yale University Press, 2004). Campbell's meticulous reconstruction of the allegorical and mythological frame of reference to the painting along with its revival of an ironic dimension to humanist *fabulae* and Mantegna's own variously encrypted signatures calls attention to the refusal to allow the text of *otium* to settle into a didactic moral calisthenics.

8 This moment of stilling recalls the concept of practical sovereignty as an "attenuated, nonmimetic relation" that inheres in spaces of ordinariness and unconscious practices developed by Lauren Berlant in "Slow Death: Sovereignty, Obesity, Lateral Agency," *Critical Inquiry* 33 (2007): 754–55, and then of "lateral agency" as an attempt to defuse the drama of subjectivation in chapter 3 of her *Cruel Optimism* (Durham, N.C.: Duke University Press, 2011). Berlant's valorization of a lateral agency that avoids mastery, future-oriented subjectivity, and optimizing technologies of self with their attendant "melodrama" of the subject resonates also with Jane Bennett's granting of an epistemological privilege or possibility to the practices of hoarders or hoarding. The focus of both, in this sense, remains trained on our encounters with an object world as mediated by the commodity form and our infrastructures. I attempt to imagine something quite distinct from this scene of active-passive consumption, which I think replays fairly traditional accounts of *otium* or deactivation even when conceived as a tactic of sorts or as, for Berlant, "a disavowal of life-affirming sovereignty." What happens, I ask, instead, if the calculus, the drama of subjectification, plays out across and with the lines of flesh as it is parceled out in animal and plant forms? On hoarding, see Jane Bennett, "Powers of the Hoard: Further Notes on Material Agency," in Cohen, *Animal, Vegetable, Mineral,* 237–69.

9 On Foucault's exhortation to finish, see Foucault, *Security, Territory, Population,* 163. For a similarly stated desire for deactivation, suspension, or inoperability, see various moments in *The Open,* when Giorgio Agamben despairs of rendering the "anthropological machine" less "bloody and lethal." Agamben, *The Open: Man and Animal,* trans. Kevin Attell (Stanford, Calif.: Stanford University Press, 2004), esp. 91–92.

Figures of suspension, idling, and deactivation are repeated throughout Agamben's work, notably in *State of Exception,* trans. Kevin Attell (Chicago: University of Chicago Press, 2005), where he speaks of a "halt[ing of] the machine" (87), and in his reading of Walter Benjamin's essay "The Critique of Violence" (61–62).

10 Leo Marx, "Pastoralism in America," in *Ideology and Classic American Literature,* ed. Sacvan Bercovitch and Myra Jehlen (Cambridge: Cambridge University Press, 1986), 45.

11 Raymond Williams, *Television: Technology and Cultural Form* (1974–75; repr., London: Routledge, 1990), 78. For a modeling of cultural forms as mediated always by the "rerun," see Derek Kompare, *Rerun Nation: How Repeats Invented American Television* (New York: Routledge, 2005).

12 Quentin Skinner, "Sir Thomas More's *Utopia* and the Language of Renaissance Humanism," in *The Languages of Political Theory in Early Modern Europe,* ed. Anthony Pagden, 123–58 (Cambridge: Cambridge University Press, 1987). This essay is revised and reprinted as "Thomas More's *Utopia* and the Virtue of True Nobility," in Quentin Skinner, *Visions of Politics* (Cambridge: Cambridge University Press, 2002), 2:213–44.

13 Quentin Skinner, "Sir Thomas More's Utopia," 132 and 131–35 more generally.

14 As Skinner writes, "the laws and customs of Utopia not only forbid *otium* and require *negotium* from everyone; they are also designed to ensure that elements of civic virtue are encouraged, praised and admired above all. Thus we learn that the Utopians are all trained in virtue." Ibid., 143. For the hugely influential model of self-cancellation, see Stephen Greenblatt, *Renaissance Self-Fashioning from More to Shakespeare* (Chicago: University of Chicago Press, 1980), 11–73. For an allied if differently staged reading in which *Utopia* produces the desired "silent body" of the middle-class consumer (input equals output), see Halpern, *Poetics of Primitive Accumulation,* 61–100. For an analysis of the tension between *otium* and *tempus* in *Utopia* as an optimizing "technology of self" allied to an ecology, see Julian Yates, "Humanist Habitats; or, Eating Well with *Utopia*," in *Environment and Embodiment in Early Modern England,* ed. Mary Floyd-Wilson and Garrett Sullivan, 187–209 (London: Palgrave Macmillan, 2007).

15 Anthony Grafton and Lisa Jardine, *From Humanism to the Humanities: Education and the Liberal Arts in Fifteenth- and Sixteenth-Century Europe* (Cambridge, Mass.: Harvard University Press, 1986). On the precarity of the liberal arts and of humanism from their inception, in Italy, see the closing sections on the fall of the humanists in Jacob Burckhardt, *The Civilization of the Renaissance in Italy,* trans. S. G. C. Middlemore (London: George Allen and Unwin, 1965).

16 On "passage techniques," see Timothy J. Reiss, *Mirages of the Selfe: Patterns of Personhood in Ancient and Early Modern Europe* (Stanford, Calif.: Stanford University Press, 2003), 469–87. On the sleeplessness of our stage of capitalism, see Jonathan Crary, *24/7: Late Capitalism and the Ends of Sleep* (London: Verso, 2014).

17 *Opus Epostolarum Des Erasmi Roterodami,* ed. P. S. Allen et al. (Oxford: Clarendon Press, 1906–58), 2, 339. Quoted also in *Complete Works of St. Thomas More,* xv. Subsequent references appear parenthetically in the text.

18 *Complete Works of St. Thomas More,* 4:38/39.

19 Jameson, "Of Islands and Trenches," 4.

20 The Yale translation renders *ocium* as "time." Moreover, throughout this passage, the preference for modern English word order means that the various conjugations of the Latin verb *prandere* in More's original text, indicating the passage of time in terms of eating and the relation to the table, are rendered instead as conjunctions. The effect is that the insistence in More's text on eating and the time devoted to food is erased.

21 Michel Jeanneret, *A Feast of Words: Banquets and Table Talk in the Renaissance,* trans. Jeremy Whitely and Emma Hughes (Cambridge: Polity Press, 1991), 114.

22 Ibid., 112. Jeanneret quotes Erasmus, *Convivium fabulosum,* in *Colloquies,* trans. C. R. Thompson (Chicago: University of Chicago Press, 1975), 257.

23 Keith Thomas, *Man and the Natural World: Changing Attitudes in England, 1500–1800* (Harmondsworth, U.K.: Penguin Books, 1984), 250–51.

24 Samuel Pepys, *The Diary of Samuel Pepys,* eds. Robert Latham and William Matthews (Berkeley: University of California Press), 8:336. Unless otherwise indicated, subsequent references appear parenthetically in the text.

25 Terry Gifford, *Pastoral* (New York: Routledge, 1999), 16.

26 Williams, *Television,* 13.

27 The phrase "making up people" comes from Ian Hacking, "Making Up People," in *Reconstructing Individualism: Autonomy, Individuality, and the Self in Western Thought,* ed. Thomas C. Heller, Morton Sosna, and David E. Wellberry, 222–36 (Stanford, Calif.: Stanford University Press, 1986).

28 Agamben, *The Open,* 63. Subsequent references appear parenthetically in the text.

29 Wolfe, *Before the Law,* 27.

30 Jameson, "Of Islands and Trenches," 19.

31 Derrida, *Beast and the Sovereign,* 1:321.

32 Ibid., 439.

33 For Derrida the "prodigious archive" in question was the "intact cadaver of an insect surprised by death, in an instant, by a geological or geothermal catastrophe, at the moment at which it was sucking the blood of another insect, 54 million years before humans appeared on earth." Jacques Derrida, "Typewriter Ribbon: Limited Ink (2) ('Within Such Limits')," in *Material Events: Paul de Man and the Afterlife of Theory,* ed. Tom Cohen, Barbara Cohen, J. Hillis Miller, and Andrej Warminski (Minneapolis: University of Minnesota Press, 2001), 331.

34 Uexküll, *A Foray into the Worlds of Animals and Humans,* 52.

35 Robert N. Watson, *Back to Nature: The Green and the Real in the Late Renaissance* (Philadelphia: University of Pennsylvania Press, 2006), 66.

For a similar reading of pastoral, keyed this time to the rise of the machine, see Jonathan Sawday, *Engines of the Imagination: Renaissance Culture and the Rise of the Machine* (New York: Routledge, 2007), 294–309.

36 Raymond Williams, *The Country and the City* (Oxford: Oxford University Press, 1975), 9–12, 13–45. Unless otherwise indicated, all subsequent references appear parenthetically in the text.

37 For an allied modeling of genre as rhizome, see Wai Chee Dimock, "Genre as World System: Epic and Novel on Four Continents," *Narrative* 14, no. 1 (2006): 85–101.

38 Williams, *Television,* 49–50.

39 Paul de Man, *The Resistance to Theory* (Minneapolis: University of Minnesota Press, 1987), 11.

40 Paul Alpers demotes Williams to a footnote and then to an in-text appearance only as a source of a "mis-application" of terms to Ben Jonson's "To Penshurst," in *What Is Pastoral?* (Chicago: University of Chicago Press, 1996), 60, but treats Empson at length (37–43). Terry Gifford wonders whether (or not) Empson really is the butt of Williams's annoyance in two references in *Country and the City*; see Gifford, *Pastoral,* 9–10.

41 Morton, *Ecology without Nature,* 31ff. The sense of mimesis as predatory and pro-active rather than as copy derives from Taussig's *Mimesis and Alterity* and Cohen's *Ideology and Inscription.*

42 David M. Halperin, *Before Pastoral: Theocritus and the Ancient Tradition of Bucolic Poetry* (New Haven, Conn.: Yale University Press, 1983), 211. Subsequent references appear parenthetically in the text. Thanks go to Carolyn Dinshaw for sending me back to Halperin. On the origins of pastoral, see also Pandian, "Pastoral Power in the Postcolony."

43 William Berg, *Early Virgil* (London: Athalone Press, 1974), 13–14. Quoted in Halperin, *Before Pastoral,* 186.

44 John Fletcher, *The Faithful Shepherdess,* ed. Florence Ada Kirk (New York: Garland, 1980), 15–16.

45 Montrose, "Of Gentlemen and Shepherds," 427.

46 Ibid., 434–35.

47 Fernand Braudel, *Capitalism and Material Life, 1400–1800* (New York: Harper and Row, 1967), 68.

48 Gifford cites the work of anthropologist Michael Herzfeld, *The Poetics of Manhood: Contest and Identity in a Cretan Mountain Village* (Princeton, N.J.: Princeton University Press, 1985) and also conversations with the "current ethnographer and two young shepherds from the village." See Gifford, *Pastoral,* 13–15.

49 Halperin, *Before Pastoral,* 85–117, esp. 96–97.

50 Rowell, "A Few Peculiar Primates," 65–66.
51 Ibid., 69.
52 Ibid., 70.
53 On this notion of composition and prospecting, see Bruno Latour, "An Attempt at a Compositionist Manifesto," *New Literary History* 41, no. 3 (2010): 471–90. For a revaluing of the adjective *interesting,* see Stengers, *Power and Invention,* 83–87. On the word *interesting* as enabling a mechanism that allows analysts to "toggle between aesthetic and non-aesthetic judgments," and so to do (or, perhaps, appear to do) ideological work (positively and negatively), see Sianne Ngai, *Our Aesthetic Categories: Zany, Cute, Interesting* (Cambridge, Mass.: Harvard University Press, 2012), 116–17.
54 Rowell, "A Few Peculiar Primates," 69.
55 On the discursive function of the "true," see Michel Foucault, "The Discourse on Language," in *The Archaeology of Knowledge,* 215–37. On the initial difficulty Rowell faced in placing work, see Vinciane Despret, *Quand le loup habitera avec l'Agneau* (Paris: Editions du Seuil, 2002), 164–65, and Despret, "Sheep Do Have Opinions," in Latour and Weibel, *Making Things Public,* 360–68.
56 Bruno Latour, "A Well-Articulated Primatology: Reflections of a Fellow-Traveller," in Strum and Fedigan, *Primate Encounters,* 368.
57 Ibid., 381.
58 Ibid., 367.
59 Ibid., 374.
60 Ibid., 372. See also Uexküll, *A Foray into the Worlds of Animals and Humans,* 44–52.
61 Rowell, "A Few Peculiar Primates," 66.
62 Ibid.
63 For a plotting of cultural studies and the wake of poststructuralism as the extension of "Universal History" to all manner of beings and an attempt to incorporate all within a single narrative, see Kerwin Lee Klein, "In Search of Narrative Mastery: Postmodernism and the People without History," *History and Theory* 34, no. 4 (1995): 275–98.
64 The key text on "falsification" as a ground for identifying scientific practice remains Karl Popper, *The Logic of Scientific Discovery* (New York: Harper and Row, 1968). Falsification remains an important rubric for Stengers's cosmopolitics, particularly in the essays gathered in *Power and Invention* and *The Invention of Modern Science,* trans. Daniel W. Smith (Minneapolis: University of Minnesota Press, 2000).
65 Derrida, *Of Grammatology,* 24.

66 Hans-Jörg Rheinberger, *Toward a History of Epistemic Things: Synthesizing Proteins in the Test Tube* (Stanford, Calif.: Stanford University Press, 1997), 184. As Rheinberger offers, he is obviously working through and with Derrida's *Of Grammatology.*

67 Despret, "Sheep Do Have Opinions," 360, and also *Quand le loup habitera avec l'Agneau,* 137–38, 164–65, and 202–5. For Donna Haraway's commentary on Despret and Rowell as oriented to Agamben's figure of the "open," see *When Species Meet,* 33–35. Sarah Franklin also treats Rowell's research in *Dolly Mixtures,* 196–98. Philip Armstrong makes good use of Rowell also in *Sheep,* 39–40 and 49–50. See also the passing but significant commentary by McHugh in *Animal Stories,* 159–62.

68 Despret, "Sheep Do Have Opinions," 361–62. Subsequent references appear parenthetically in the text.

69 Rowell, "A Few Peculiar Primates," 367. See also T. E. Rowell and C. A. Rowell, "The Social Organization of Feral *Ovis Aries* Ram Groups in the Pre-Rut Period," *Ethology* 95 (1993): 213–32, and T. E. Rowell and C. A. Rowell, "Till Death Do Us Part: Long-Lasting Bonds between Ewes and Their Daughters," *Animal Behaviour* 42 (1991): 681–82.

70 Rowell, "A Few Peculiar Primates," 368.

71 Despret, *Quand le loup habitera avec l'Agneau,* 208. Unless otherwise indicated, references appear parenthetically in the text. For a translation of this section of the book regarding the raven, see Vinciane Despret, "The Enigma of the Raven," trans. Jeffrey Bussolini, *Angelaki* 20, no. 2 (2015): 57–72. This special issue of *Angelaki* devoted to the work of Vinciane Despret, edited by Brett Buchanan, Matthew Chrulew, and Jeffrey Bussolini, is an invaluable aid to readers of her work and includes important translations of key essays and commentaries from the editors, translators, and fellow travelers, such as Donna Haraway.

72 See, e.g., the work of Bernd Heinrich, *Ravens in Winter* (New York: Vintage Books, 1991), and Heinrich, *Mind of the Raven* (New York: HarperCollins, 2000).

73 Isabelle Stengers, "Who Is the Author?," in *Power and Invention,* 171.

74 Despret continues this work through a variety of different publications, most recently in *Que diraient les animaux, si... on leur posait les bonnes questions?* (Paris: La Découverte, 2012), which extends the reach of the inquiry to consider artistic endeavors and new media alongside scientific practice.

75 Despret, *Quand le loup habitera avec l'Agneau,* 276–77. The French reads as follows: "Si le prophète a eu raison de nous faire confiance, le 'nous' que désigne cette confiance reste encore à constuire. Ce 'nous' qui

enrole de plus en plus d'humains et de non-humains dans une histoire et dans un monde communs reste encore à négocier, progressivement. Avec confiance et vigilance; avec de bonnes et de mauvaises habitudes. Hériter de la prophétie, c'est aussi hériter de ce qu'ont appris ceux qui ont commencé à la mettre en oeuvre: ce qui compte pour nous compte autrement pour ceux que nous invitons à constuire ce 'nous.'

"Alors? Alors, le loup habitera peut-être avec l'agneau mais l'agneau ne dormira sans doute que d'un oeil. Si Isaïe nous proposait la confiance, il nous rappelait en même temps, avec cette sage précaution, la vigilance: n'oublions pas de demander à l'enfant de les conduire par la main."

76 "'Eating Well,' or the Calculation of the Subject: An Interview with Jacques Derrida," in *Who Comes after the Subject?,* ed. Eduardo Cadava, Peter Connor, and Jean-Luc Nancy (New York: Routledge, 1991).

77 Despret, *Quand le loup habitera avec l'Agneau,* 277.

78 Bruno Latour, *The Pasteurization of France,* trans. Alan Sheridan and John Law (Cambridge, Mass.: Harvard University Press, 1988), 194. Martin Heidegger, "Letter on Humanism," in *Basic Writings,* ed. David Farrell Krell (San Francisco: Harper San Francisco, 1992), 234.

79 Topsell, *Historie of Four-Footed Beastes,* 629–30.

80 For this Rabelais-inspired version of Ovid's myth, see Sir John Harington, *A New Discourse on a Stale Subject, Called the Metamorphosis of Ajax* (1596), ed. Elizabeth Story Donno (London: Routledge and Kegan Paul, 1962), 67–68.

3. INVISIBLE INC. (TIME FOR ORANGES)

1 Sir John Peyton to the Privy Council, October 5, 1597. Historical Manuscript Commission, *Calendar of Manuscripts of the Most Ho. the Marquis of Salisbury, K.G., etc. Preserved at Hatfield House* (London: HMSO, 1883) (hereinafter HMC), 8:417–18.

2 Leonard Mascall, *Of Spots and Stains* (London, 1588), A3r.

3 Rowell, "A Few Peculiar Primates," 70.

4 Despret, *Quand le loup habitera avec l'Agneau,* 266–67.

5 Michael Marder, *Plant-Thinking: A Philosophy of Vegetal Life* (New York: Columbia University Press, 2013), 9.

6 Ibid., 10.

7 Jacques Derrida, *Glas,* trans. John P. Leavey Jr. and Richard Rand (Lincoln: University of Nebraska Press, 1986), 15. Marder, *Plant-Thinking,* 112 and 10. On the structure of living on or "*survivance,*" see Derrida, *Beast and the Sovereign.* In the course of these seminars, and in particular

in the coreading of Daniel Defoe's *Robinson Crusoe* alongside Martin Heidegger's *The Fundamental Concepts of Metaphysics,* the structure of *survivance* comes to comprehend questions of archiving, the archive, the book. See esp. Derrida, *Beast and the Sovereign,* 2:119–46. For an excellent consideration of plants in relation to concepts of "life," "territory," and the "animal" in Foucault, Derrida, and Deleuze and Guattari, see Jeffrey T. Nealon, *Plant Theory: Biopower and Vegetable Life* (Stanford, Calif.: Stanford University Press, 2015).

8 On a desire to move beyond detective fiction and discovery, see Michel Serres, *Biogea,* trans. Randolph Burks (Minneapolis, Minn.: Univocal, 2010), 19. Serres offers readers the structure of a detective story tied to the appearance of a dead body in a river but aims to deactivate or cancel out this genre in favor of the presentation of a *milieu* or mesh of plant, animal, mineral actors or, better still, an understanding that it makes no sense to separate out the actors that make up *Biogea.* On singularity or totality writing, see Paul Augé, *In the Metro,* trans. Tom Conley (Minneapolis: University of Minnesota Press, 2002), 60. On the archive as a repertoire to be performed, see Diana Taylor, *The Archive and the Repertoire: Performing Cultural Memory in the Americas* (Durham, N.C.: Duke University Press, 2003). I am grateful to Ayanna Thompson for sending me to Taylor.

My modeling of an "event" as some *thing* that changes or alters by and through our performance of it remains indebted to the work of Bruno Latour and actor network theory. Latour's *Aramis, or the Love of Technology,* trans. Catherine Porter (Cambridge, Mass.: Harvard University Press, 1996), provides one source of inspiration for this chapter; Iain Pears's *An Incident of the Finger Post* (London: Riverhead Books, 1998) provides another.

9 On recusant life in sixteenth-century England, see Hugh Aveling, *Northern Catholics: The Catholic Recusants of the North Riding of Yorkshire 1558–1790* (London: Geoffrey Chapman, 1966); John Bossy, *The English Catholic Community 1570–1850* (London: Darton, Longman, and Todd, 1975); and Christopher Haigh, *Reformation and Resistance in Tudor Lancashire* (Cambridge: Cambridge University Press, 1975). On anti-Catholic prejudice, see Peter Lake, "Anti-Popery: The Structure of a Prejudice," in *Conflict in Early Stuart England,* ed. Richard Cust and Ann Hughs, 81–210 (London: Longman, 1989), and Arthur F. Marotti, ed., *Catholicism and Anti-Catholicism in Early Modern English Texts* (London: St. Martin's Press, 1999). On the network of sites and the practice of Jesuits and missionary priests in England, see Michael Hodgetts, *Secret*

Hiding-Places (Dublin: Veritas, 1986), and Julian Yates, *Error, Misuse, Failure: Object Lessons from the English Renaissance* (Minneapolis: University of Minnesota Press, 2003).

10 "An Acte against Jesuites, Semynarie Priestes and such other Sundrie Persons," in *Statutes of the Realm* (London, 1819), 4:706–7.

11 The phrase "seedman of treason" was in widespread use, at least among the authorities. It appears, for example, in a letter to Elizabeth I written by Richard Topcliffe, chief priest hunter to the crown, dated 1588. See Henry Foley, *Records of the English Province of the Society of Jesus* (London: Burns and Oates, 1878), 2:355–56.

12 Gramsci, *Prison Letters,* 51–54. The Italian reads as follows: "Lo scrivere mi è anche diventato un tormento fisico, perché mi dànno degli orribili pennini, che grattano la carta e domandano un'attenzione ossessionante alla parte meccanica dello scrivere. Credevo di poter ottenere l'uso permanente della penna e mi ero proposto di scrivere i lavori ai quali ti ho accennato; non ho però ottenuto il permesso e mi dispiace insistere. . . .

"Questo ti spiega come passo il tempo, quando non leggo; ripenso a tutte queste cose, le analizzo capillarmente, mi ubbriaco di questo lavoro bizantino. . . .

"Ecco, vedi; un altro oggetto di analisi molto interessante: il regolamento carcerario e la psicologia che matura su di esso da una parte, e sul contatto coi carcerati, dall'altra, tra il personale di custodia. Io credevo che due capolavori (dico proprio sul serio) concentrassero l'esperienza millenaria degli uomini nel campo dell'organizzazione di massa: il manuale del caporale e il catechismo cattolico. Mi sono persuaso che occorre aggiungere, sebbene in un campo molto piú ristretto e di carattere eccezionale, il regolamento carcerario, che racchiude dei veri tesori di introspezione psicologica." Gramsci, *Lettere dal carcere,* 41–42.

13 Sir John Peyton to the Privy Council, October 5, 1597, HMC, 417–18.

14 Peyton was appointed in June 1597, replacing Sir Richard Berkeley. He would be replaced by Sir William Waad. John Gerard, *The Autobiography of an Elizabethan,* trans. Philip Caraman (New York: Longmans, Green, 1951), 114.

15 Royal Armouries, MS I. 243. Reproduced in Anna Keay, *The Elizabethan Tower of London: The Haiward and Gascoyne Plan of 1597,* London Topographical Society Publication 158 (London: Society of Antiquaries and Historic Royal Palaces, 2001), 58.

16 MS Eng Hist. E. 195, Bodleian Library, University of Oxford. Reproduced also in Keay, *Elizabethan Tower of London,* 61.

17 As Peter Lake and Michael Questier make clear, while the Tower could on occasion come the closest to a maximum security prison that Tudor prison ecology may have had, it was frequently if not always possible for well-connected prisoners to arrange to move around within its precincts. In October 1588, the Earl of Arundel had managed to bribe servants and officials to allow him to hear Mass. Attempts to improve security at the Tower, such as Peyton conducted, were intermittent and constant initiatives and at the mercy of a system that was "privatized and decentralized" (188) and without a central authority that would follow through on proposals. In 1582, for example, the Privy Council had specified "exact measures to be taken to ensure security . . . even specifying the shape and structure of all the windows" (196). Lake and Questier, *The Anti-Christ's Lewd Hat: Protestants, Papists, and Players in Post-Reformation England* (New Haven, Conn.: Yale University Press, 2002), 187–269.

18 Gerard, *Autobiography of an Elizabethan,* 116, and Stonyhurst MS A. v. 22, 158. Gerard's memoir came to England in the seventeenth century, when the Jesuits were expelled from France and its territories and the College of St. Omer was translated to Stonyhurst, Lancashire, where it became Stonyhurst College, now a private Catholic school. The original manuscript is thought to be lost, but the text survives at Stonyhurst in two copies: an incomplete seventeenth-century version and a fuller eighteenth-century copy. The manuscript appears in a very literal translation that sought to capture something of Early Modern English in *The Life of John Gerard,* trans. John Morris (London: Burns and Oates, 1881), and then in a modernized but free translation by Philip Caraman. Throughout, I use Caraman's translation, supplementing it with the Latin original where matters of sense and lexical specificity seem most crucial.

19 Gerard, *Autobiography of an Elizabethan,* 119, and Stonyhurst MS A. v. 22, 162–63.

20 Kristie Macrakis offers a redaction of Gerard's story in *Prisoners, Lovers, and Spies: The Story of Invisible Ink from Herodotus to Al Qaeda* (New Haven, Conn.: Yale University Press, 2014), 41–45, and notes that, curiously, Gerard's faith in orange juice seems misplaced. In laboratory conditions, "all citrus react in the same way" (43, 306).

21 John Baptista Porta, *Natural Magick* (London, 1658), 341.

22 Hugh Platt, *The Jewell House of Art and Nature* (London, 1594), 13.

23 For examples of recipes in manuscripts, see MSS V. a. 159 fol. 66 and V. a. 140 fols. 14v–15r, Folger Shakespeare Library.

24 Ben Jonson, *Volpone, or the Fox,* ed. Alvin B. Kernan (New Haven, Conn.: Yale University Press, 1962), 2.1.68–73.

25 Thomas Middleton, *Thomas Middleton: The Collected Works,* ed. Gary Taylor and John Lavagnino (Oxford: Clarendon Press, 2007), 1.1.308–10.

26 *Calendar of State Papers, Domestic Series, of the Reign of James I, 1603–1610,* nos. 241–46. The letters are reproduced also in Foley, *Records of the English Province of the Society of Jesus,* 4:102–7, and in J. Travers, *Gunpowder: The Players behind the Plot* (Kew, U.K.: National Archives, 2005). The most clearheaded paleographical analysis of this confused archive comes from Paul Wake in "Writing from the Archive: Henry Garnet's Powder-Plot Letters and Archival Communication," *Archival Science* 8, no. 2 (2005): 69–84. Wake describes the way the *State Papers* rationalizes but also scrambles the order and content of the letters and offers them as an example of the way the archive itself is already a structuring phenomenon that encrypts the circumstances that gave rise to the letters themselves. For a wider consideration of secret communications within a culture of letter writing, see James Daybell, *The Material Letter in Early Modern England: Manuscript Letters and the Culture and Practices of Letter-Writing, 1512–1635* (London: Palgrave Macmillan, 2012), 148–74, esp. 167–68.

27 Christopher Marlowe, *Doctor Faustus,* ed. Roma Gill (New York: W. W. Norton, 1965), 2.1.62–81.

28 On dematography, see Jeffrey Kahan, *Shakespiritualism: Shakespeare and the Occult, 1850–1950* (New York: Palgrave Macmillan, 2013), 163.

29 Lowell Gallagher, "Faustus's Blood," *English Literary History* 73, no. 1 (2007): 10.

30 Abraham Cowley, *The Collected Works of Abraham Cowley,* vol. 2, part 1, *The Mistress,* ed. Thomas O. Calhoun, Laurence Heyworth, and J. Robert King (Newark: University of Delaware Press, 1993), 28–29. I am grateful to Liza Blake for sending me to Cowley's poem.

31 On deinscription and for a rich rendering of the way the figure of the Book of nature shaped early modern imaginations, see James Bono, *The Word of God and the Languages of Man: Interpreting Nature in Early Modern Science and Medicine,* vol. 1 (Madison: Wisconsin University Press, 1995).

32 Juliet Fleming, *Graffiti and the Writing Arts of Early Modern England* (Philadelphia: University of Pennsylvania Press, 2001), 20.

33 On "ash" as a figure of archival loss, see Derrida, *Archive Fever,* 100–101. In the last paragraph of the postscript, Derrida writes, "We will always wonder what, in this *mal d'archive,* he [Freud] may have burned. We will always wonder, sharing with compassion in this archive fever, what have burned of his secret passions, of his correspondence, or of his 'life.'

Burned without limit, without remains, and without knowledge. With no possible response, be it spectral or not, short of or beyond suppression, on the other edge of repression, originary or secondary, without a name, without the least symptom, and without even an ash." For an earlier, equivalent caution, see also Michel de Certeau in "Reading as Poaching," where the remnants of the tactical operations of other readers (annotations or marks on books) serve as evidence merely of the presence of the other, of "life," or of "lives" lived in ways that cannot be assimilated to any set of codes and that endure merely and maximally *now* as residues that enliven a book. De Certeau is scrupulous in this regard and names such traces "a relic," "a remainder," "a sign of their erasure." De Certeau, *The Practice of Everyday Life,* trans. Steven Rendall (Berkeley: University of California Press, 1984), 1:165–76 and 35.

34 Marder, *Plant-Thinking,* 33. It seems worth pointing out here the way, coincidentally, the action of lemon juice on parchment, their "sympathy" in an early modern lexicon, serves for Graham Harman and a foundational text in object-oriented ontology as a way of approaching the way every and any *thing* exists in and of and for itself and endures quite happily in the absence of human witnessing. Bypassing post-Kantian philosophy, here he advocates for a return to metaphysics that resurrects a theory of substances. For Harman, we can describe the shape and color of a lemon, its smell and taste, but in the absence of other entities that cause notes or aspects of lemon being to presence, we can only allude to, say, the effect of lemon juice as an acid that weakens paper (rags) or parchment (sheep skin), causing the area marked to oxidize more quickly than the surrounding white space. If it seems that Harman is still speaking a post-Kantian language of relationality, assembly, or network—the lemon morphing in and by its joining or attachment to other things—he is eager to point out that "while all of this suggests that the notes or qualities of [orange] belong to the domain of relationality . . . it is only that the [paper] or parchment which can prehend acid-emission as a distinct feature of [orange] being are now present: indeed the [paper] or parchment *makes* it a separate feature." *Things* endure in and of themselves and exteriorize their potentialities by their encounter with other things. For Harman, then, the figure of a network, contexture, infrastructure, traces the encounters between *things.* It exists then, in my terms, as a dead-alive fossil of their phenomenalization: an archive. Harman, *Tool-Being,* 263–64.

35 On writing in the period as twinned with cutting and the coarticulation of pen and knife, see Goldberg, *Writing Matter,* 57–109. On pens more

generally and the use of the word to delimit a field of techniques that included painting, drawing, and writing, which opens this discussion to the elaborated forms of women's text(ile) production, see Susan Frye, *Pens and Needles: Women's Textualities in Early Modern England* (Philadelphia: University of Pennsylvania Press, 2010), 1–30. On erasable tablets in the period, see Peter Stallybrass, Roger Chartier, J. Franklin Mowery, and Heather Wolfe, "Hamlet's Tables and the Technologies of Writing in Renaissance England," *Shakespeare Quarterly* 55 (2004): 379–419. On whitewashing, see Fleming, *Graffiti,* 73–78. On palimpsesting (scraping) as a partial but incomplete erasure that produces what we now call palimpsests or polytemporal texts, see Jonathan Gil Harris, *Untimely Matter in the Time of Shakespeare* (Philadelphia: University of Pennsylvania Press, 2009), 1–25.

36 Philip Caraman calls the narrative "a private account of the adventures written for his fellow Jesuits and perhaps, in the first place, novices under his direction." Gerard, *Autobiography of an Elizabethan,* xvii. Alice Dailey offers also that the memoir marks Gerard's attempt to craft an explanation of the way his career in England was not able to correspond to the available tropes and genres of the martyrdom narrative, necessitating his production of a new genre that hybridizes memoir, hagigography, spiritual autobiography and the how-to book. Dailey, *The English Martyr from Reformation to Revolution* (Notre Dame, Ind.: University of Notre Dame Press, 2012), 1–2, 196–206. On Catholic spiritual biography or life writing, see also Molly Murray, *The Poetics of Conversion in Early Modern English Literature* (Cambridge: Cambridge University Press, 2009), 36–51.

37 H. H. Cooper, "Clandestine Methods of the Jesuits in Elizabethan England as Illustrated in an Operative's Own Classic Account," Central Intelligence Agency, *Studies Archive Indexes* 5, no. 2 (1993), https://www.cia.gov/library/center-for-the-study-of-intelligence/kent-csi/vol5no2/html/v05i2a12p_0001.htm.

38 Stonyhurst MS A. v. 22, 158–59.

39 For a treatment of this episode as part of a larger pedagogical function of the text sensitive to the way knowing the codes of invisible ink built or reinforced a sense of community, see Anne M. Myers, "Father John Gerard's Object Lessons: Relics and Devotional Objects in Autobiography of a Hunted Priest," in *Catholic Culture in Early Modern England,* ed. Ronald Corthell, Frances E. Dolan, Christopher Highley, and Arthur F. Marotti (Notre Dame, Ind.: University of Notre Dame Press, 2007), 216–35.

40 William Shakespeare, *Much Ado about Nothing*, ed. F. H. Mares (Cambridge: Cambridge University Press, 2003), 2.1.217–24.

41 For a useful overview of the status of the *Exercises* as a set of routines for managing or modeling or attuning somatic process, see Philip Endean, "The Spiritual Exercises," in *The Cambridge Companion to the Jesuits,* ed. Thomas Worcester (Cambridge: Cambridge University Press, 2008), 52–67. On the *Exercises* as a shaping force in Early Modern English writing, see Louis Martz, *The Meditative Poem: An Anthology of Seventeenth-Century Verse* (New York: New York University Press, 1963).

42 On the Tower as a contested site of inscription, see Elizabeth Hanson, *Discovering the Subject in Early Modern England* (Cambridge: Cambridge University Press, 1998), 24–54; Lake and Questier, *Anti-Christ's Lewd Hat,* 187–269; and Dailey, *English Martyr,* 197–98.

43 For this version of the poem, see Gerard Kilroy, *Edmund Campion: Memory and Transcription* (Aldershot, U.K.: Ashgate, 2005), 153; on the graffiti as source of consolation, see also Myers, "Father John Gerard's Object Lessons," 229–30.

44 For a modeling of human users of technological platforms as "wetware," see Doyle, *Wetwares.* For a radicalization of animal recruitment through the offers of sweetness from fruits and color and nectar from flowers, see Doyle's exploration of altered states of consciousness attained through psychedelic drug use in *Darwin's Pharmacy: Sex, Plants, and the Evolution of the Noösphere* (Seattle: University of Washington Press, 2011).

45 Another strategy would be to follow Hélène Cixous, who leaves the orange whole in *Vivre l'orange* (To live the orange), although she admits that "*oran/je*" was "the first word I cut." In living the orange, Cixous tropes her earlier "*je vois l'orange,*" opting for the tactile contiguity of the touch or caress over the violence of the realm of sight, the orange always remaining both intact and multiple even as it is performed. The same is true for Ponge, I think, though "L'orange" focuses on the consumption of a single, historical orange. Cixous, *L'heure de Clarice Lispector, précédé de Vivre l'orange,* trans. Ann Liddle and Sarah Cornell (Paris: Des femmes, 1989).

46 Francis Ponge, *Selected Poems,* trans. C. K. Williams and John Montague (Winston-Salem, N.C.: Wake Forest University Press, 1994), 23. The French reads as follows:

"Comme dans l'éponge il y a dans l'orange une aspiration à / reprendre contenance après avoir subi l'épreuve de l'expression. / Mais où l'éponge réussit toujours, l'orange jamais: car ses cellules ont / éclaté, ses tissus se sont déchirés. Tandis que l'écorce seule se / rétablit mollement dans sa forme grâce à son élasticité, un liquide / d'ambre s'est répandu, accompagné de rafraîchissement, de parfums / suaves, certes,—mais souvent aussi de la conscience amère d'une / expulsion prématurée de pépins.

"Faut-il prendre parti entre ces deux manières de mal supporter / l'oppression ?—L'éponge n'est que muscle et se remplit de vent, / d'eau propre ou d'eau sale selon : cette gymnastique est ignoble. / L'orange a meilleurs goût, mais elle est trop passive,—et ce sacrifice / odorant... c'est faire à l'oppresseur trop bon compte vraiment.

"Mais ce n'est pas assez avoir dit de l'orange que d'avoir rappelé sa / façon particulière de parfumer l'air et de réjouir son bourreau. Il / faut mettre l'accent sur la coloration glorieuse du liquide qui en / résulte et qui, mieux que le jus de citron, oblige le larynx à s'ouvrir / largement pour la prononciation du mot comme pour l'ingestion / du liquide, sans aucune moue appréhensive de l'avant-bouche dont / il ne fait pas hérisser les papilles.

"Et l'on demeure au reste sans paroles pour avouer l'admiration / que suscite l'enveloppe du tendre, fragile et rose ballon ovale dans cet / épais tampon-buvard humide dont l'épiderme extrêmement mince / mais très pigmenté, acerbement sapide, est juste assez rugueux pour / accrocher dignement la lumière sur la parfaite forme du fruit.

"Mais à la fin d'une trop courte étude, menée aussi rondement / que possible,—il faut en venir au pépin. Ce grain, de la forme d'un / minuscule citron, offre à l'extérieur la couleur du bois blanc de / citronnier, à l'intérieur un vert de pois ou de germe tendre. C'est en / lui que se retrouvent, après l'explosion sensationnelle de la lanterne / vénitienne de saveurs, couleurs, et parfums que constitue le ballon fruité / lui-même,—la dureté relative et la verdeur (non d'ailleurs / entièrement insipide) du bois, de la branche, de la feuille: somme / toute petite quoique avec certitude la raison d'être du fruit" (22–25).

47 Sianne Ngai, "The Cuteness of the Avant-Garde," *Critical Inquiry* 31, no. 4 (2005): 832. See also Ngai, *Our Aesthetic Categories,* 91. In addition to Ngai, my reading of the end of Ponge's poem owes a debt to Helen Deutsch's inspiring reading of the poem as one in a succession of anecdotal fragments produced by the orange as itself an anecdotal entity. See Deutsch, "Oranges, Anecdotes, and the Nature of Things," *SubStance* 38, no. 1 (2009): 31–54.

48 De Man, *Resistance to Theory,* 44. Ngai draws on de Man's "Autobiography and Defacement" in *The Rhetoric of Romanticism* (New York: Columbia University Press, 1984), 67–81.

49 Marder, *Plant-Thinking,* 115, quoting Ponge, *Selected Poems,* 71. And on Ponge's aesthetic as a form of flower writing, see Claudette Sartiliot,

Herbarium Verbarium: The Discourse of Flowers (Lincoln: University of Nebraska Press, 1993), 85–115.

50 Ibid., 116.

51 Ibid., 114.

52 The word *orange* would exist then as another transspecies pidgin or ethonym, as elaborated by Kohn in *How Forests Think,* 199. In *Darwin's Pharmacy,* Richard Doyle offers a wild extension of Louis Althusser's figure of ideological hailing "appropriate to the experience of ecodelic interconnection." He names this experience "the transhuman interpellation." Whether or not you subscribe to Doyle's positing of instances, as in psychedelics, in which we may be "hailed by the whole," his model remains true to a general, if nonlinear and erring, recruitment of animal actors to aid in the reproductive processes of plants.

53 Jacques Derrida, *Signésponge | Signsponge,* trans. Richard Rand (New York: Columbia University Press, 1984), 8. For his commentary on the orange that reprises this notion of self-grounding mimesis or the mimesis of mimesis, see 65–66.

54 For this formulation of "Eucharistic blood and flesh of Christ" as claiming "the absolute right of appropriation of the mute body of the plant," see Marder, *Plant-Thinking,* 33.

55 Nealon, *Plant Theory,* 107.

4. GOLD YOU CAN EAT (ON THEFT)

1 William Pietz, "Afterward: How to Grow Oranges in Norway," in Spyer, *Border Fetishisms,* 246. Subsequent references appear parenthetically in the text. See also Paul R. Krugman and Maurice Obstfeld, *International Economics: Theory and Policy*, 7th ed. (New York: HarperCollins, 2005), 3. Krugman and Obstfeld are now in their ninth reprinting.

2 William Pietz, "The Problem of the Fetish, 1," *Res* 9 (1985): 5–17; Pietz, "The Problem of the Fetish, 2," *Res* 13 (1987): 23–45; and Pietz, "The Problem of the Fetish, 3a," *Res* 16 (1988): 105–24.

3 Pietz, "Problem of the Fetish, 1," 7.

4 The choice is never between having a "fetish" or not but between competing fetishes, some bad, some good, some too good to be true. See Peter Stallybrass's unpacking of Marx's use of the word in "Marx's Coat," 184. Here it's worth noting the way Pietz's modeling of the fetish as the result of a crisis or contest of values anticipates Bruno Latour's modeling of fact and fetish as broken halves of a preoriginary "factish,"

as he names it. See Bruno Latour, *On the Modern Cult of the Factish Gods,* trans. Catherine Porter and Heather MacLean (Durham, N.C.: Duke University Press, 2010). For a rousing reading of Pietz as enabling us to think about forms of association routed through the object world, see Nicholas Thorburn, "Communist Objects and the Values of Printed Matter," *Social Text 103,* 28, no. 2 (2010): 1–32. For an allied observation, see Michael Taussig's estimation that what Pietz "does for us with his genealogizing is restore certain traces and erasures and weave a spell around what is, socially speaking, at stake in *making.*" Michael Taussig, *The Nervous System* (New York: Routledge, 1992), 118–19.

5 Abbie Hoffman, *Steal This Book* (1971; repr., New York: De Capo Press, 2002).

6 This phrase comes from John Gerard's memoir preserved as Stonyhurst MS A. v. 22, 158. See also Gerard, *Autobiography of an Elizabethan,* 116.

7 For this particular notion of life as in excess or surplus, see Massumi, *What Animals Teach Us,* 10–15. Antonio Negri's affirmative, expansive, autonomist Marxist formulation of labor and life as resistant to biopower is also, obviously, an influence here. See Negri, *Time for Revolution* (London: Bloomsbury, 2013), and, with Michael Hardt, *Empire* (Cambridge, Mass.: Harvard University Press, 2001), *Multitude,* and *Commonwealth* (Cambridge, Mass.: Belknap Press, 2011). It is sometimes hard when starting down this path not to become rapturous with respect to what quickly becomes an immanent sense of resistance. In what follows, I am more inclined to point to the extreme neutrality of a liking or inclination as precisely something that the techniques of biopower might be said to manage and distribute. This is not, however, to say that such inclinations might not form the basis of progressive modes of association or radically enlarged senses of the common. For one modeling of civil society as a zone that actively manages forms of life as mundane as a liking, see Miguel Vatter, *The Republic of Living: Biopolitics and the Critique of Civil Society* (Brooklyn, N.Y.: Fordham University Press, 2014).

8 Johannes Baptista Ferrarius, Frater, *Hesperides, Sive de Malorum Aureorum Cultura et Usu* (1646), book 1, 33. Ferrarius cites this tradition as part of his survey of Hesperidean myths. The translations that follow are slightly adapted from the version made by Lily Thompson Hawkinson, "An Introduction to and Notes on the Translation of *Hesperides, Sive de Malorum Aureorum Cultura et Usu*" (MA thesis, Claremont Colleges, 1936), 33–34. Philip Armstrong offers the clearest explanation of this confusion or condensation of terms in *Sheep,* 72–74.

9 Ferrarius, *Hesperides,* book 1, 2: "Ego Hercule fortunatior totum Hesperidium pomarium, modico inclusum volumine, tuum in obsequium trasnfero." See also Hawkinson, "An Introduction," 2. The only extended critical treatment of Ferrarius comes from art historian David Freedberg, who approaches the volumes as a trove of images related to citrus lore compiled at the origin of empirically driven natural science. See Freedberg and Enrico Baldini, *Citrus Fruit* (London: Harvey Miller, 1997), and, in passing, Freedberg, *The Eye of the Lynx: Galileo, His Friends, and the Beginnings of Natural History* (Chicago: University of Chicago Press, 2002). This collection of orange-colored writers might include Samuel Tolkowsky, *Hesperides: A History of the Use and Culture of Citrus Fruits* (London: J. Bale and Currow, 1938); John McPhee, *Oranges* (New York: Farrar, Strauss, and Giroux, 1966); Cixous, *L'heure de Clarice Lispector*; Pierre Laszlo, *Citrus: A History* (Chicago: University of Chicago Press, 2007), and Clarissa Hyman, *Oranges: A Global History* (London: Reaktion Books, 2013).

10 Ferrarius, *Hesperides,* book 1, 1: "pomaria eduli auro"; and with regard to the bees, "Vides, ut Apes, florum arbitrae prudentissimae aurei mali argenteo flore porissimum delectantur." See also Hawkinson, "An Introduction," 6 and 10.

11 Ferrarius, *Hesperides,* book 1, 8–9. See also Hawkinson, "An Introduction," 22–24.

12 Georges Bataille offers the great solar orange, the sun, as the figure of this pure expenditure: "The source and essence of our riches are given in the effulgence of the sun, who dispenses riches—without a counterpart. Men were aware of this long before [modern science] measured this unceasing prodigality. They linked the splendor [of the sun] to the gesture of he who gives without receiving." Georges Bataille, *The Accursed Share* (1949), trans. Robert Hurley (New York: Zone Books, 1988), 35.

13 Ferrarius, *Hesperides,* book 1, 9: "Ut que adeo in hortensi rapina posse temperare, Herculeum est." See also Hawkinson, "An Introduction," 23.

14 Ibid., book 4, 417: "pugnacissima Dianae Amazon." See also Hawkinson, "An Introduction," 94.

15 Ibid., book 4, 418: "Nam vulnerata erupit ex arbore subitus curor. . . . Hei mihi! . . . Leonilla notae voces indicio matrem fortuito sagittae casu triectam rata, amore partier ac dolore volucris, in auxilim complexumque properavit." See also Hawkinson, "An Introduction," 96.

16 Ibid., book 4, 418: "fructus . . . abortiuos." See also Hawkinson, "An Introduction," 98.

17 As Timothy Morton observes, "Bataille implicitly includes ecology in thinking about economics: 'Should we not, given the constant development of economic forces, pose general problems that are linked to the movement of energy on the globe?'" Morton, *Ecology without Nature,* 109, and Bataille, *Accursed Share,* 20.

18 On the exchange of properties between person and object by their joining, see Bruno Latour, *Aramis,* 61. The *Oxford English Dictionary* traces *orange-woman* to 1748, though references appear much earlier. The most famous example would be the comic actress Nell Gwyn (1650–87), who immortalized the figure while simultaneously rendering her ambiguous. Oranges and orange girls play a crucial role in William Wycherley's *The Country Wife* (1675), in which squeezing an orange or having one squeezed serves as a double entendre. Part 4 of Richard Head's *The English Rogue* (London, 1671) contains a description of "Fair Oranges—Finc Lemmons—a cunning slut, who by a fifteen years practice, had got her Trade to her finger ends" (209–10). John Tatham's *The Rump* (1660) includes the character of Priscilla, who goes everywhere with her "basket of oranges," and act 4, scene 1 of the anonymous *The Hectors* (1656) includes the stage direction "Enter an orange woman with oranges." Sir Walter Besant's novel *The Orange Girl* (1899) celebrates the ubiquitous figure of the child regatress/go-between.

For a study of how orange-women marked a shift in women's work in England's metropolis, see Natasha Korda, *Labors Lost: Women's Work and the Early Modern English Stage* (Philadelphia: University of Pennsylvania Press, 2011), 216.

Orangemen, traditionally supporters of the House of Orange, are something quite different—even as the linguistic transfer that gives them their name is consistent with the naming of orange-women. The relationship between the House of Orange and oranges is entirely coincidental. As Tolkowsky offers, "there is no other connection than that which exists between any two homonyms, namely a purely accidental similarity of sound." "Close to the left bank of the river Rhône," he elaborates, "lies the capital of the department known as Vaucluse, one of the oldest towns in France." When this area formed an independent principality "in the days of Charlemagne," one of its rulers is "found mentioned as the first prince of Aurenja, as it was called in the Provençal tongue." *Auranja* also happens to be the earliest word in Provençal for the fruits of the orange tree. Tolkowsky traces the philology of the word's passage, charting how the oranges make their impression on different languages. Tolkowsky, *Hesperides,* 227–30.

19 On this altered sense of scale, see Timothy Morton, *Ecology without Nature,* 20–21, and, more recently, his *Hyperobjects.* For a more depressing but allied series of observations, see Cohen et al., *Theory and the Disappearing Future.*

20 According to the *Oxford English Dictionary* and the *Middle English Dictionary.*

21 See Arthur Golding's rendering of the story of Atalanta in *Ovid's Metamorphoses, the Arthur Golding Translation* (1567), ed. John Frederick Nims (Philadelphia: Paul Dry Books, 2000).

22 Tolkowsky, *Hesperides,* 168–70. On the great variety of names for citrus fruits, see chapter 1, "The Home of the Genus Citrus," 1–38. Laszlo's *Citrus* covers much of the same ground more schematically, but with a rewarding focus on the chemical properties of orange-derived substances. For a discussion of the morphology of citrus and its derived products, see S. V. Ting and Russell L. Rouseff, *Citrus Fruits and Their Products: Analysis and Technology* (New York: Marcel Dekker, 1986).

23 Edward Norgate, *Miniatura or the Art of Limning,* ed. Jeffrey M. Muller and Jim Murrell (New Haven, Conn.: Yale University Press, 1997), 59.

24 Ibid., 63.

25 Ibid., 59–60.

26 Nicholas Hilliard, *Nicholas Hilliard's the Art of Limning,* ed. Arthur Kinney and Linda Bradley Salamon (Boston: Northeastern University Press, 1983), 21.

27 Henry Peacham, *The Art of Drawing with a Pen, and Limning with Water Colours* (London, 1606), and Peacham, *The Compleat Gentleman* (London, 1622). Peacham published numerous revised editions of his *Compleat Gentleman,* enlarging the sections on drawing and painting and so ennobling these activities.

28 On this model of *poiesis,* see Bruno Latour, *Pandora's Hope: Essays on the Reality of Science Studies* (Cambridge, Mass.: Harvard University Press, 1999), 281.

29 Deleuze and Guattari, *A Thousand Plateaus,* 2:411–12.

30 On the efficacy of gemstones and minerals in this tradition, see Valerie Allen, "Mineral Virtue," in Cohen, *Animal, Vegetable, Mineral,* 123–52, and Kellie Robertson, "Exemplary Rocks," ibid., 91–121.

31 Aristotle, *De Anima,* 418.b.14–20. Quoted in Bruce R. Smith, *The Key of Green: Passion and Perception in Renaissance Culture* (Chicago: University of Chicago Press, 2009), 58. Aristotle's color spectrum included the following: "black | gray | deep blue | leek-green | violet | crimson | yellow | white." See John Gage, *Color and Culture: Practice and Meaning*

from Antiquity to Abstraction (Berkeley: University of California Press, 1993), 12–13.

32 Serres, *Conversations on Science,* 57–62, and Haraway adapting Alfred North Whitehead in *When Species Meet,* 7.

33 On the ethical and political functions of politeness in scientific and ecological practice, see Stengers, *Cosmopolitics I.*

34 Jim Long, *The Munsell Book of Color,* 3rd ed. (London: Fairchild, 2011), and Long, *Munsell® Soil Color Book,* rev. ed. (2009).

35 Latour, *Pandora's Hope,* 59.

36 On "circulating reference," ibid., 24–79.

37 Tens of thousands of oranges and lemons are recorded for ports such as London and Bristol; see, e.g., details from the *London Port Book, 1567/8,* describing duty collected on "40 thou[sand] oranges and lemons"; like quantities in Brian Dietz, *The Port and Trade of Early Elizabethan London Documents* (London: London Record Society, 1972), 85; and similar instances in *Documents Illustrating the Overseas Trade of Bristol in the Sixteenth Century,* ed. Jean Vanes (Bristol: Bristol Record Society, 1979), 144–49. Trade with areas beyond London and the key ports occurred more sporadically. "Even oranges and lemons came rather from London than from Spain direct," observes one historian, who goes on to cite the unusual arrival of "20,000 oranges and 1,000 lemons" at Norwich in 1581. N. J. Williams, *The Maritime Trade of the East Anglian Ports, 1550–1590* (Oxford: Clarendon Press, 1988), 176. Trade became more difficult, obviously, at times of conflict with Spain, when such imports were listed as contraband. See R. H. Tawney and E. Power, *Tudor Economic Documents* (London: Longmans, Green, 1951), 2:81.

38 Oranges and lemons appear as a frequent item in the bills of Henry VIII and subsequent monarchs and were a particularly frequent item during fish days. See *A Collection of Ordinances and Regulations for the Government of the Royal Household Made in Divers Reigns from King Edward III to King William and Queen Mary* (London: Society of Antiquaries, 1790), 175–76, and also *The Star Chamber Dinner Accounts . . . during the Reigns of Queen Elizabeth I and King James I of England,* ed. André L. Simon (London: Wine and Food Society, 1959), 30. For a more general assessment of the status of such luxuries, see Keith Wrightson, *Earthly Necessities: Economic Lives in Early Modern Britain* (New Haven, Conn.: Yale University Press, 2000).

For more exact pricing, see Jane Whittle and Elizabeth Griffiths, *Consumption and Gender in the Early Seventeenth-Century Household: The World of Alice Le Strange* (Oxford: Oxford University Press, 2012), 104.

Whittle and Griffiths note that "between April 1619 and October 1621, the Le Stranges bought 8d worth of oranges (probably just a few fruits) [and] nine lemons, which cost 8d each."

More anecdotally, one of the dialogues in John Florio's *Florio's Second Fruites* (London, 1591) contains a description of an elaborate fruit course including oranges and lemons (62–63). Oranges and lemons were also a food for special occasions; see their inclusion in "a brief note of the greatest charge spent at the . . . marriage . . . of dowghters Ann and Dorothie . . . for meats" in *A Book of Several Transactions of the Kays of Woodsome from Queen Elizabeth's Days to 1642,* Folger MS W.b. 482 MS Add 348.

On their use as projectiles, see Michael Hattaway, *English Popular Theater: Plays in Performance* (London: Routledge, Kegan, and Paul, 2004), 46, where he notes "orange and beer sellers plied their trade before the play began."

39 On this point, see Arjun Appadurai, "Introduction: Commodities and the Politics of Value," in *The Social Life of Things: Commodities in Cultural Perspective* (Cambridge: Cambridge University Press, 1986), 41–42.

40 While lauded for their excellence as a sauce for meat, along with myriad other uses, Edward Muffett's sixteenth-century dietary includes "citrous and lemons" among "dry meates" suitable for the "over-moistured," but possibly unsuitable for other temperaments. He is particularly wary of citrons that are not "altered by sugar . . . and parboiling." Edward Muffet (elsewhere appears as Moffett), *Health's Improvement: or, Rules Comprizing and Discovering the Nature, Method, and Manner of Preparing All Sorts of Food* (circa 1595) (London, 1655), 34–35. On eating habits generally and the changing role of fruit in diet, see Joan Thirsk, *Food in Early Modern England: Phases, Fads, Fashions, 1500–1760* (London: Hambledon Continuum, 2007), 294–303.

41 L. Anne Wilson, *The Country House Kitchen, 1600–1850* (London: Sutton, 1998), 4–5. On the gentry's fascination with orange and lemon trees, see 56–57, and on the use of "glass houses," see 100–103. On gardens and gardening in practice and as they marked an English imaginary, see also Rebecca Bushnell, *Green Desire: Imagining Early Modern English Gardens* (Ithaca, N.Y.: Cornell University Press, 2003).

On the continent, the engineer Salomon de Caus, whose name in England is associated with the machinery of Court Masques, was the most famous designer of orangeries, which, like the masque, constitute a technological superlative, requiring the latest in hydraulic engineering to ensure that the oranges survive the winter. De Caus, *New and Rare Invention of Water-Works* (in French, 1615), trans. John Leak (London, 1701).

42 Key treatises include the following: Jean de La Quintinye, *Instruction Pour Les Jardins Fruitiers et Potagers, Avec La Traite, suivy de quelques Reflections sur l'Agriculture* (Amsterdam, 1642); Nicholas Bonnefons, *The French Gardiner* (Paris, 1658); Henry Van Oosten, *The Dutch Gardiner or the Complete Florist... Together with a Particular Account of the Nursing of Lemon and Orange Trees in Northern Climates* (London, 1711).

43 S. Commelyn, *The Belgick, or Netherlandish Hesperides. That Is the Management, Ordering, and Use of the Limon and Orange Trees, Fitted to the Nature and Climate of the Netherlands. Made English by G. V. N.* (London, 1683), 2–3.

44 The difficulty of growing oranges and lemons was not something that plagued northern Europe alone. Tolkowsky diagnoses similar problems centuries earlier as Roman gardeners sought to domesticate citrus to Italy. Based on his study of mosaics and other media, Tolkowsky ventures that many of the orange trees depicted were representational hybrids, grafting citron flowers onto orange trees for want of a living model. Tolkowsky, *Hesperides,* 100–103.

45 Platt, *Jewell House of Art and Nature,* 5.

46 Quoted in Mea Allen, *The Tradescants: Their Plants, Gardens, and Museum, 1570–1663* (London: Michael Joseph, 1964), 45–46.

47 This meeting is recorded by William Brenchley Rye, *England as Seen by Foreigners in the Days of Elizabeth and James the First* (London: John Russell Smith, 1865), 118–19.

48 Anthony Low, *The Georgic Revolution* (Princeton, N.J.: Princeton University Press, 1985), remains an excellent account to be read.

49 McRae, *God Speed the Plough,* 146.

50 Gerard de Malynes, *Lex Mercatoria* (London, 1622), 84. I am grateful to Brad Ryner for this reference

51 Ibid., 85.

52 [Smith], *A Discourse of the Commonweal of This Realm of England.*

53 Ralph Austen, *A Dialogue or Familiar Discourse and Conference betweene the Husbandman and the Fruit-Trees* (Oxford, 1676), 56.

54 John Paston to Sir John Paston (1470), letter LXXXII, in *The Paston Letters 1422–1509,* ed. John Gardiner (London, 1904), 4:183.

55 George Cavendish, *The Life and Death of Cardinal Wolsey* (1557), ed. Richard S. Sylvester and Davis P. Harding (New Haven, Conn.: Yale University Press, 1962), 25. On a similarly charged orange become pomander as memorial, see the story of Charles I's execution and the orange that lives on as Royalist relic as reported by Patricia Fumerton in *Cultural Aesthetics: Renaissance Literature and the Practice of Ornament* (Chicago: University of Chicago Press, 1991), 1–28.

56 Shakespeare, *Much Ado about Nothing,* 4. 1. 25–37.

57 On Herbert's prosthetic poetics, see David Glimp, "Figuring Belief: George Herbert's Devotional Creatures," in Judith H. Anderson and Joan Pong Linton, *Go Figure: Energies, Forms, and Institutions in the Early Modern World,* 112–31 (New York: Fordham University Press, 2011).

58 Shakespeare, *Much Ado about Nothing,* note to 4. 1. 27.

59 R. E. R. Madelaine, "Oranges and Lemans: *Much Ado about Nothing,* IV, 1, 31," *Shakespeare Quarterly* 33, no. 4 (1982): 491–92. On the wider associations of oranges and lemons in the period, see Robert Palter, *The Duchess of Malfi's Apricots and Other Literary Fruits* (Columbus: University of South Carolina Press, 2002), 384–481.

60 References to these practices are common, especially in Restoration drama and prose. For an earlier example, see Francis Beaumont and John Fletcher, *Philaster; or Love Lies a Bleeding,* in which the King warns Meg that "all the Court shall hoot thee through the court, / Fling rotten oranges, make ribald Rhymes." Beaumont and Fletcher, *Philaster; or Love Lies a Bleeding,* ed. Suzanne Gossett (London: Arden Early Modern Drama, 2009), 2. 1. 425–26.

61 Roland Barthes, *A Lover's Discourse: Fragments,* trans. Richard Howard (New York: Hill and Wang, 1978), 110–11. Helen Deutsch treats Barthes in passing in her wonderfully nuanced essay "Oranges, Anecdotes, and the Nature of Things," *SubStance* 38, no. 1 (2009): 31–54, which came to my attention after I was deep into my own orange endeavors. Her sense of the orange as keyed to the anecdote has proved immensely useful to my thinking and has taught me to understand the way many of the textual instances that I read mark attempts to render the orange as event.

62 Francis Bacon, *The Essays or Counsels Civil and Moral on the Wisdom of the Ancients,* ed. S. W. Singer (London, 1857), 322–23.

63 Deutsch, "Oranges, Anecdotes," 34.

64 *Ovid's Metamorphoses: The Arthur Golding Translation of 1567,* ed. John Frederick Nims (Philadelphia: Paul Dry Books, 2000), lines 778–97. Golding's translation sticks remarkably closely to Ovid's. See also *Ovid's Metamorphoses,* ed. William S. Anderson (Norman: University of Oklahoma Press, 1927), book 10, lines 598–680. In Ovid's *Metamorphoses,* emphasis is placed on this doubt, which translators read as evidence of his singular fascination with Atalanta's psychology. The Latin reads, "Inque latus campi, quo tardius illa redirect, / Iecit ab oliquo nitidum iuvenaliter aurum. / An preteret, virgo via est dubitare." *Ovid's Metamorphoses,* book 10, lines 674–76.

65 Derrida, *Signésponge | Signsponge,* 102–3. You can obviously eat gold—gold

leaf is an adornment for festival foods in many cultures. On this use and for a reckoning of the mineral's phenomenological heft, see Graham Harman, "Gold," in *Prismatic Ecology: Eco-theory beyond Green*, ed. Jeffrey Jerome Cohen (Minneapolis: University of Minnesota Press, 2013), 106–23.

66 Iona Opie and Peter Opie, eds., *The Oxford Dictionary of Nursery Rhymes,* 2nd ed. (Oxford: Oxford University Press, 1997), 398.

67 Ibid.

68 Nurse Lovechild, *Tom Thumb's Pretty Song Book* (1744; repr., Glasgow: Lumsden and Sons, 1813), 24. On Playford's *Dancing Master* (1665), see Opie and Opie, *Oxford Dictionary of Nursery Rhymes,* 399.

69 Edward Ravenscroft, *The London Cuckolds* (1682). Additional street cries for golden, ripe, and fine oranges are recorded in Charles Hindley, *A History of the Cries of London* (London: Reeves and Turner, 1881).

70 See frontispiece in Hindley, *A History of the Cries of London.*

71 Graham Harman, "On Vicarious Causation," in *Collapse II,* ed. R. Mackay (Falmouth, U.K.: Urbanomic, 2007), 171–205.

72 Susan Stewart, *On Longing: Narratives of the Miniature, the Gigantic, the Souvenir, the Collection* (Durham, N.C.: Duke University Press, 1993), 23.

73 Maria Edgeworth, *The Little Dog Trusty; the Orange Man; and the Cherry Orchard; Being the Tenth Part of Early Lessons* (1801; repr., Los Angeles, Calif.: William Andrews Clark Memorial Library and the Augustan Reprint Society, 1990), 27. Subsequent references appear parenthetically in the text.

74 George Orwell, *The Complete Works of George Orwell,* vol. 9, *1984,* ed. Peter Davison (London: Secker and Warburg, 1984), 3. Subsequent references appear parenthetically in the text.

75 The absence of oranges in putative futures marks speculative fiction of this period—a testament to their novelty and seasonality (they were frequently stocking stuffers for Christmas) and also a memory of wartime rationing. See, e.g., Josella's question to Bill in *The Day of the Triffids* as the two say good-bye to the London that was and is no longer: "We lay back in two superbly comfortable armchairs . . . the plutocratic-looking balloon with the puddle of unpriceable brandy was mine. Josella blew out a feather of smoke and took a sip of her drink. Savoring the flavor, she said: 'I wonder whether we shall ever taste fresh oranges again . . . Okay, shoot.'" John Wyndham, *The Day of the Triffids* (1951; repr., New York: Doubleday, 2003), 76.

76 Karen Tei Yamashita, *The Tropic of Orange* (Minneapolis, Minn.: Coffee House Press, 1997), 11. Subsequent references appear parenthetically in the text.

77 On the mythological life of Atlas as world-bearer as indexed to questions of technology and what he calls the "project of metaphysical globalization," see Peter Sloterdijk, *Globes: Spheres II,* trans. Wieland Hoban (South Pasadena, Calif.: Semiotext(e), 2014), 45–134.

78 Ursula K. Heise, *Sense of Place and Sense of Planet: The Environmental Imagination of the Global* (Oxford: Oxford University Press, 2008), 201.

79 For a sense of how elaborate these networks are, see, as one example, Shane Hamilton, "The Political Ecology of Frozen Concentrated Orange Juice," *Agricultural History* 77, no. 4 (2003): 557–81.

5. BREAD AND STONES (ON BUBBLES)

1 On the need to maintain the future as strategically blank, see Karl Marx, "The Eighteenth Brumaire of Louis Bonaparte," in *The Marx and Engels Reader,* 2nd ed., ed. Robert C. Tucker (New York: W. W. Norton, 1978), 595.

2 Tim Ingold, *The Life of Lines* (London: Routledge, 2015), 43.

3 Ibid., 45.

4 Ibid., 43–45.

5 The word Michel Serres uses to describe such structures is *circumstance*—literally, from the Latin, the way *things* stand around. A "circumstance" designates an eddy of negentropy amid the flux, a momentary stabilization or form that endures for a time. "Circumstance" marks the "place [also] where writing emerges as the mnemonic preserver," writes Serres, "and so things and words are negentropic tablets . . . [that] escape, for as long as the code is memorized, the irreversible flux of dissolution." Serres, *The Birth of Physics,* trans. Jack Hawkes (Manchester, U.K.: Clinamen Press, 2000), 148.

On the automorphism and so the animation of matter, see Michel Serres, *Rome: The Book of Foundations,* trans. Felicia McCarren (Stanford, Calif.: Stanford University Press, 1991), 80–85, and Serres, *Statues: The Second Book of Foundations,* trans. Randolph Burks (London: Bloomsbury, 2015), 172–78.

For a book-length examination of stone's lithic materiality and animation, see Jeffrey Jerome Cohen, *Stone: An Ecology of the Inhuman* (Minneapolis: University of Minnesota Press, 2015).

6 Serres, *Rome,* 80.

7 Ibid., 81.

8 Steven Connor writes well about leavening and fermentation specifically as a way of thinking about the process of transmutation in *The Making*

of Air: Science and the Ethereal (London: Reaktion Books, 2010), 312–25.

For a similar sense of bread making as figuring the "non-self-identity that affects everything" that Jacques Derrida names "dissemination," see Scott Cutler Shershaw's *Bread* (London: Bloomsbury, 2016), 7–9. Shershaw's excellent "object lesson" book arrived while I was in the final throes of copy editing this chapter. I regret therefore that it has not been possible to engage with his frequently allied observations as fully as I might have liked.

One particular resonance between our projects lies in our shared sense of the way bread convokes beings. Indeed, I am particularly indebted to Shershaw's book for its illuminating discussion of the challenges posed by capturing the essence of the Greek word *epiousios* in the Lord's Prayer so as to capture its sense of "something like 'necessary' or 'needful,' or perhaps 'continual'" (128–30), as opposed simply to "quotidian." As Shershaw observes, the translation "daily" belongs to William Tyndale's early-sixteenth-century translation (1526) of Matthew 6:11 and Luke 11:3, "Give us this day our daily bread," which goes on to influence the King James Version. Shershaw reminds that in the Roman Catholic Douai–Rheims Bible, the line in Matthew is rendered as "Give us this day our supersubstantial bread," which I think Shershaw very rightly asks readers to consider an attempt to capture what is understood to be the enduring fullness or plenitude of this gift of physical and spiritual sustenance (128).

It is this material–semiotic heft to bread (and stone), this sense of permanence or the givenness of the given, that drives my reading of bread as the product of and a figure for infrastructure.

9 Peter Sloterdijk, *Spheres,* vol. 1, *Bubbles,* trans. Wieland Hoban (Cambridge, Mass.: MIT Press, 2011), 53.

10 For the coinage "microbiopolitics," see Heather Paxson, "Post-Pasteurian Cultures: The Microbiopolitics of Raw Milk Cheese in the USA," *Cultural Anthropology* 23, no. 1 (2008): 17. Paxson grounds the term in a "parallel history" that yokes Bruno Latour's tracking of "microbial life into the very constitution of the social field" against Foucault's account of the rise of biopolitics in the nineteenth century through the "fashioning of new categories of persons to facilitate the statistical measurement and rational management of populations, largely via sex and reproduction" (17–18). Paxson reads the discourses that arise from competing alliances between cheese makers and "raw" as opposed to pasteurized milk as symptomatic of changing attitudes to microbial life. See also Latour, *Pasteurization of France.*

11 Haraway, *When Species Meet,* 17. On the "companion species," taking the dog-human relation as privileged, see also Haraway, *Companion Species Manifesto.*

12 Sloterdijk, *Bubbles,* 45–46.

13 Ibid., 32–33. There are of course two contradictory creation stories cohabiting in Genesis: Genesis 1:1–2:3, in which the world, plants, and animals all predate the coappearance of Adam and Eve, and Genesis 2:4–25, in which Adam precedes the animals and then the creation of Eve.

14 Ibid., 36–37.

15 Ibid., 53.

16 Ibid., 39.

17 Peter Sloterdijk, *You Must Change Your Life* (New York: Polity Press, 2014).

18 Luce Irigaray, *The Forgetting of Air in Martin Heidegger,* trans. Mary Beth Mader (Austin: University of Texas Press, 1999), 3. Subsequent references appear parenthetically in the text. Carla Mazzio uses Irigaray to excellent effect in "The Making of Air: *Hamlet* and the Trouble with Instruments," *South Central Review* 26, no. 1–2 (2009): 153–96.

19 Walter Benjamin, *Walter Benjamin, Selected Writings,* vol. 2, part 2, *1931–1934,* trans. Rodney Livingstone et al., ed. Michael W. Jennings, Howard Eiland, and Gary Smith (Cambridge, Mass.: Belknap Press of Harvard University Press, 1999), 674. Subsequent references appear parenthetically in the text. See also Benjamin, *Gesammelte Schriften,* ed. Rolf Tiedermann (Frankfurt, Germany: Suhrkampf, 1972), vol. 4, part 1, 414–15.

20 Sloterdijk, *Bubbles,* 18.

21 "Cigarettes," we know, because Richard Klein's still wonderful book says so, "are sublime." In it he tracks the cigarette as index, as emitter and obliteration of time effects through the underside of modernity, sensitive to the trance that Benjamin inhabits, which relies on the "connection between smoking and writing." "Like writing," Klein observes, "smoking belongs to that category of action that falls between the states of activity and passivity—a somewhat embarrassed condition." Klein, *Cigarettes Are Sublime* (Durham, N.C.: Duke University Press, 1993), 35. For a use of the scene of smoking as an object lesson in thinking through what he calls the "*faire faire*" of an "actor network" that distributes different entities so as to produce agentive effects, see Bruno Latour, "Factures/Fractures: From the Concept of Network to the Concept of Attachment," *Res* 20 (1999): 1–20.

22 I borrow the phrase "meta-munchies" from the virtuoso reading of "Hashish in Marseilles" by Tom Cohen as one instance of Benjamin's "mock-descriptive archaeo-graphematics of cities (Marseilles, Naples, Moscow)" in *Ideology and Inscription,* 221–37. My debts to Cohen will be obvious in what follows. On Benjamin's reworking of his travel experiences over the course of his life, see Momme Brodersen, *Walter Benjamin: A Biography,* trans. Malcolm R. Green and Ingrida Ligers, ed. Martina Dervis (London: Verso, 1996), 33–34.

For an anthology of Benjamin's writings on hashish sensitive to the altered state of consciousness the experience offers, see Walter Benjamin, *On Hashish,* ed. Howard Eiland (Cambridge, Mass.: Belknap Press of Harvard University Press, 2006). For a review of the philosophical moorings to his trials, see Gary Shapiro, "Ariadne's Thread: Walter Benjamin's Hashish Passages," in *High Culture: Reflections on Addiction and Modernity,* ed. Anna Alexander and Mark S. Roberts (Albany, N.Y.: SUNY Press, 2002), 59–74. On the rhetoric of the drug high, see Jacques Derrida, "The Rhetoric of Drugs," trans. Michael Israel, ibid., 19–45.

23 The German here reads as follows: "Und im Haschisch sind wir genießende Prosawesen höchster Potenz." Benjamin, *Gesammelte Schriften,* vol. 4, part 1, 415.

24 The sentence comes from an unspecified text by Johannes Vilhelm Jensen in Benjamin, *Selected Writings,* vol. 2, part 2, 679.

25 The German here reads as follows: "Während der Satz bei Jensen für mich darauf hinauskam, daß die Dinge so sind, wie wir ja wissen, durchtechnisiert, rationalisiert, und das Besondere steckt heute nur noch in Nüancen, war die neue Einsicht durchaus anders. Ich sah nämlich nur Nuancen: diese jedoch waren gleich." Benjamin, *Gesammelte Schriften,* vol. 4, part 1, 415.

26 Roland Barthes, *The Neutral: Lecture Course at the Collège de France (1977–1978),* trans. Rosalind E. Krauss and Denis Hollier (New York: Columbia University Press, 2002), 11. For Barthes, the word *nuance* becomes one of the defining orientations that his lecture course devoted to the neutral, or, as he prefers "display[ing] a series of neutrals," explores. Benjamin's "Hashish in Marseilles" appears once in the text to illustrate the way drugs, for Benjamin, exist as one instance of living "unproductively." Love, for Barthes, refers to another such state of being.

27 For an allied series of observations on Benjamin's sense of what he calls a "'generalized text'—one that would accommodate a 'city' no

less than a verbal structure"—and the way this manifests as an ongoing interrogation of "modes of signifying," see Samuel Weber, *Benjamin's-abilities* (Cambridge, Mass.: Harvard University Press, 2008), 227–28. Cohen argues that, for Benjamin, the "'drug' is also a counter-poison to the undeclared stupor or stupefaction and political mortification of the senses, the 'mimetic faculty' run aground." Cohen, *Ideology and Inscription,* 236. For Cohen, Benjamin's rhetoric of drugs comes indexed to an allohuman sense of linguistic materiality, so much so that it becomes a sublime, wild allegory. I am inspired by Cohen's virtuoso renderings but would revise his conclusions so that Benjamin's animal, vegetable, and mineral metamorphoses register not the mimetic faculty "run aground" so much as working overtime, caught in an ongoing series of transformations, expending itself as it is unable to settle on one object as it wants always to become more.

28 In German, this passage reads as follows: "Ich vertiefte mich in das Pflaster vor mir, das durch eine Art Salbe, mit der ich gleichsam darüber hinfuhr, als eben dieses Selbe und Nämliche auch das Pariser Pflaster sein konnte. Man redet oft davon: Steine für Brot. Hier diese Steine waren das Brot meiner Phantasie, die plötzlich heißhungrig darauf geworden war, das Gleiche aller Orte und Länder zu kosten." Benjamin, *Gesammelte Schriften,* vol. 4, part 1, 415.

29 The German here reads as follows: "'Barnabe' stand auf einer Elektrischen, die vor dem Platze, an dem ich saß, kurz hielt. Und mir schien die traurig-wüste Geschichte von Barnabas kein schlechtes Fahrtziel für eine Tram ins Weichbild von Marseille." Ibid.

30 The German here reads as follows: "Es schmeichelte mir der Gedanke, hier in einem Zentrum aller Ausschweifung zu sitzen, und mit 'hier' war nicht etwa die Stadt, sondern der kleine, nicht sehr ereignisreiche Fleck gemeint, auf dem ich mich befand." Ibid.

31 The German here reads as follows: "Und wenn ich dieses Zustands mich erinnere, möchte ich glauben, daß der Haschisch die Natur zu überreden weiß, jene Verschwendung des eignen Daseins, das die Liebe kennt, uns—minder eigennützig—freizugeben. Wenn nämlich in den Zeiten, da wir lieben, unser Dasein der Natur wie goldene Münzen durch die Finger geht, die sie nicht halten kann und fahren läßt, um so das Neugeborene zu erhandeln, so wirft sie nun, ohne irgend etwas zu hoffen oder erwarten zu dürfen, uns mit vollen Händen dem Dasein hin." Ibid., vol. 4, part 1, 416.

32 For a radicalization of Benjamin's drug trance as sublime access to an unreduced ecology of being, see Doyle, *Darwin's Pharmacy.*

33 See Walter Benjamin, *The Writer of Modern Life: Essays on Charles Baudelaire,* trans. Rodney Livingstone, ed. Michael W. Jennings and Howard Eiland (Cambridge, Mass.: Belknap Press of Harvard University Press, 2006), 36. On the figure of the collector and collections, see Walter Benjamin, "Eduard Fuchs, Collector and Historian," in *Selected Writings,* 3:260–302; "Unpacking My Library: A Talk about Collecting," ibid., vol. 2, part 2, 486–93, where Benjamin concludes by comparing books to stones—"I have erected before you one of his dwellings with books as the building stones"; and Benjamin, *The Arcades Project,* trans. Howard Eiland and Kevin McLaughlin (Cambridge, Mass.: Belknap Press of Harvard University Press, 2002).

34 Walter Benjamin, "On the Mimetic Faculty," in *Selected Writings,* vol. 2, part 2, 720.

35 Benjamin Franklin, *Benjamin Franklin's Autobiography,* ed. Joyce E. Chaplin (New York: W. W. Norton, 2012), 28. Unless otherwise indicated, subsequent references appear parenthetically in the text.

36 On "biscuit" or "bisket bread" and its transformation from soldier and sailor fare into a highly refined and enriched comestible in the late seventeenth century, see Thirsk, *Food in Early Modern England,* 109–10. It is unclear which version Franklin requests.

37 E. P. Thompson, "The Moral Economy of the Crowd," in *Customs in Common* (New York: New Press, 1991), 189. Originally published in 1971, the essay was revisited, largely to answer detractors and critics, in "The Moral Economy Reviewed," ibid., 259–351. On food fantasies and the psychic and social associations to other forms of bread in a European context, see Piero Camporesi, *Bread of Dreams: Food and Fantasy in Early Modern Europe* (Cambridge: Polity Press, 1989).

38 Thompson, "Moral Economy of the Crowd," 194.

39 Thompson, "Moral Economy Reviewed," 261 and, esp., 269.

40 On attempts to negotiate these conflicting imperatives and on the regulation of food markets more generally, see Karin J. Friedman, "Victualling Colonial Boston," *Agricultural History* 47 (1973): 189–205; Phyllis Whitman Hunter's broader contextual study of the market in *Purchasing Identity in the Atlantic World: Massachusetts Merchants, 1670–1780* (Ithaca, N.Y.: Cornell University Press, 2001), esp. 102–6; and the story of a bakers' strike in Simon Middleton's "How It Came That the Bakers Bake No Bread: A Struggle for Trade Privileges in Seventeenth-Century New Amsterdam," *The William and Mary Quarterly* 58, no. 2 (2001): 347–72.

41 On Franklin's experiments with diet, see *Benjamin Franklin's Autobiography,* 20–25. On the "errata of my life," ibid., 25. On the co-imbrication

of Franklin's ethical precepts and print media, see Peter Stallybrass, "Benjamin Franklin: Printed Corrections and Erasable Writing," *Proceedings of the American Philosophical Society* 150, no. 4 (2006): 553–67.

42 On the pseudonymity of Franklin's voice as he negotiates the tensions that derive from the republican statesman's impossible legitimation by the voice of the "people," see Michael Warner, *The Letters of the Republic: Publication and the Public Sphere in Eighteenth-Century America* (Cambridge, Mass.: Harvard University Press, 1990).

43 See, e.g., J. A. Leo Lemay's reading of the scene of arrival as reference to his future and so current success in *The Life of Benjamin Franklin* (Philadelphia: University of Pennsylvania Press, 2006), 1:222–34.

44 Lemay provides maps of the city in his biography, accepting Franklin's invitation, ibid.

45 Jacques Lezra, "Enough," *Political Concepts: A Critical Lexicon* 3, no. 3 (2014): 1–7, http://www.politicalconcepts.org/enough-jacques-lezra/. Lezra offers his characteristically dazzling commentary on the word enough as a word that indexes the political as such. On the origins of the word as keyed to questions of moral economy in the early modern Atlantic world, see Hillary Eklund's *Literature and Moral Economy in the Early Modern Atlantic: Elegant Sufficiencies* (Farnham, U.K.: Ashgate, 2015). As Eklund observes in the concluding cast to her book, "defining 'enough' is an ongoing rhetorical, ethical, and ideological process" (194–95) in which we are still obviously engaged, though doing so today requires that we wean ourselves off the notion that there exist "empty spaces" in the world from which value may simply be extracted.

46 Ibid., 1–2.

47 Gerard, *Autobiography of an Elizabethan,* 116, and Stonyhurst MS A. v. 22, 158. Unless otherwise indicated, all subsequent references will be to Caraman's translation supplemented with the Latin original, as needed.

48 As Regina Schwartz observes, the key word here might be "efficacy." Beyond a signifying system, sacramental "rites make something happen" even in the absence of the Host. Schwartz's parses this efficacy taking the Eucharist as her example. Schwartz, *Sacramental Poetics at the Dawn of Secularism* (Stanford, Calif.: Stanford University Press, 2008), 6–10.

49 These bills are reproduced in *Tower Bills, 1591–1685 with Gatehouse Certificates, 1592–1603,* ed. J. H. Pollen, Catholic Record Society Miscellanea IV (London: Arden Press, 1907), 232–34. These bills combine the costs for what they itemize as "dyett, keeper, fuell, washing."

50 The Latin here reads as follows: "esta ibi cibus constitutus diversus pro diversis gradibus hominum in quo genere ipsi religiosum statum non currant sed quod humanum est et minoris esse debet, id pluris faciunt." Stonyhurst MS A. v. 22, 158–59.

51 Diane Purkiss, "Crammed with Distressful Bread? Bakers and the Poor in Early Modern London," in *Renaissance Food from Rabelais to Shakespeare,* ed. Joan Fitzpatrick (Farnham, U.K.: Ashgate, 2010), 18.

52 Ibid., 19, quoting Marx, *Capital,* 198.

53 *The Assize of Bread* (London, 1597), A3v.

54 Ibid., A4r.

55 The assize has, understandably, attracted a good deal of scrutiny from economic historians, and there remains significant debate as to its efficacy and the possibility that the serial revisions of the regulations sought to identify and respond to an error in some of its provisions. The foundational study of the statute along with the Worshipful Company of Bakers remains Sylvia Thrupp, *A Short History of the Worshipful Company of Bakers* (London, 1936), and of the assize itself in isolation, F. J. Nicholas, "The Assize of Bread in London during the Sixteenth Century," *Economic History* 2 (1930–33): 324–27. Subsequent reevaluations have come from Alan S. C. Ross, "The Assize of Bread," *Economic History Review, New Series* 9, no. 2 (1956): 332–42, and James Davis, "Baking for the Common Good: A Re-assessment of the Assize of Bread in Medieval England," *Economic History Review* 58, no. 3 (2004): 465–502.

On enforcement of assizes for beer and brewing (and on occasion baking) regulations and punishments meted out in the period, see Judith Bennett's *Ale, Beer, and Brewsters in England: Women's Work in a Changing World, 1300–1600* (Oxford: Oxford University Press, 1996), esp. 103–5, and Thrupp, *A Short History,* 25–26; on regatresses, 41–42; and on the frequency of searches for delinquent breads, 50–51. And for an attention to brewing and baking as it marked the representation of ordinary women on stage and page, Natasha Korda, *Shakespeare's Domestic Economies: Gender and Property in Early Modern England* (Philadelphia: University of Pennsylvania Press, 2002), and Korda, *Labor's Lost: Women's Work and the Early Modern English Stage* (Philadelphia: University of Pennsylvania Press, 2011).

For an overview of the food webs of medieval and early modern England, see, among others, Christopher Dyer, *Standards of Living in the Later Middle Ages: Social Change in England c. 1200–1520* (Cambridge: Cambridge University Press, 1989), and Keith Wrightson, *Earthly Necessities*

(New Haven, Conn.: Yale University Press, 2000). For an excellent introduction to London's world of government and regulation, see Caroline M. Barron, *London in the Later Middle Ages: Government and People, 1200–1500* (Oxford: Oxford University Press, 2004).

56 Elizabeth David, *English Bread and Yeast Cookery* (Harmondsworth, U.K.: Penguin Books, 1977), 92. David quotes *The Brewer's Book* (Norwich, 1468–69). The same phrase is found used colloquially in early modern England. On this usage in the context of arguments over the exchange of yeast between neighbors and charges of interfering with the process of fermentation in witchcraft trials, see Barbara Rosen, *Witchcraft in England, 1558–1619* (Amherst: University of Massachusetts Press, 1969), esp. 142–43. The most detailed description of bread making in the period, geared to household kitchens, comes from Gervase Markham, *The English Housewife* (1615), chapter 8, "Of the Office of the Brew-House, and the Bake-House, and the Necessary Things Belonging to the Same."

For an archaeological description of a bakehouse and shop belonging to the baker Robert Hollier (1624), see *The London Surveys of Ralph Treswell,* ed. John Schofield, London Topographical Society Publication 1358 (London: Society of Antiquaries and Historic Royal Palaces, 1987), 85–87. Modern accounts of medieval and early modern baking practices can be found in numerous sources. One of the best remains Peter Brears, *All the King's Cooks: The Tudor Kitchens of King Henry VIII at Hampton Court Palace* (London: Souvenir Press, 1999), 39–45.

57 On the making of this microscopic and so microbial world, see Latour, *Pasteurization of France,* and on the multiplication of "Pasteurians and hygienists" as a parasitic cascade because microbes are everywhere, 38–40 esp.

On the personification of such microbial practices in the likeness of the fairy or changeling Puck in William Shakespeare's *A Midsummer Night's Dream,* see Robert N. Watson, "The Ecology of the Self in *A Midsummer Night's Dream,*" in *Eco-Critical Shakespeare,* ed. Lynne Bruckner and Daniel Brayton, 33–56 (Farnham, U.K.: Ashgate, 2011).

On failed fermentation (of bread or beer) as one of the charges frequently leveled against witches, see Rosen, *Witchcraft in England.*

58 Moffett, *Health's Improvement,* 235. The key study of early modern dietaries or "regimes of health" is Ken Abala, *Eating Right in the Renaissance* (Berkeley: University of California Press, 2002), 194–201, esp. regarding the social differentiation that marked the "bread nexus."

59 Moffett, *Health's Improvement,* 242. On the ambiguous value of fermentation in the Judeo-Christian tradition, see Shershaw, *Bread,* 53–63.

60 Thirsk, *Food in Early Modern England,* 230 and 230–35 more generally.

61 I am suggesting that, for Gerard, the dry Mass effects something on the order of what Jean-Luc Marion calls a "saturating" or "saturation phenomenon," a phenomenon that by appearing in and out of itself evades metaphysical grounding or, more correctly, alludes to that grounding by the absent presence or efficacy of the mimed Host, treated as if present to the cell. Gerard's gestures, the space of his cell, constitute an anchoring "event" or sacramental bubble within the confines of the Tower. See, among others, Jean-Luc Marion, *The Visible and the Revealed,* trans. Christina M. Gschwandtner et al. (New York: Fordham University Press, 2008), and Marion, *In Excess: Studies in Saturated Phenomena,* trans. Robyn Horner and Vincent Berraud (New York: Fordham University Press, 2002). For a more general overview of the phenomenology of the sacraments and, in particular, ways of negotiating the "gap" or "connection" between the "form of the sacramental symbol and the 'sacramental thing' *(res sacramenti)*—the grace of Christ," see Marion, "The Phenomenality of the Sacrament—Being and Givenness," trans. Bruce Ellis Benson, in *Words of Life: New Theological Turns in French Phenomenology* (New York: Fordham University Press, 2010), 89–102.

The bibliography on the theology of the Eucharist is, of course, voluminous. But key studies informing my thinking here include John Bossy, "The Mass as Social Institution, 1200–1700," *Past and Present* 100 (1983): 29–61; Miri Rubin, *Corpus Christi: The Eucharist in Late Medieval Culture* (Cambridge: Cambridge University Press, 1992); P. J. Fitzpatrick, *In the Breaking of Bread: The Eucharist and Ritual* (Cambridge: Cambridge University Press, 1993), whose parsing of "Eucharistic presence" is remarkable; and Lee Palmer Wandel, *The Eucharist in the Reformation: Incarnation and Liturgy* (Cambridge: Cambridge University Press, 2006), which offers a magisterial overview of differing accounts of the sacrament.

Compelling also are the accounts of sacramental poetics as they shaped literary and cultural practices, past and present; see, among others, Schwartz, *Sacramental Poetics*; Sarah Beckwith and David Aers, "The Eucharist," in *Cultural Reformations: Medieval and Renaissance in Literary History,* ed. Brian Cummings and James Simpson, 153–65 (Oxford: Oxford University Press, 2010).

For a somewhat totalizing account of the Eucharist as suprasign or "sublime object" in the period, see Catherine Gallagher and Stephen Greenblatt, *Practicing New Historicism* (Chicago: University of Chicago Press, 2000), 75–109 and 136–92. For an important critique emphasizing, by contrast, the breadth and depth to sacramental theology that offers the

Eucharist as "enmeshed in complex webs of relationships . . . the sacrament of Christian unity, the sacrament of charity, the consummation and completion of all the sacraments," see David Aers, "New Historicism and the Eucharist," *Journal of Medieval and Early Modern Studies* 33, no. 2 (2003): 241–59. Although he does not use these terms, Aers is at pains to point out the way the sacraments constitute an infrastructure and language of belief that suture signs to matter in a way that organizes the world. They constitute a material-semiotic system.

The sacraments are, in Bruno Latour's terms, an "actor network," something I now find taken up in Michael Wintroub's analysis of the "metrological work of trying to establish, maintain, and extend the faithfulness of translation—in domains as diverse as literature, politics, religion, and commerce." Wintroub, "Translations: Words, Things, Going Native, and Staying True," *American Historical Review* 120, no. 4 (2015): 1186. My work is in deep sympathy with the translational poetics Wintroub investigates.

62 Daniel Defoe, *A Journal of the Plague Year,* ed. Louis Landa (Oxford: Oxford University Press, 2010), 9. Unless otherwise indicated, subsequent references appear parenthetically in the text.

63 The most recent iteration of this response to H.F. comes from Ernest B. Gilman in his inspiring *Plague Writing in Early Modern England,* in which he tracks the advent of the plague in early modern England, indexing its fallout to the passage from a faith in political theology and providential narrative as modes of explanation to the rise of the novel. Gilman situates Defoe's *A Journal of a Plague Year* at the moment of disintegration when belief in the divine decouples from belief in the supernatural, coming to rest, instead, in a rationalized sense of civic industry. See Gilman, *Plague Writing in Early Modern England* (Chicago: University of Chicago Press, 2009), 229–43. Gilman's reading of the *Journal* as a spectral discourse trades on the logical holes in the narrative and owes debts to Jayne Elizabeth Lewis's "Spectral Currencies in the Air of Reality: *A Journal of the Plague Year* and the History of Apparitions," *Representations* 87 (2004): 82–101.

64 *The Children's Robinson Crusoe: or the Remarkable Adventures of an Englishman . . . by a Lady* (Boston: Hilliard, Gray, Little, and Wilkins, 1830), 69–70.

65 Gallagher and Greenblatt, *Practicing New Historicism,* 113–14. Gallagher and Greenblatt offer a complicating revision to E. P. Thompson's sense of the "bread nexus" in "The Moral Economy of the Crowd," though, to be fair, they seem at a loss of where to go after they motivate the

potato to destabilize Thompson's account. In my terms, Gallagher and Greenblatt's recovery of "tuber culture" (72) reveals the multispecies enmeshment of material–semiotic actors such as bread and its differently encoded parceling of nutrients and matter, the potato, the breadfruit, and other plants, such as the cassava, that Crusoe also fails to find on the island. Each of these actors challenges one particular axiomatic or settlement of the "raw" and the "cooked" and its attendant structures. See also, and inevitably, Claude Lévi-Strauss, *The Raw and the Cooked: Mythologiques,* vol. 1 (1969; repr., Chicago: University of Chicago Press, 1983).

66 Daniel Defoe, *Robinson Crusoe,* 2nd ed., ed. Michael Shinagel (New York: W. W. Norton, 1994), 86. Unless otherwise indicated, subsequent references appear parenthetically in the text.

67 Early on in his *Autobiography,* Franklin refers to Defoe's Moll Flanders, his "Cruso," and his essay on projects (26).

68 On the nineteenth-century use of Defoe's novel to inculcate the "duty of self-help," see David Blewett, *The Illustration of Robinson Crusoe, 1719–1920* (Gerrards Cross, U.K.: Colin Smythe, 1995).

69 Lydia H. Liu, "Robinson Crusoe's Earthenware Pot," *Critical Inquiry* 25 (1999): 728–57.

70 As Gilles Deleuze notes, the sense of the novel as an "'instrument of research'—which starts out from the desert island and aspires to reconstitute the origins and the rigorous order of works and conquests which happen with time . . . is twice falsified. On the one hand, the image of the origins presupposes that which it tries to generate (see, for example, all that Robinson has pulled from the wreck). On the other hand, the world which is reproduced on the basis of this origin is the equivalent of the *real*—that is, economic—world, or the world as it would be if there were no sexuality." Deleuze, *The Logic of Sense,* trans. Mark Lester, with Charles Stivale (New York: Columbia University Press, 1989), 302–3.

71 Robert Markley, *The Far East and the English Imagination, 1600–1730* (Cambridge: Cambridge University Press, 2006), 178. For Markley's inspired and synoptic treatment of the trilogy, see 177–240.

72 Deleuze, *Logic of Sense,* 318–19.

73 Derrida, *Beast and the Sovereign,* 2:75.

74 Ibid., 2:78. For an explicit return to *Of Grammatology,* 2:82–83.

75 Ibid., 2:83–85. And on the homonym lurking in the phrase "'*Qu'est une île?*'"—what is an island; what is a he, ibid., 2:3. Marc Shell puts this founding ambiguity to provocative use in his *Islandology: Geography, Rhetoric, Politics* (Stanford, Calif.: Stanford University Press, 2014).

76 Derrida, *Beast and the Sovereign,* 2:50.

77 Understandably, the appearance of this footprint is something of an icon within the history of criticism. My reading owes debts to Patrick Brantlinger's reading of the scene as a derangement of "bourgeois rationality" in *Crusoe's Footprints: Cultural Studies in Britain and America* (London: Routledge, 1990), 1–3; Srinivas Aravamudan's phenomenologically nuanced reading in *Tropicopolitans: Colonialism and Agency, 1688–1804* (Durham, N.C.: Duke University Press, 1999), 71–76; and Robert Folkenflik's survey of misreadings of the scene in "*Robinson Crusoe* and the Semiotic Crisis of the Eighteenth Century," in *Defoe's Footprints,* ed. Robert N. Manquis and Carl Fisher, 98–125 (Toronto: University of Toronto Press, 2009).

78 Michel Tournier, *Friday,* trans. Norman Denny (Garden City, N.Y.: Doubleday, 1969), 77. The French here reads as follows: "Cette fois il ne se refusa pas la joie de se faire du pain. . . . Une fois de plus, il replongeait ainsi dans l'élément à la fois materiel et spiritual de la communauté humaine perdue. Mai si cette première panification le faisait remonter, par toute sa signification mystique et universelle, aux sources de l'humain, elle comportait aussi dans son ambiguité des implications toutes individuelles celles-là—cachées, intimes, enfouies parmi les secrets honteux de sa petite enfance—et promises par là même à un épanouissement impévu dans sa sphère solitaire." Michel Tournier, *Vendredi, ou les Limbes du Pacifique* (Paris: Gallimard, 1967), 68.

79 Italics appear in the English translation. The French reads as follows: "Je le sais maintenant, j'imaginais d'étranges épousailles entre la miche et le mitron, et je rêvais même d'un levain d'un genre nouveau qui donnerait au pain une saveur musquée et comme un fumet de printemps." Tournier, *Vendredi, ou les Limbes du Pacifique,* 69.

80 The French reads as follows: "Il y vit le signe que la voie végétale n'était peut-être qu'une dangereuse impasse." Ibid., 102.

81 The French reads as follows: "ses jambes prenaient appui sur le roc, massives et inébranlables comme des colonnes." Ibid., 205.

82 Deleuze, *Logic of Sense,* 320.

83 Claude Lévi-Strauss, *Tristes Tropiques,* trans. John Weightman and Doreen Weightman (New York: Atheneum, 1974), 334 and 330.

84 Ibid., 334.

85 For a medium-specific reading of Defoe's fading ink, see Lothar Müller, *White Magic: The Age of Paper,* trans. Jessica Spengler (Cambridge: Polity Press, 2014), 104–7.

86 These are Psalms 13 and 52 in Vulgate numbering, 14 and 52, respectively, in the Hebrew Bible and in Reformation Bibles.

87 V. A. Kolve, *Telling Images: Chaucer and the Imagery of Narrative II* (Stanford, Calif.: Stanford University Press, 2009), 235 and 237. As Kolve offers, "in his *Summa theologiae,* ca. 1265–68, where the second question—'whether there is a God'—is answered by magisterial proofs of His existence," Aquinas addresses his "argument explicitly to the psalter fool. But . . . he is brought forward as a straw man, a striking way to launch the proof" (225).

88 Ibid., 226.

89 For a positive reclaiming of the figure of the idiot, of the overly idiomatic, and so of idiocy as a self-enclosed, self-predicating assertion of nonknowledge or not knowing, which slows down our ability to make decisions, see Stengers, "Cosmopolitical Proposal."

90 Haraway, *When Species Meet,* 245.

91 Sophia Roosth, "Screaming Yeast: Sonocytology, Cytoplasmic Milieus, and Cellular Subjectivities," *Critical Inquiry* 35 (Winter 2009): 332.

92 Ibid., 338–39.

93 Ibid., 350.

94 Anna Lowenhaupt Tsing, *The Mushroom at the End of the World: On the Possibility of Life in Capitalist Ruins* (Princeton, N.J.: Princeton University Press, 2015).

95 Ibid., 139.

ERASURE

1 Tolkowsky describes the fourteenth- and fifteenth-century Italian pictorial tradition in *Hesperides,* 151–55, as does McPhee in *Oranges,* 78–80. The painting in my mind's eye is, perhaps, Titian's *The Last Supper* or *Supper at Emmaus* (1557–64), intended for the Escorial; or perhaps it is Leonardo da Vinci's mural in the refectory at the monastery of Santa Maria delle Grazie (1495–98).

2 On such fashions, see John L. Varriano, *Tastes and Temptations: Food and Art in Renaissance Italy* (Los Angeles: University of California Press, 2009), 105, and also his short essay "At Supper with Leonardo," *Gastronomica* 8, no. 1 (2008): 75–79, in which he ventures that the oranges garnish a dish of eels. I am grateful to Stephen J. Campbell for these references.

Index

Aardman Animation, 55–57, 86–89. *See also* Serta mattresses; *Shaun the Sheep*; *Timmy Time*
Abla, Ken, 338n58
acedia (sloth), 93–96. *See also* idling/idleness; *otium*
actor, theater: as screen, 5–6; as sound box, 4–5; as wetware, 1–11. *See also* Cade, Jack; Dick the Butcher
actor network theory (ANT), 14, 23–26, 340n61; and archive fever, 288n46; and differently scaled actors, 18–19; and dropped actors, 25–26; and the humanities, 25–26; and model of events, 312n8. *See also* Latour, Bruno
Acumen (journal), 47
Aers, David, 340n61
aesthetics: as discourse on sense data, 14, 25, 46, 289n53; as mode of attack, 219–20
Agamben, Giorgio, 286n34, 305n9, 310n67; and anthropological machine, 106–11; and *otium* archive, 108–11; and state of exception / suspension, 107–8
Agnus Dei, 7, 51–52, 109–10, 121
Ahmed, Sara, 1
air: as groundless ground of metaphysics, 229–30. *See also* breath; bubbles; fermentation
Ajax (mythological figure), 131
Alaimo, Stacy, 24
Allen, Valerie, 324n30
Allet, Sir John (Lord Mayor of London), 198
allure: as captivation by things, 30, 196–208, 211
Alpers, Paul: *What Is Pastoral?,* 92, 113, 308n40
Althusser, Louis, 320n52
America: as an idea, 240–43
Anidjar, Gil, 295n35
animal/s (as category), 26–29; artificial/electric, 78–86; co-making with ethologists, 119–25; communication, 16–17; companion, 17; empathy for, 76–84; and factory farming, 17; health insurance for, 17; naming of the Middle Ages and Renaissance, 303n96; questions of reaction and response, 14, 16, 23, 81; recruitment of by plants, 318n44; sense perception as multispecies contact zone, 190–91; sympathy, 1–2. *See also* Derrida, Jacques; fable, animal;

Haraway, Donna; human; multispecies; sheep; Wolfe, Cary
animation: differing states of, 15–16; as an effect of writing, 16; hierarchies of, 286n38
anthropocene (as concept / naming practice), 21–22
anthropology (multispecies), 119, 284, 283n18
anthropomorphism, 36–42. *See also* zoomorphism
anthropo-zoo-genesis / genetic practices, 2–3, 10–11, 13–14, 27–29, 54, 72, 95, 117, 137, 223–25, 260; as co-making of animal and plant actors, 13–14; as practice of coauthoring, 119–224; and tropes/tropology, 12–14, 20, 137, 224, 260. *See also* Despret, Vinciane; Rowell, Thelma
antiquarianism, 212–15; as fetish work, 210–12
Appadurai, Arjun, 326n39
apples, golden, 30, 166–69, 179–90, 192, 204–8, 210. *See also* Atalanta; Hercules; oranges
Arcadia, 116
Archer, John Michael, 282n15
archive, 1–30; as constellation of animal, plant, and mineral matter, 7–8; as contact zone, 15, 27; human body as, 12; humanities modeled as, 24–26; as multispecies impression, 1–2, 10–11, 13–14, 175–76; as repertoire to be performed, 140–41; shifting ontology of, 141. *See also* flesh: material-semiotic; metaplasm; writing
Arden, John (escapee from Tower of London), 136, 160–63. *See also* Gerard, John; Tower of London
Aristotle, 304n4, 324n31; on color, 194
Armstrong, Philip, 297n61, 321n8
art: and affective labor, 63–68; modernist work of, 173; technics versus "Nature," 205–6
Arundel, Earl of (prisoner in the Tower of London), 314n17
Ascham, Roger: *The Schoolmaster,* 299n71
askesis, 22
assemblage (additive paradigms), 14
Assize of Bread, The, 248–52, 255–56, 337n55
Atalanta (mythological figure), 192, 204–8
Atlas (mythological figure), 186, 217–20, 330n77
Auerbach, Erich: *Mimesis,* 13–14. *See also* figures/figuration
Augé, Paul, 312n8
Austen, Ralph: *A Dialogue or Familiar Discourse and Conference betweene the Husbandman and the Fruit-Trees,* 200–201

Babington Plot, the, 161
baboons, 121, 126
Bacon, Sir Francis, 158; *New Atlantis,* 291n3; *On the Wisdom of the Ancients,* 204–8
bakers and baking: and bread nexus, 239–40; as figure of process, 225–27; as kinaesthetic, 225–26; kneading dough and sexuality, 262–63; use of ale barm in making of, 247–48; Worshipful Company of Bakers, 248–52,

257. *See also Assize of Bread, The*; beer and brewing; bread; fermentation; leaven; yeast
Barad, Karen, 24
Barnabas, Saint, 236, 243
Barrick Gold, 183
Barthes, Roland: on the neutral/nuance, 233, 333n26; on orange/s, 203–4, 328n64; on theft, 290n59
Bataille, Georges, 109, 322n12
beachcombing, as method, 14
Beaumont, Francis: *Philaster: or Love Lies Bleeding*, 328n60
Beckwith, Sara, 339n61
Becon, Thomas: *The Jewell of Joy*, 60
beer and brewing, 248–49, 337n55. *See also* bread; fermentation; leaven
bees, 3–4, 131
Benjamin, Walter: on collecting, 335n33; *Hashish in Marseilles*, 223, 230–36, 250, 333n22; *One Way Street*, 108; "On the Mimetic Faculty," 235; *Passagenwerk*, 235
Bennett, Jane, 24, 305n8
Bennett, Judith, 337n55
Beresford, M. W., 60–63
Berkeley, Sir Richard (Lieutenant of the Tower of London), 158, 313n14
Berlant, Lauren, 305n8
Besant, Sir Walter: *The Orange Girl*, 323n18
bestiality: accusations of, 77–78. *See also* sex, interspecies
Bible, Hebrew, 104, 108
biblical references: Genesis 1:1–2, 332n13; Genesis 4:25, 228–30; Isaiah 11:6, 125–31, 139; John 1:29, 51–52; Leviticus 20:15, 78; Luke 11:3 (The Lord's Prayer), 331n8; Matthew 4:3, 267; Matthew 6:11 (The Lord's Prayer), 331n8; Psalms 13 and 52, 265–69; Psalm 23, 72, 301n84
Big Sheep (theme park), 84–89
Bill of Rights, 243
biopolitics, 17–20, 27–28, 54, 59, 95–96, 109–10, 129–30, 137–39, 142, 224; affirmative, 185, 321n7; history of, 55–56, 109–10; as reduced form of ecology, 20; and sheep/shepherding, 28–29, 129–31, 291n4; as value neutral, 41; and zoopolitics, 18, 177. *See also* blood; Christianity; *otium*; pastoral; pastoral care; sheep; shepherds and shepherding
biopower, 27, 36
Birch, Thomas: *The History of the Royal Society*, 51
Black Sheep (film), 55
Blade Runner (film) 47–48
blazon/blazoning, 68–70, 79
blood: 294nn33–34, 294n35, 320n54; of Jesus as universal substrate, 52; transfusion, history of, 48–55, 59, 62; and writing, 62, 151–52, 165
Boehrer, Bruce, 300n72
Bogost, Ian, 289n53
Bonner (warder at the Tower of London): as archival artifact/chimera, 165–67; as figure for tourists at the Tower of London, 167–68; as hybrid, 145–46; illiterate, 160–61; as informant,

158–59; legal identity of, 145; and liking for oranges, 145, 159–69, 183–85, 196, 206–8, 244–45; twinned with gold and oranges, 169. *See also* Gerard, John; Peyton, Sir John; Tower of London
Bono, James, 315n31
boredom: as fatigue, 92, 96; profound, 105–11. *See also* Heidegger, Martin; idling/idleness; labor: and leisure; *otium*; sleep
Bosman, Anston, 279n2
Boston, Colonial, 236–37, 239
Bowden, Peter, 296n46
Boyle, Robert, 51
Braddocks (safe house), 158
Braidotti, Rosi, 24
Bradford, William: *Of the Plymouth Plantation,* 77–78, 302nn85–87
Brantlinger, Patrick, 342n77
Braudel, Fernand, 115–16, 308n47
bread, 30–31, 223–69; bread nexus, 239–40; daily, 30, 245–52, 257–58; as figure for infrastructure/security, 30–31, 224–25; as fossilized breath, 226–27; hieroglyphical character of, 249–50; as index of culture, 256–57; kneading dough and sexuality, 262–64; labor in production of, 247–48; and Last Supper, 271–73; regulation of in medieval and Renaissance London, 248–50; taste of, 250; unleavened, 227, 237, 335n36. *See also* air; *Assize of Bread, The*; bakers and baking; beer and brewing; bubbles; Eucharist; fermentation; leaven; yeast
breath: divine, 128–29, 228–29, 251; human, 223–69; regulation of, 257–58. *See also* air; bubbles; yeast
Brewer's Book, The, 338n56
British Army Corporal Manual, 142
Brueghel, Pieter the Elder, 264
bubbles, 30–31, 223–69; as figure of enough/sufficiency, 242; as morpho-immunological figure, 228–29. *See also* air; bread; Sloterdijk, Peter; yeast
Burckhardt, Jacob, 306n15
Burke, Peter, 304n6
Butler, Judith, 288n47

Cade, Jack (character in *Henry VI, Part 2*), 1–11, 20–26, 31, 36, 37–42, 73–74, 91, 127; becoming sheep, 10–11; as idiot, 9–10; as mimetic operator, 8; as screen, 5–7; self-narrating, 282n16. *See also* actor, theater; idiot/idiocy
Calhoun, Joshua A., 280n3, 301n80
Callois, Roger, 283n20
Calthorp, Elizabeth, 201–3
Campbell, Stephen J., 304n7
Campbell, Timothy, 27
Campion, Edmund, 165–66
Camporesi, Piero, 335n37
capitalism, industrial, 37–42, 218–20; and fetishism of the commodity, 30, 173–77, 179–83, 205–6; and gains of trade, 180–83, 200. *See also* Marx, Karl; Pietz, William
Carew, Sir Francis, 198
Cartelli, Thomas, 282n16
cash nexus, 237–38
catechresis, 210. *See also* figures/figuration

Catholics/Catholicism: legislation against in Renaissance England, 142; mission to convert Elizabethan England, 141–42. *See also* Jesuits
Caus, Salomon de, 326n41
Cecil, Sir Robert (Lord Chancellor of England), 77
ceramics: divine creator as ceramist, 228–30; in *Robinson Crusoe,* 258–59
Certeau, Michel de, 316n35
Chakrabarty, Dipesh, 285n27
Chaplin, Charlie: as Little Tramp, 37–42
character *(ethos),* 1–2, 6. *See also* actor, theater; *ethopoeia*
Charles I, 327n55
Chartier, Roger, 281n9, 317n35
Chen, Mel Y., 286n38
chimpanzees, 120–21
choreography, 190–91, 214–15; ontological, 11, 118, 125, 283n17
Christianity: and biopolitics, 36–42; and blood, 51–55; blood of Jesus as universal substrate, 52; and meekness, 70; mutuality and obligation, 59–60; and pastoral care/power, 53, 141–42; as sheep-like, 76–77; and universalism, 7, 76. *See also* biopolitics; Eucharist; sheep; transubstantiation
CIA (Central Intelligence Agency), 157, 317n37
Cicero Marcus Tullius: *De Natura,* 93–96; *De Officiis,* 67, 97–98
cigarettes, 332n21
CinemaScope (magazine), 72
Cixous, Hélène, 318n45
Clement's Danes, Saint (church), 209, 213
Clement's Eastcheap, Saint (church), 209
cloning, 46–48, 50–55, 293n23. *See also* Franklin, Sarah; sheep
Close Shave, A (film), 55
Coe, Sue: *Ship of Fools,* 55
Coga, Arthur, 51–55, 57, 109, 294nn33–34
Cohen, Jeffrey Jerome, 284n20, 330n5
Cohen, Tom, 288n45, 308n41, 333n22, 334n27
Coke, Edward (Attorney General of England), 149–50, 158, 160
Colebrook, Claire, 288n45
collecting, 212–14, 335n33
College of St. Omer, 157
color: fixity, 193; materiality of, 192–96. *See also* Munsell Code
Comelyn, S.: *The Belgick of Netherlandish Hesperides,* 197–98
commons and commonwealth, 2–11, 58–63, 128–31, 238, 241–43, 272–73; the Commonwealth Men, 60, 67, 85–88; and cosmopolitics, 24
Connor, Steven, 330n8
Constitution, U.S., 243
counting (as *logos* and logistics), 21, 28–29, 72–73, 83–84, 91–92, 131, 137–39, 292n14, 303n1; and biopolitics, 36–42; counting sheep, 55–58; as disanimating, 303n98; resignified as scene of encounter, 87–88. *See also* biopolitics; shepherds and shepherding
Cowley, Abraham: "Written in the Juice of Lemmon," 153–55, 172.

See also ink, invisible; writing: secret
cows and calves, 51–2; and Middle Passage, 292n12; and origins of capital, 40, 291n5
Crane, Susan, 300n72
critic, literary: expertise of refigured in posthumanities, 24–26
criticism, nuclear, 287n45
cuts/cutting, 31, 288n50. *See also* Dick the Butcher; decision; erasure

Dailey, Alice, 317n36
Dasein, 79–84. *See also* finitude: human; Heidegger, Martin
David, Elizabeth, 338n56
Daybell, James, 315n26
decision, madness of, 21–24, 229, 242–43, 271. *See also* cuts/cutting; Dick the Butcher
Defoe, Daniel: bread making episode, 257–59; ceramics in, 258–59; *Journal of a Plague Year,* 252–56; *The Life and Strange Surprising Adventures of Robinson Crusoe,* 256–59, 342n77
Delanda, Manuel, 284n26
Deleuze, Gilles: on the axiomatic, 284n26; and Félix Guattari, on nonorganic life, 194; on Robinsonade, 262–64, 341n70; on time, 35–36
dermatography (skin writing), 315n28
Derrida, Jacques: on allure, 207; on the animal as a concept, 16, 83, on archives and archive fever, 16, 286n34, 315n33; on bare life, 83, 109; biologistic continuum, 15–16; on biopolitics, 109; on divine incarnation, 109; on flowers, 135, 140–41; on the infra-human, 15; on insects in amber, 307n33; on ipseity, 260–61; on the madness of decision, 288n50; on mimesis, 320n53; on Francis Ponge, 175; and response, 14–16, 302n90; on *Robinson Crusoe,* 260–62; on *survivance,* 311n7
Descartes, René: concepts, 78–84, 300n72, 302n89
Descola, Phillipe, 283n18
desire, as "empty set," 27
Despret, Vinciane: on anthropo-zoo-genetic practices, 19–20, 138–39, 287n41, 310n74; *Quand le loup habitera avec l'agneau,* 125–31; on Thelma Rowell's sheep-observing, 119, 122–25
detective fiction (genre), 141
Deutsch, Helen, 204–8, 319n47, 328n61
Dick the Butcher (character in *Henry VI, Part 2*): as personification of decision/cutting/erasure, 10–11, 21–22, 26, 31, 37–42, 229, 271. *See also* cuts/cutting; decision; erasure
Dick, Philip K.: *Do Androids Dream of Electric Sheep?,* 47, 78–84, 91, 118, 302n89; on herbivores and carnivores, 80–82; on relation between Deckard and Descartes, 76–84; Voight Kampf test in, 80–81
Dimock, Wai Chee, 308n37
disorientation, as a mode of reading, 1–2

DNA, 176
dogs, 42
Dolly the sheep, 46–48, 293n23. *See also* cloning; Franklin, Sarah; Roslin Institute
donkeys, 42
Doyle, Richard, 281n7, 320n52, 334n32
drugs: vegetal efficacy of, 230–36. *See also* Benjamin, Walter; Doyle, Richard; plants

ecology, 14, 177; and disaster, 287n45
economy, moral, 232–43. *See also* commons and commonwealth; enough, concept of; Thompson, E. P.
Edgeworth, Maria: "The Orange Man," 211–12
Edward VI, 60
E.E.C. (European Economic Community), 85
Eklund, Hillary, 282n16, 336n45
ekphrasis, 102–6, 108. *See also* figures/figuration; idyll; media ecology; pastoral
Elizabeth I, 77
Ellison, Joan Jarvis: *Shepherdess,* 301n77
Empson, William, 92, 113, 118, 121, 303n2, 308n40
enclosure, 48; as coding of land and labor, and cowriting of sheep and humans, 7–8, 59–60, 85–86; critique of, 57–63; and Scottish land clearances, 63–68. *See also* More, Sir Thomas; multispecies; sheep; writing
endo-colonization, 50
Enlightenment, the, 181
enough, concept of, 236–43, 246, 257
Enterline, Lynn, 281n7
epic, 114–18
epistemology, historical, 194–95
Erasmus, Desiderius, 98–99
erasure, 22–23, 31, 271–73. *See also* cuts/cutting; decision; Dick the Butcher
eschatology, 126–31, 138–39. *See also* cuts/cutting; decision; exception/suspension, state of
Esposito, Roberto, 286n36
ethology, 92–93
ethonym, 72
ethopoeia (ethos), 6, 27, 281n7. *See also* figures/figuration; *prosopopoeia*; tropes/tropology
Etter, David Rent, 241–42
Eucharist, 52–55, 128, 227–28, 181, 234, 259; as appropriation of plant being, 176, 320n54; as biopolitical technique, 54; convoking function of, 30; efficacy of, 336n48; and poetics, 294n34; as sublime object, 339n59; theology of, 339n61; and xenotransfusion, 51–55. *See also* Christianity; Mass, the; transsignification; transubstantiation
exception/suspension, state of, 106–11, 125–31, 305n9. *See also* Agamben, Giorgio; eschatology

fable, animal, 20, 287n42; and political theology, 125–31
factory farming, 17
failure, risk of: in knowledge making, 120–21

Farge, Arlette, 284n25
Febvre, Lucien, 279n3
Feerick, Jean E., 282n16
Fella, Thomas, 72–76
fermentation, 31, 338n59. *See also* bakers and baking; bread; bubbles; yeast
Ferrarius, Joannes Baptista: *Hesperides sive de Malorum Aureorum Cultura et Usu,* 185–91; as multispecies archive, 188
fetish/fetishism, 30, 199–200; border fetishism, 181–82; as concept-problem, 179–83; and DIY sacramental objects, 181; good and bad fetishes, 68, 182, 320n4; 299n70; and work, 210–12. *See also* allure; capitalism, industrial; Pietz, William
figures/figuration, 13–15. *See also* Auerbach, Erich; *catechresis*; *ethopoeia*; metaplasm; *proposopoeia*; tropes/tropology
finitude: human, 12–15; multiple forms of, 16
Fleming, Juliet, 154–55, 317n35
Fleming, Thomas (solicitor general), 158
flesh: material-semiotic, 1–11; as substrate for biopower, 2, 7, 17–18, 26–27, 36–42, 53–54, 235–36, 286n36, 305n8. *See also* archive; Esposito, Roberto; Merleau Ponty, Maurice; writing
Fletcher, John: *The Faithful Shepherdess,* 115; *Philaster: or Love Lies Bleeding,* 328n60
florilegium (book of flowers), 30, 186–88, 190–91
Florio, John: *Florio's Second Fruites,* 326n38
flowers: and sacrifice, 18, 140–41, 172. *See also florilegium*; plants
Folkenflik, Robert, 342n77
food/eating: eating well, 27, 208; markets in colonial America, 335n40; and meals / Last Supper / Passover Seder, 271–72; and the munchies, 232; the raw and the cooked, 10–11, 256–57; as scene of animal recruitment by plants, 170–77; species hierarchy and, 10–11; status of messianic banquet, 128–31; surplus of in ethological protocols, 123–24; and *Utopia,* 98–102; webs in medieval and Renaissance England, 337n55. *See also* bread
fool: medieval iconography of, 265–69. *See also* idiot/idiocy
form/formalism, 24–26, 35–36
Foucault, Michel, 36–42, 39–41, 87–88, 91, 291n4, 292n14, 300n72, 305n9, 309n55
Franklin, Benjamin, 236–43, 335n41
Franklin, Sarah, 48, 298n62, 293n23
Freedberg, David, 322n9
Freeport McMoRan, 183
Freud, Sigmund, 287n44
Friedman, Karin J., 335n40
fruit: and sacrifice, 140–41. *See also* plants
Fudge, Erica, 279n2, 300n72, 303n96
Fumerton, Patricia, 327n55
fungi, 26–30, 223–69; and conservation, 12. *See also* bakers and baking; bread; bubbles;

fermentation; leaven; yeast
future/futurity, 217–20, 273, 330n1, 336n43

Gallagher, Catherine, 340n65
Gallagher, Lowell, 152
gardens and gardening, 326n41; ambivalent value of in Renaissance England, 100–102
Garnet, Henry (Jesuit superior), 149–50
genome: human, as archive, 12; micromanipulation of, 47, 50–55. *See also* endo-colonization; Haraway, Donna
genre: infragenre, concept of, 13–14; as rhizomatic, 35–36, 308n37
geometry, 261
Georgic (mode), 184, 199–201
Gerard, John: and bread in the Tower, 243–52; genre of, 314n18, 317n36; and prison escape from Tower of London, 141–42, 145–47, 157–69, 175, 184–85; *Narratio Johannis Gerardi,* 133–77; and production of sacramental bubble in cell, 251–52; Tower bills for, 336n49
Gifford, Terry, 105, 116, 308n40
Gilman, Ernest B., 340n63
Gimzeski, Jim, 268
Glimp, David, 296n48, 328n57
goats, 42
Goethe, Johann Wolfgang von: *The Sorrows of Werther,* 203–4
gold, 30; edible, 328n65. *See also* oranges
Goldberg, Jonathan, 282n14, 302n85, 316n35
Golden Age, the, 111, 115
Golding, Arthur: *Ovid's Metamorphoses,* 206–8
Goldsworthy, Andy, 63–69
graffiti, 165, 244–45, 255, 318n43
Grafton, Anthony, 306n15
Gramsci, Antonio: on prison psychology, 135, 142; on writing in prison, 142; and zoomorphism, 35, 42–46, 91. *See also* life; skin
Granger, Thomas, 77–78, 91, 302n85
Greenblatt, Stephen J., 306n14, 340n65
Griffin, Dr. Harry, 47
Griffiths, Elizabeth, 325n38
ground, concept of, 224–26, 235–36, 268–69. *See also* bread; infrastructure; Ingold, Tim; stone
Gunpowder Plot, 149, 160, 244
Gustafsson, Laura, 303n101
Gwyn, Nell, 323n18

Haapoja, Terika, 303n101
Hacking, Ian, 307n27
Hales, John: *The Causes of Dearth,* 61
Halperin, David, 113–18
Halpern, Richard, 67, 306n14
Hamilton, Shane, 330n79
Hamlet (Shakespeare), 42
Hanson, Elizabeth, 318n42
Haraway, Donna, 1, 11–13, 21, 24, 124, 40–41, 124–25, 227–28, 267; and Catholicism, 295n37; on companion species relating, mess-mates, and multispecies, 11–14, 54–55; on species

discourse, 20–21, 287n44
Hardt, Michael, 302n88, 321n7
Harington, Sir John: *A New Discourse upon a Stale Subject: The Metamorphosis of Ajax,* 131, 311n80
Harman, Graham, 285n32, 289n54, 316n34, 329n65
Harris, Jonathan Gil, 317n35
Hartington, John, Jr., 283n18
Harvey, Elizabeth, 279n2
Hattaway, Michael, 326n38
Hatton, Sir Christopher, 77
Head, Richard: *The English Rogue,* 323n18
Hegel, Georg Wilhelm Friedrich, 122, 309n63
Heidegger, Martin, 106–11, 130–31, 311n78
Heise, Ursula K., 82, 218
Helmreich, Stefan, 283n18
Henrietta Maria, Queen, 102, 153
Henry the Fifth: as ideologeme, 2, 10, 39, 53
Henry VI, Part 2 (Shakespeare): controversies over editing, 282n13; sources for the Jack Cade sequence, 8; zoomorphic play in, 1–11, 36, 127, 271–72. *See also* Cade, Jack; enclosure; innocent lamb; multispecies; skin
Hentschell, Roze, 295n43
Herbert, George: *The Temple,* 201–3, 328n57
Hercules: as figure for writing about oranges, 178, 185–91, 197–98, 217. *See also* Hesperides, Garden of; oranges
Herzfeld, Michael, 308n48
Hesperides, Garden of, 185–91. *See also* Hercules; oranges
heterotopia/heterospace, 4, 183, 184–85. *See also* Pietz, William
Highfield, Roger, 293n23
Hilliard, Nicholas: *The Art of Limning,* 192–93
Hiltner, Ken, 304n3
Hippomenes (mythological figure), 192, 204–8
hoarding, 305n8
Hoffman, Abbie: *Steal This Book,* 183
Holbein, Hans, 100–102, 105
Hollywood, 86
holocene, 22
Holsinger, Bruce, 279n2
Homer: *The Odyssey,* 114
hospitality, concept of, 19–20, 22, 25–26, 31, 195–96. *See also* Derrida, Jacques; enough, concept of; politeness; trust, cultivation of; Wolfe, Cary
human animal: as an archive, 12, 140–41, 150–52; as belated, 22–23, 27; and cowriting with sheep, 28–29, 36–42, 48–55, 88–89, 91–92; and handedness, 93–96, 97–98, 101–2; and idleness, 98–102; and infra-human, 15; and mimesis, 82–84; and parasitism, 13–16; and recruitment by plants, 157, 169–77; as screen, 16–17, 24–26; as a sheep-being, 73–74
humanism, Renaissance: and concepts of "profit," 68–69; and making up people, 105–6; and *otium*, 97–102; as rewritten by Thelma Rowell's sheep observing protocols, 119–25; and scripts for labor and leisure,

96–102. *See also* commons and commonwealth; More, Sir Thomas; *negotium*; *otium*
humanities: as archival contact zone, 24–26; precarity of, 306n15; and the sciences, 24–25
hunting: as a method, 14
hydraulics, 326n41

iconography, 93–96
identification, transversal, 5–7; canalized by differences in scale, 15, 19–20
ideology, 112–13
idiot/idiocy: as slowing down, 9–10, 21, 26, 271–73, 282n14. *See also* Cade, Jack; Stengers, Isabelle
idling/idleness, 73–76, 91, 93–96, 106–11. *See also* boredom; Heidegger, Martin; *otium*
idyll/idyllic, 97–98, 102–6; *eidyllion*, 105. *See also* media ecology; pastoral
infrastructure, 7, 12, 219–20, 223–69; as keyed to multispecies writing, 12, 31, 235–36; as routinization, 23–24. *See also* bread; ground, concept of; stone; writing
Ingold, Tim, 225–26, 243
ink, invisible: 28–29, 136, 146–57; as co-optation of and exhaustion of plant efficacy, 155–57, 176–77; as eventalization of media, 150–51; explanations of in Renaissance England, 148–49; as frustration of meaning, 151–55; literary references to in Renaissance England, 152–55; and network building, 160–62; recipes for, 148; techniques for using, 148–49
innocent lamb: as figure for multispecies archive, 1–11, 21. *See also* Cade, Jack; Dick the Butcher; enclosure; *Henry VI, Part 2*; sheep
insects: and camouflage rhetoric, 12
interest/interesting (as orientation), 118, 309n53. *See also* Ngai, Sianne; Stengers, Isabelle
interpellation, transhuman, 174. *See also* Althusser, Louis; Doyle, Richard
inventory, as method, 284n25. *See also* Perec, Georges
Irigaray, Luce, 229–30
Irishness, stereotypes of, 256–57
islands/islandology, 341n75
Italy, 42–46; idea of, 42; 1920s–30s prison ecology, 43

James I, 198–99
Jameson, Frederic, 66, 99, 108
Jardine, Lisa, 306n15
Jason (mythical hero): and the Golden Fleece, 52–53
Jeanneret, Michel, 99–100
Jensen, Johannes Vilhelm, 233
Jesuits, 141–42; mission, 157–58. *See also* Catholicism; Garnet, Henry; Gerard, John; Loyola, Saint Ignatius
Jesus Christ, 109–10
Johnson, Barbara, 289n55
Jonson, Ben: *Volpone,* 149
Justice, Steven, 281n9

Kay, Sarah, 279n2
Kett's Rebellion, 62

Kilroy, Gerard, 318n43
kinesthesia/kinesthetics, 225–26. *See also* bakers and baking; prison/s; walking
King, Dr. Edmund, 51–55. *See also* Royal Society; xenotransfusion
Kirksey, Eben, 283n18
Klein, Kerwin Lee, 309n63
Klein, Richard, 332n21
kleptomania, 14. *See also* theft
Kohn, Eduardo, 283n18, 284n27, 320n52
Kolve, V. A., 266–67
Kompare, Derek, 305n11
Korda, Natasha, 323n18, 337n55
Krugman, Paul, 179–80, 181, 200

labor, 7, 33–89, 91–131, 179–221; affective, 79, 302n88; Atlas as figure for, 217–20, 330n77; of bakers, congealed, 247–50; disputes, 53–55; and fetish work, 210–12; labor power, 36–42; labor time, 179–82, 185; and leisure, 92–102, 185; oranges as a waste of, 199–201; plants as self-laboring, 179–221; sheep as summoned to tell the truth about, 28–29, 35–60; of shepherding, 68–78. *See also* Atlas; Franklin, Benjamin; Hercules; Marx, Karl; sheep; shepherds and shepherding
Lake, Peter, 314n7
lamentation / lamentable things, 1–11, 22, 271–72. *See also* Butler, Judith; Cade, Jack
Lancaster, House of, 191
language: and possibility/ resistance, 211–12; as relation to exteriority/technicity, 4, 14–15
lapidaries, 194
Laszlo, Pierre, 324n22
Latimer, Hugh: *Sermon on the Plough,* 58–59
Latin/Latinity, 10, 51–55
Latour, Bruno, 24, 25–26, 119–21, 130, 193, 195–96, 288n46, 289n54, 309n53, 312n8, 320–21n4, 323n18, 324n28, 331n10, 331n21, 338n57, 340n61
laundry: as repository of chemical knowledge, 136–37
law, 3–4; and lyric poetry, 289n55
leaven: as evidence of the divine, 226–27, 248–50. *See also* bakers and baking; bread; fermentation; yeast
Lemay, J. A. Leo, 336nn43–44
lemons: and homonym “leman,” 153, 203, 328n59; and secret writing, 146–58. *See also* ink, invisible; oranges; writing: secret
Leonardo da Vinci, 272–73, 343n1
Leonilla (mythological figure): figure for exchange of properties between humans and oranges, 186–88, 202
Leroi-Gourhan, André, 15
letters and letter-writing, 42–46, 135–36, 149–50
Lévi-Strauss, Claude, 264–65, 272, 341n85
Lewis, Jayne, 340n63
Lezra, Jacques, 242–43, 336n45
life: bare, 42–46, 142, 184–85, 286n34; *bios* and *zoë,* 17–18, 76; as emerging from our notions of an archive, 13–15; forms of,

14–16; grievable, 288n47; and living well, 67; as surplus, 321n7. *See also* biopolitics
liking, structure of, 30, 143, 169–77; as ideological, 219–20; as multispecies contact zone, 171–77. *See also* allure; Bonner; fetish/fetishism; Pietz, William
limning, practice of, 156, 192–96. *See also* Hilliard, Nicholas; miniatures; Oliver, Isaac; Teerlinc, Lavinia
literacy, differential, 3–9, 10; in secret writing, 160–61
literary study: scripts for, 12
Liu, Lydia, 259
livestock, 37, 40, 45–46, 67–68, 85–86, 88–89, 291n5. *See also* capitalism, industrial; cows; human animal; sheep
llamas, 42
Loyola, Saint Ignatius: *Spiritual Exercises,* 158–60, 163–65, 244–46, 250–51, 318n41. *See also* Garnet, Henry; Gerard, John; Jesuits
Luhmann, Niklas, 288n46

Macdonald, Russ, 282n13
Macrackis, Kristie, 314n20
Madelaine, R. E. R., 328n59
Malynes, Gerard de: *Lex Mercatoria,* 199–200
mammals, 128
Man, Paul de, 112, 172–73, 289n55, 303n2, 319n48
Mandeville, Sir John, 301n82
Mantegna: *Pallas Expelling the Vices from the Garden of Virtue,* 94–96, 106, 304n7
Marder, Michael, 140–41, 156, 173–75
Marin, Louis, 67, 294n34
Marion, Jean Luc, 339n61
Markham, Gervase: *Cheap and Good Husbandrie,* 70; *The English Housewife,* 338n56
Markley, Robert, 259, 341n71
Marlowe, Christopher: *Dr. Faustus,* 150–52
Marquez, Gabriel Garcia, 215
Marseilles, 231–36
Martin, Henri Jean, 279n3
martyrdom, 153–55
Marx, Karl: *Das Capital,* 36–37, 63–68, 112, 173, 179, 183, 247, 330n1; *Eighteenth Brumaire of Louis Napoleon,* 223–24; Marxism, autonomist, 321n7. *See also* capitalism, industrial; Gramsci, Antonio; Hardt, Michael; Negri, Antonio
Marx, Leo, 96
Mary, Queen of Scots, 149
Mascall, Leonard, 68–69, 73, 79–80, 136–37
Mass, the, 53–55, 245; *Missa Sicca* (dry Mass), 164, 158–64, 250–51, 339n61. *See also* Catholicism; transubstantiation
Massumi, Brian, 284n24, 321n7
matsutake mushrooms, 268–69
matter, 35–36, 226–27; and metaphor, 4–5, 190–91; and vital materialism, 24
matzah, 227
Mazzio, Carla, 332n18
McHugh, Susan, 302n85
media ecology, 6–7, 12–13, 16, 24–25, 27, 84, 95, 117–18, 124,

139–40, 235, 240; broadcast, 95–98; ethics and politics rooted through, 23. *See also* pastoral; television; translation, modes of
Meillasoux, Quentin, 285n32
memory, 1–11. *See also* skin; touch, sense of
Menely, Thomas, 290n58
Merchant of Venice, The (Shakespeare), 291n5
Merleau Ponty, Maurice, 286n36
metaphysics, 24, 229–30
metaplasm, 1–11, 13–15, 45–55, 64, 72, 113, 282n13. *See also* archives; Haraway, Donna; tropes/tropology
microbiopolitics, 31, 227, 331n10
Middle Passage, 292n12, 303n98
Middleton, Thomas: *A Game at Chess,* 149
Midsummer Night's Dream, A (Shakespeare), 338n57
mimesis, 46, 82–84, 113–15, 171–77, 251, 308n41, 320n53
miniatures: portrait, 156, 192–96
mise en abime, 173
Modern Times (film), 37–42
monasteries, dissolution of, 59–60
Montrose, Louis A., 115, 304n3
More, Sir Thomas: *Utopia,* 36, 57–59, 63, 65–85, 76, 85–89, 98–102, 105–6, 123–25
Morton, Timothy, 24, 113, 285n32, 289n53, 323n17, 324n19
Mowery, J. Franklin, 317n35
Much Ado About Nothing (Shakespeare), 164, 201–3
Mudrew, Craig, 296n44
Muffett, Edward, 326n40, 338n58
Müller, Lothar, 342n85
multispecies, 2, 7, 11–21, 24; as *catechresis,* 20, 271–73; and cowriting of different entities, 19–20, 126–31; and enclosure, 7–8, 35–78; and Eucharistic poetics, 227–28; as neutral, 19–20, 271–73; as an orientation, 272; rival, 7–8, 18, 35–78; writing machines, 28–29, 88, 126–31, 138. *See also* Haraway, Donna; species, discourse of
Munsell Code, 195–96
Murakami, Haruki: *Wild Sheep Chase, Dance, Dance, Dance,* 298n63
Murray, Molly, 317n36
Myers, Anne M., 317n39

NAFTA (North American Free Trade Agreement), 216
Nature (journal), 46–47
Nealon, Jeffrey, 177, 312n7
negotium (business / daily affairs): humanist concept of, 88, 97–111, 124. *See also* humanism, Renaissance; More, Sir Thomas; *otium*
Negri, Antonio, 302n88, 321n7
Newton, Sir Isaac, 195
New York: colonial, 236
Ngai, Sianne, 172–73, 309n53, 319n47
Nicholson, Marjorie Hope, 51–52
Noah's Ark, 125, 303n98
non-conventional entities: Brian Massumi's concept of, 13
Northumberland, Duke of, 61
Norway, 179–83, 198
nostalgia, 210, 214–15
nuance (the neuter/neutral), 250

object-oriented ontology, 316n34
Obstfeld, Maurice, 179–80, 181, 200
occupatio, 206–8. *See also* figures/figuration; tropes/tropology
Odyssey, The, 114
Oliver, Isaac, 192–93
ontology 12; flat, 14, 26; ontological exhaustion of plants in human uses, 176–77; as projective not categorical, 20–21, 82–84
Orange, House of, 179–80, 323n18
orangemen, 323n18
orange/s, 26–30, 133–220; as archival entity, 173–75; as a color, 191–96; and ecology, 188–90; as ethonym, 320n52; and the fetish of the commodity, 30, 169–77, 196–208; as homonym for sheep in Greek, 185; and industrialized farming, 330n79; as localized surplus, 205–8; material semiotic transfer with human animals, 140–41, 190–91, 201–4; morphology of, 204; as a multiplicity and event, 29, 141, 172–73, 207–8; as reproductive technology of a plant, 169–77; as rhetorical entity, 174–75; in speculative fiction, 329n75; and time/temporality, 28, 188–90, 195, 202–3; as tropic weapon, 215–20
"Oranges and Lemons" (nursery rhyme), 208–15
oranges in medieval and Renaissance England: and amatory poetry, 153–58, 328n59; as anti-Platonic fruit, 202–3; as a challenge to Georgic, 199–202; as a commodity, 197; as a corruptible, 196–97, 202; cultivation of, 179–80, 197–201, 324n42, 326n41, 327n2; desire for, 27, 203–8; as edible gold / golden apples, 30, 166–69, 184, 188, 207–8; as fetish, 200–201; as foodstuff, 136, 326n40; as foreign, 196–201; and gender, 202–7; and iconography of the Last Supper, 271–73; imports and prices of, 196, 325nn37–38; liking for code for Spanish sympathies, 164, 197–98; and *otium,* 199–200; as projectiles, 197, 326n38, 328n60; and secret writing, 146–57, 314n20, 317n39; as stain removers, 136–37; taxidermied, 201–3; as technologized, 326n41. *See also* lemons; plants
orange women, 190, 323n18
Orwell, George: *1984,* 212–15
otium, 26–29, 67–68; and anthropological machine, 28–29, 93–96; and creaturely indistinction, 28–29, 92–96, 118; drug, 234; and *negotium,* 88–89, 306n14; and oranges, 199–200; rezoned by Thelma Rowell and comparative ethological studies of sheep, 119–25; and sheep, 92–93, 137–38; and utopian thinking, 92–96. *See also* biopolitics; boredom; humanism, Renaissance; More, Sir Thomas; sheep; sleep
Ovid: *Metamorphoses,* 50, 131, 204–8, 328n64; *Remedia Amoris,* 94

pain/suffering, 1–2, 6–7
palimpsest: as figure of multitemporality, 317n35
Pandian, Anand, 292n7
parasitism, 6–7, 12, 128–31, 229. *See also* Serres, Michel
parchment, 1–11, 26, 36, 146, 156, 271–72, 279n3, 280n4, 280n6, 316n34. *See also* archive; sheep; writing
Parton, Dolly, 47
passion fruit, 218–20
Pasteur, Louis, 130–31, 149
Paston Letters, The, 201–3
pastoral, 28–29, 91–131; and biopolitics, 36–42; and history of technology, 114–16; literary history of, 89, 111–18; and metaplasm, 113; and miniaturization, 104–6; as mixed media, 95–96, 102–6, 117–18; as multispecies contact zone, 117–18; and *otium*/leisure, 92–96, 113–18; and pastoralism, 114–16. *See also* sheep; shepherds and shepherding
pastoral care, 137–39. *See also* biopolitics; Christianity; shepherds and shepherding
Patterson, Annabel, 281n12
Paxson, Heather, 331n16
Peacham, Henry: *The Complete Gentleman,* 193–94
Pears, Iain: *An Incident of the Finger Post,* 312n8
Pepys, Samuel, 102–6, 111
Perec, Georges, 284n25
perversion, structure of, 266–69. *See also* Deleuze, Gilles
Peyton, Sir John (Lieutenant of the Tower of London), 135–69, 183–84, 313n14
phenomenology, 110–11, 190
Philadelphia, Colonial, 256–63
physicalism, 116–17
pidgins, transspecies, 284n27
Pietz, William, 179–83, 218–20, 299n70, 321n4
pigs, 126
plants, 26–30, 133–220; and being, 138–39; and headless discourse, 140; and *otium,* 93–111; and proto-writing, 174–76; and secret writing, 152–58, and sex, 263–64; as substrate, 131, 224; temporality of, 139–41. *See also* biopolitics; Marder, Michael; oranges; *otium*
Platt, Sir Hugh, 148, 198
Plautus, Titus Maccius, 20, 287n42
Playford's *The Dancing Master*, 209
Plymouth Plantation, 77–78
poiesis (invention, making), 14, 16, 22, 174, 194, 288n46, 324n28
polar bears, 126
politeness, 19–20, 122–23, 195, 324n27, 325n33. *See also* hospitality, concept of; trust, cultivation of
Ponge, Francis, 202, 170, 172–73; *Le parti pris des choses,* 170–77
Popper, Karl, 309n64
population, 36–37. *See also* biopolitics; sheep
porcupines, 281n11
Porta, Giambattista Della: *Magiae Naturalis,* 148
posthuman/posthumanism, 21–24, 31
posy/flower, 154–55

Portugal, 196
praeteritio, 271. *See also* figures/ figuration; tropes/tropology
primatology, 118–22
prison/s, 42–46, 247; as kinetic dead zone, 44; object world in, 142–43; in Renaissance London, 158. *See also* Gerard, John; Gramsci, Antonio
Privy Council, 143, 158, 160, 162,
profit, humanist conception of, 72, 77, 200–201, 239, 299n71
prosopopoeia, 6–7, 172–73, 190, 289n55. *See also ethopoeia*; figures/figuration; Man, Paul de; metaplasm; trope/tropology
Pugliatti, Paola, 281n12
Purkiss, Diane, 247
Puttenham, George: *The Arte of English Poesy,* 115

Questier, Michael, 314n17
questions and questioning, form of: and the "human," 6, 271–72; as *praeterito* or retarding effect, 21, 31. *See also Dasein*; Gramsci, Antonio; Heidegger, Martin

Rabelais, François, 311n80
Raber, Karen, 63, 297n61
race, 40–41
rationality, 254–55, 259–63, 268
ravens, 125, 169, 310n71
Ravenscroft, Edward: *The Lover's Cuckolds,* 209
Reed, Ronald, 279n3, 280n6
Reiss, Timothy, 98, 306n16
reptiles, 128
Rheinberger, Hans-Jörg, 122
rights, discourse of, 40
risk management, 22; and shepherding, 36–37, 70–71
Robertson, Kellie, 324n30
Rockford Files, The (television show), 96
Robinsonade (genre), 257–63; and creaturely indistinction, 257; as perverse or psychotic, 260. *See also* Deleuze, Gilles
Roosth, Sophia, 268
Rosen, Barbara, 338n56
Roslin Institute, 46–48, 293n23
Rowell, Thelma, 29, 88–89, 92, 118–25, 129–31, 137–39, 177; epistemological status of observational protocol, 122–23; and fictionalizing strategies, 120–22; concept of "gregarious-long-lived-vertebrates," 118, 124, 138, 177; and multispecies coauthoring, 119–25; and rezoning of *otium,* 94–96, 119–25; twenty-third bowl and pastoral, 123–25. *See also* biopolitics; Despret, Vinciane; Franklin, Sarah; Haraway, Donna; sheep
Royal Socicty, 51–55, 67, 109–10, 200–201
ROYGBIV (mnemonic), 191–92
Rudy, Katherine M., 280n6
Ryder, Michael L., 280n4, 297n60

sacraments, the, 52–55. *See also* Christianity; Eucharist; transubstantiation
Satiliot, Claudette, 319n49
Sawday, Jonathan, 308n35
scale, 15, 17–19, 28, 157, 224, 226–27, 324n19

Schneider, Jane, 295n43
Scotland, 63–68; Museum of, 47–48
Scott, Ridley: *Blade Runner,* 47, 78
Schwartz, Regina, 336n48
Sedgwick, Eve Kosofsky, 279n2
self, as circling motion, 259–61
sensorium, human, 12, 42–46
Serres, Michel, 12, 19–20, 22, 36, 226–27, 287nn41–42, 288n50, 312n8, 330n5
Serta mattresses, 56–58, 63, 67, 86. *See also* Aardman Animation; *Shaun the Sheep*; *Timmy Time*
sex, interspecies, 81–82, 302n85
Seymour, Sir Edward, Duke of Somerset, 60–63, 67
Shadwell, Thomas: *The Virtuoso,* 48–55
Shannon, Laurie, 300n72, 303n98
Shapin, Steven, 294n31
Shaun the Sheep (television show), 55, 86–89. *See also* Aardman Animation; Serta mattresses; *Timmy Time*
sheep, 1–11, 26–29, 35–89, 91–131, 137–40, 224, 271–73; as anti-narcissism / divine gift, 11, 75–78; and biopolitics, 28–29; black sheep as figure, 37–39; blood, figurative compatibility with human, 51–55; breeds, 46–48, 69; census of in Renaissance England, 60–63; as destructive, 65, 296n61; diseases, 298n62; domestication of, 297n60; and ethology, 119–25; figured as homicidal, 57–63; as historical subjects, 18–19, 27, 84, 122, 139; history of, 299n72; human animals and, as coeval multiplicities, 7–8, 137–38; as interchangeable units, 87–88; in Japan, 298n63; and labor, 36–42; as most useful animal in Renaissance England, 68–69; and pastoral, 114–16; perspective of, 303n100; as plant-like, 40, 57, 76–77, 137, 139; and practical sovereignty, 92–94; as rhetorical creatures, 123–25; and Thelma Rowell, comparative ethological studies and co-authoring, 29, 88–89, 92–93, 119–31; and service industry, 84–89; and sheepishness, 291n3, 297n61; and sheepy reading, 83–84; talking, 36–37; taxes on in Renaissance England, 60–63; as tool of colonization, 63–68, 298n65. *See also* biopolitics; pastoral; shepherds and shepherding
Shell, Marc, 291n5, 341n75
shepherds and shepherding, 36–42, 46, 70–78, 116, 141–42; as ethical discourse, 72, 300n77; as figure for *homo sacer,* 117–18; scenography of, 131, 219–20; shepherd effects, re-signified, 129–31; shepherd of being, 130–31
Shershaw, Scott Cutler, 331n8, 338n59
shipwreck, 252–61
Shoah, the, 292n12
Shrank, Cathy, 296n51
Shukin, Nicole, 17–18, 286n38, 291n4
skin, 1–11, 12–13, 42–46, 151–52,

271. *See also* Cade, Jack; memory
Skinner, Quentin, 97, 306n14
sleep, 46, 55–58, 91–92, 304n4; and *otium,* 98–102, 110–11; as plant-like, 82, 89. *See also* counting; idling/idleness; leisure; *otium*
Sloterdijk, Peter, 223, 227–30, 231–32, 260, 330n77
smell, sense of, 110–11. *See also* sensorium, human
Smith, Bruce R., 194, 301n84, 324n31
Smith, Thomas: *A Discourse of the Commonweal,* 199–200
sovereignty, practical, 92–94. *See also* Berlant, Lauren
Spivak, Gayatri Chakravorty, 290n56
Squier, Susan, 294n23
Srinivas, Aravmudan, 342n77
Stallybrass, Peter, 299n70, 317n35, 320n4, 336n41
Steel, Karl, 300n72
Stengers, Isabelle, 24, 27, 125–26, 282n14, 284n25, 290n57, 325n33
Stubbe, Dr. Henry: *The Plus Ultra Reduced to a Non Plus,* 51–52
surplus (value), 183–84, 185. *See also* kleptomania; life; theft
Spain, 196
species, discourse of, 11–21, 17–18, 20–21, 40–41, 125, 137–39, 229, 286n38; 302n85. *See also* multispecies
stardom, discourse of, 86–87
Star Trek (television show), 96
stone, 223–69; and automorphism, 12; paving, 30–31, 233–36; woolly, 63–68
subjectivity: collective historical ("we"), 2, 8–9, 125–31; as multiple, 228–30
Sweetgrass (film), 70–72
symbiosis, 12, 19–20, 22, 219, 287n41. *See also* parasitism; Serres, Michel
synesthesia, 3–4. *See also* sensorium, human

tables/table-talk as genre, 65–68, 99–102; and posthumanism, 31, 272–73; as tabulations, 272–73
Tatham, John: *The Rump,* 323n18
Taylor, Diana, 312n8
Taylor, John (the Water Poet), 301n81
Taussig, Michael, 321n4
taxonomy, 128–31
technology/technicity: history of, 205–6, 228–31, 260–62; technoscience and theology, 53–54; televisual, 96–102; and transport, 102–6
Teerlinc, Lavinia, 192
telephony, 21–22
television, 55–57, 96–97; and remote control, 97; and reruns, 305n11. *See also* media ecology; Williams, Raymond
Thames River, 161
theft, 27–28, 30, 46, 182–83, 191, 219–20, as method, 14. *See also* Barthes, Roland; Hoffman, Abbie; kleptomania; Pietz, Bill
Theocritus: *Idylls,* 113–18, 124
Thirsk, Joan, 58, 250, 335n36
Thomas, Keith, 102, 303n96
Thompson, Charis, 283n17
Thompson, E. P., 239–40, 335n37, 340n65

Thorburn, Nicholas, 321n4
Thrupp, Sylvia, 337n55
ticks (insects), 107–11
time/temporality, 8–9, 14, 35–36; and the anthropocene, 21–22; calendar, 103; as captured in bread making, 226–27; liturgical, 103–6; as multiple and asynchronous, 28, 195; and *otium,* 95–96; secular, 103–6; and topology, 28; vegetal, 135–77
Timmy Time (television show), 55, 86. *See also* Aardman Animation; Serta mattreses; *Shaun the Sheep*
Titian: *Nymph and Shepherd,* 108–9, 272–73, 343n1
Tolkowsky, Samuel, 323n18, 327n44
Tom Thumb's Pretty Song Book, 209
Topcliffe, Sir Richard (priest hunter), 158, 244, 313n11
Topsell, Edward: *The History of Foure-Footed Beasts,* 60, 69, 281n11
torture, 158–59, 244
touch, sense of, 1–11, 271–73, 318n45
tourism, 84–89, 102–6
Tournier, Michel: *Friday,* 262–64
Tower of London, 135–77, 183–85, 243–52, 314n17, 318n42; bills for prisoners in, 336n49; contemporary exhibitions in, 146–47, 166–69; escape from, 29–30; food in, 245–47; graffiti in, 164–65, 245, 318n43; survey of in 1598, 143–45; warders' terms of service in, 143–45
Tradescant, John, 198
transhumance, 298n65. *See also* shepherds and shepherding
translation, modes of, 12, 83–84, 95–96, 199. *See also* animal/s; media ecology
transsignification (Louis's Marin's concept of), 52–55
transubstantiation, 51–52, 158–60, 236. *See also* Eucharist
tropes/tropology (as method), 3–5; as anthropo-zoo-genetic switches, 12–14; as expertise offered in the humanities, 24–26; as orientation to history, 14, 26–28; as over and under-producing, 14. *See also* figures/figuration
trust, cultivation of, 19–20. *See also* Despret, Vinciane; hospitality, concept of; Haraway, Donna; politeness; Stengers, Isabelle
Tsing, Anna Lowenhaupt, 268–69, 283n18; on storytelling as method, 284n25
Tulp, Nicolaes, 292n8
Tusser, Thomas: *Five Hundred Points of Good Husbandry,* 199
TV Guide (magazine), 47

Uexküll, Baron Jakob Johann von, 19, 107–11, 120
Ulysses (mythological hero), 131
Umwelt (environment), 19, 110–11, 190
unemployment (downsizing): scripts for, 55–58
universalism, 125–31; and discourse of species, 20–21
utopia/utopianism, 46, 48, 84–89, 105–11, 124–25; and ewe-topia, 84–89; as genre, 67–68. *See also* More, Sir Thomas

Varriano, John, 343n2
Vatter, Miguel, 321n7
Vaux, Anne (Patron of Henry Garnet), 149–50
ventriloquism, 26–27
Venus (goddess), 192, 207–8, 219
vermiculation, 39
Vickers, Brian, 91, 93–96
Virgil, 113, 199
Vives, Juan: *Dialogues,* 250

Waad, Sir William (Lieutenant of the Tower of London), 149, 158, 313n14
Wake, Paul, 315n26
Wakefield, Battle of, 191
Wakelin, David, 280n3
walking, 225–26
Walpole, Hugh (Jesuit priest), 251, 255; and oratory as sacral space, 164–68
Warner, Michael, 336n42
Warren, Roger, 282n13
Watson, Robert N., 111, 338n57
Weber, Samuel, 334n27
wellness, discourse of, 98
wetware, 5, 16, 39, 242, 255, 281n8, 318n44. *See also* Doyle, Richard
whitewashing: of walls in Renaissance English houses, 317n35
Whittle, Jane, 325n38
Wilkes, Rhoda, 292n5, 295n38, 301n77
Williams, Raymond, 96–97, 105–6, 111–12, 123, 308n40
Wills, David, 285n30
Wilmut, Ian, 293n23
Wilson, L. Anne, 326n41
Wilson, Thomas: *A Discourse upon Vsurye,* 296n51
Wintroub, Michael, 340n61
witches and witchcraft, 338nn56–57
Wolfe, Cary, 14–18, 22–26, 107–8
Wolfe, Heather, 317n35
Wolsey, Cardinal, 201–3
wolves, 42
wool, 5, 7–8, 46, 48–50, 58–60, 63–66, 71–76, 93, 117, 136–37, 296n46, 301n81. *See also* sheep
Wren, Sir Christopher, 50, 209
writing: apparatus of in medieval and Renaissance England, 1–11, 73–74, 156–57, 176, 317n35; and cutting, 282n14, 316n35; and exteriority, 14–16; and general or generative text, 333n27; as horizon/contact zone with other entities, 16; human, as subset of a common relation to coding, 12–13; inhuman (coding/inscription), 11–13, 283n20; insurgent, 7–9, 281n9; as magical, 150–57, 176; as parasitic on animals, plants, minerals, 6–7, 11–12; plant, 139–41, 173–77; secret, 154–55; sheepy, 84–89; and survival, 12–16
Wunderkammer, 181
Wycherley, William: *The Country Wife,* 323n18
Wyndham, John: *The Day of the Triffids,* 329n75

xenotransfusion, 48–55. *See also* biopolitics; blood

Yachnin, Paul, 300n72
Yamashita, Karen Tei: *The Tropic of Orange,* 215–20, 223–24

yeast *(saccharomyces cerevisiae)*, 26–31, 223–69; absence of as sacramental void, 251–52, 256–65; as evidence of the divine, 249; screaming, 267–68; and sexuality, 263; and temporality, 28. *See also* bakers and baking; bread; fermentation; leaven

yersina pestis (bubonic plague), 58, 252–56

York, House of, 191

zooarchaeology, 280n3, 297n60

zoomorphism, 1–11, 36–42. *See also* anthropo-zoo-genesis

zoopolitics, 18, 110, 169–77

(continued from page ii)

29 NEOCYBERNETICS AND NARRATIVE
Bruce Clarke

28 CINDERS
Jacques Derrida

27 HYPEROBJECTS: PHILOSOPHY AND ECOLOGY AFTER THE END OF THE WORLD
Timothy Morton

26 HUMANESIS: SOUND AND TECHNOLOGICAL POSTHUMANISM
David Cecchetto

25 ARTIST ANIMAL
Steve Baker

24 WITHOUT OFFENDING HUMANS: A CRITIQUE OF ANIMAL RIGHTS
Élisabeth de Fontenay

23 VAMPYROTEUTHIS INFERNALIS: A TREATISE, WITH A REPORT BY THE INSTITUT SCIENTIFIQUE DE RECHERCHE PARANATURALISTE
Vilém Flusser and Louis Bec

22 BODY DRIFT: BUTLER, HAYLES, HARAWAY
Arthur Kroker

21 HUMANIMAL: RACE, LAW, LANGUAGE
Kalpana Rahita Seshadri

20 ALIEN PHENOMENOLOGY, OR WHAT IT'S LIKE TO BE A THING
Ian Bogost

19 CIFERAE: A BESTIARY IN FIVE FINGERS
Tom Tyler

18 IMPROPER LIFE: TECHNOLOGY AND BIOPOLITICS FROM HEIDEGGER TO AGAMBEN
Timothy C. Campbell

17 SURFACE ENCOUNTERS: THINKING WITH ANIMALS AND ART
Ron Broglio

16 AGAINST ECOLOGICAL SOVEREIGNTY: ETHICS, BIOPOLITICS, AND SAVING THE NATURAL WORLD
Mick Smith

15 ANIMAL STORIES: NARRATING ACROSS SPECIES LINES
Susan McHugh

14 HUMAN ERROR: SPECIES-BEING AND MEDIA MACHINES
Dominic Pettman

13 JUNKWARE
Thierry Bardini

12 A FORAY INTO THE WORLDS OF ANIMALS AND HUMANS, *WITH* A THEORY OF MEANING
Jakob von Uexküll

11 INSECT MEDIA: AN ARCHAEOLOGY OF ANIMALS AND TECHNOLOGY
Jussi Parikka

10 COSMOPOLITICS II
Isabelle Stengers

9 COSMOPOLITICS I
Isabelle Stengers

8 WHAT IS POSTHUMANISM?
Cary Wolfe

7 POLITICAL AFFECT: CONNECTING THE SOCIAL AND THE SOMATIC
John Protevi

6 ANIMAL CAPITAL: RENDERING LIFE IN BIOPOLITICAL TIMES
Nicole Shukin

5 DORSALITY: THINKING BACK THROUGH TECHNOLOGY AND POLITICS
David Wills

4 BÍOS: BIOPOLITICS AND PHILOSOPHY
Roberto Esposito

3 WHEN SPECIES MEET
Donna J. Haraway

2 THE POETICS OF DNA
Judith Roof

1 THE PARASITE
Michel Serres

Julian Yates is professor of English and material culture studies at the University of Delaware. He is the author of *Error, Misuse, Failure: Object Lessons from the English Renaissance* (Minnesota, 2003), which was a finalist for the Modern Language Association Best First Book Prize, and *What's the Worst Thing You Can Do to Shakespeare?* (with Richard Burt, 2013). He is editor of *Object Oriented Environs* (with Jeffrey Jerome Cohen, 2016).